X-ray Imaging of the Soil Porous Architecture

Sacha Jon Mooney • Iain M. Young •
Richard J. Heck • Stephan Peth
Editors

X-ray Imaging of the Soil Porous Architecture

Springer

Editors
Sacha Jon Mooney
School of Biosciences
University of Nottingham
Nottingham, UK

Iain M. Young
Biological and Environmental Science and Engineering Division
King Abdullah University of Science and Technology
Thuwal, Saudi Arabia

Richard J. Heck
School of Env. Sciences
University of Guelph
Guelph, ON, Canada

Stephan Peth
Inst. für Bodenkunde
Bodenkunde Kassel
Witzenhausen, Hessen, Germany

ISBN 978-3-031-12175-3 ISBN 978-3-031-12176-0 (eBook)
https://doi.org/10.1007/978-3-031-12176-0

This Springer imprint is published by the registered company Springer Nature Switzerland AG
The registered company address is: Gewerbestrasse 11, 6330 Cham, Switzerland

Foreword

Soils are literally fundamental to all human civilisations and underpin all terrestrial ecosystems. As such, their importance to the Earth system, and all life thereon and therein—including humans, of course—is paramount. Soils function by virtue of their spatial organisation, and they are arguably unique systems in terms of both the diversity of their mineral and organic constituents and in the way these are arranged in four-dimensional space (i.e. three dimensions over time) over many orders-of-magnitude. Soil structure is the term traditionally used to describe and conceptualise these spatial characteristics, but arguably *soil architecture* is a more apposite term since it explicitly integrates living entities in the framework and encourages consideration of soils as integrated ecological systems. And it is pore networks, manifest as extraordinarily complex multi-scale labyrinths, which are one of the most essential features of soil systems. This is because they represent the *inner space* of the soil system, in and through which all gases, liquids, solutes and organisms are bound, reside, move, react, transform and more besides. There is a curious irony in that these pore systems are effectively defined by where the solid phases of the soil are *absent*.

One of the major challenges in studying soil architecture, which has certainly constrained progress in these terms since the onset of pedology, as a well-found scientific discipline, is that they are (generally) friable and (certainly) opaque to the unaided human sensory experience. One can only progress so far in quantifying and understanding the origins and consequences of soil architecture by direct visual observation, nor even with light or electron microscopes and modifying geologists' or histologists' approaches based on carefully spatially-preserved thin-sections.

Enter Godfrey Hounsfield with his visionary (pun intended) invention and development of X-ray Computed Tomography in a medical context, with the emergent means to non-destructively image many of the key constituents of soil, and there was then a means to overcome what was previously considered as an intractable challenge. This opened new frontiers to explore soil systems, and over the past 40 years the tomographic tools that have evolved for application in soil science have been revolutionary. These encompass multi-disciplinary hardware, software and conceptual (i.e. modelling) engineering endeavours, allied to imaginative experimental systems. This laudable volume provides a comprehensive description and synthesis of all these aspects of the science and art of X-ray Computed Tomography as applied to soil systems. It reveals often astonishing new views of the underworld and sets the scene for the exciting future of this powerful approach to understand, and therefore effectively manage, the critical soil resources on which we depend.

Emeritus Professor of Soil Ecology
University of Nottingham, Nottingham,
UK
e-mail: karl.ritz@nottingham.ac.uk

Karl Karl

Acknowledgements

The editors acknowledge the preliminary discussions and editorial support from Craig Sturrock, Brian Atkinson and other researchers at the Hounsfield Facility, University of Nottingham who helped with ideas and suggestions for this book. SJM is grateful to Emma, Dylan and Lily for their support and encouragement during the development of this book.

Contents

Chapter 1
40 Years of X-ray CT in Soil: Historical Context

Iain M. Young, Sacha J. Mooney, Richard J. Heck, and Stephan Peth

1.1 Introduction

In December 1973 Godfrey Hounsfield published his seminal paper on transverse axial scanning, describing a methodology that could non-destructively analyse a human head (Hounsfield, 1973). Two years previously, a patient's head had been scanned in what was the first system available to hospitals developed by Hounsfield and his research partners. Thereafter, an explosion in the use of Computed Tomography (CT) systems for medical purposes was seen in most western countries, and today Tomography systems (including MRI), in general, and CT systems, in particular, are some of the most widely used techniques in hospitals around the world. Two decades ago, a CT exam would take more than 30 minutes. Now, it is possible to collect high-resolution images in 2 seconds, along with vast improvements in detector hardware and associated software, whilst reducing any dose by up to 80%.

In 1979, Hounsfield and Allan Cormack won the Nobel Prize in Physiology and Medicine "*for the development of computer assisted tomography*". Less than 3 years

I. M. Young (✉)
Biological and Environmental Science and Engineering Division, King Abdullah University of Science and Technology, Thuwal, Kingdom of Saudi Arabia
e-mail: iain.young@kaust.edu.sa

S. J. Mooney
School of Biosciences, University of Nottingham, Sutton Bonington Campus, Leics, UK
e-mail: Sacha.Mooney@nottingham.ac.uk

R. J. Heck
School of Environmental Sciences, University of Guelph, Guelph, ON, Canada
e-mail: rheck@uoguelph.ca

S. Peth
University of Hannover, Institute of Soil Science, Soil Biophysics Group, Hannover, Germany
e-mail: peth@ifbk.uni-hannover.de

S. Jon Mooney et al. (eds.), *X-ray Imaging of the Soil Porous Architecture*,
https://doi.org/10.1007/978-3-031-12176-0_1

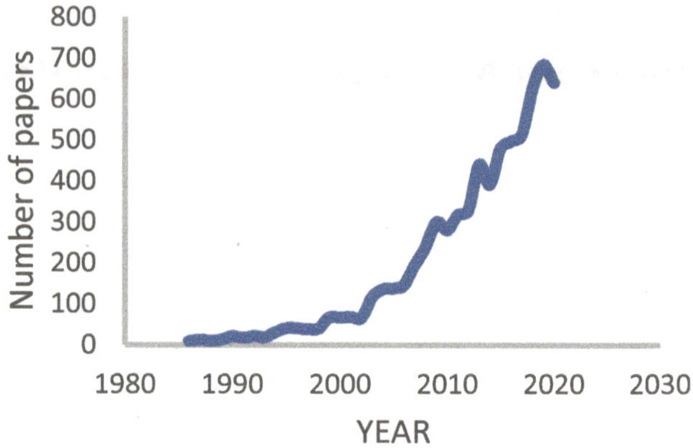

Fig. 1.1 Publications using X-ray CT in the soil sciences

later, in a short technical note, the first X-ray microtomography system (μCT) produced an image of the internal structure of a snail's shell (Elliott and Dover, 1982), with a 15 μm spatial resolution, opening up the possibility of relatively inexpensive (at least compared to medical CT systems) lab-based scanners for a wide range of medical and non-medical applications.

Petrovic et al. (1982), in a first for soil science, using a fourth generation medical CT system, successfully quantified bulk density changes in soil, and a year later in Perth, Western Australia, using a bespoke laboratory gamma-ray system, Hainsworth and Aylmore (1983) determined the spatial distribution of water in a soil column. A new dawn in investigative technology had arrived for soil scientists who could now observe structural elements of soil in a non-destructive fashion.

Figure 1.1 shows the published papers (using Scopus and the search terms "tomography and soil or plant"—28th Jan 2021). A slow rise in publications between 1986–1995 (195 publications) through to 672 papers published between 1996–2000, with an obvious upswing in 2003. Between 2016–2020 we see an exponential rise in publications (2963). This is probably related to an increase in the development and accessibility of benchtop high-resolution X-ray CT systems (hereafter referred to as X-ray CT), with much of this development driven by the oil, mining and engineering industries interests in porous media.

Whilst many of the challenges of using an X-ray CT for soil systems research map onto those of other systems, soil brings peculiar and complex problems not seen in any other opaque architectures, including the human body. Many such examples are presented in this book. It is important to recognise, for instance, that the use of CT for medical purposes is generally focussed on divergence from the norm in human bodies. So, abnormalities such as bone fractures, emergence of dense tumours or changes in the density of lungs are daily occurrences and most medical systems are driven to produce clearer and faster images for such architectures, with the lowest radiation doses possible. Often these features are readily detectable both in terms of contrast and resolution. Similarly, in engineering where defects in prescribed designs are an important focus. However, in soil, where compositional and spatiotemporal heterogeneities are inherent across multiple scales, where complex geometries of structures exist over space and time, in generally unsaturated conditions within complex organo-mineral constructs that shrink and swell, change is a constant. It is, however, within this complex architecture of soil that the many macro-, meso- and micro-communities live and imprint their own activity, requiring accurate observations and analysis, that present us with problems that are of many orders of magnitude more difficult to deal with.

As hardware has advanced, so has associated software. Due to the complexity of soil however, the problems of image processing and analysis of the soil-plant-microbe complex remain a great challenge (Chaps. 4 and 5). Whilst much has been achieved, the reality is that the problems related, for example, to the segmentation of unsaturated organo-mineral complexes, that comprise soil, have so far not been solved to the extent that automated processes can be used across soil types and X-ray CT hardware. Much improvement has been made, however, to isolate and quantify root systems (see Chap. 9) in various soil textures despite limits of spatial resolutions and some progress has been made observing organic matter in soil (Chap. 10). The greatest progress has been made in the analysis of soil porosity (Chaps. 6, 7, 8), and associated metrics (see Feeney et al., 2006). This has advanced to the extent that we can now scan an undisturbed soil core and from the captured 3-D architecture simulate water flow, retention and gas flow. The linkages of model simulations and 3-D pore space architecture are highlights of the advancement of the use of X-ray CT in soils (Chap. 11).

This book deals with the great challenges of using X-ray CT for soils and looks in detail at the recent developments in X-ray CT applications in soil research. The associated opportunities and problems are described, from a range of researchers covering different fields, including tips on the best way forward (Chaps. 3 and 4) and, in some cases, differing views on the same subject; and provides practical approaches to some limiting problems in using and/or choosing hardware systems (Chap. 2). We hope that this book will be of interest to the soil scientists undertaking their first forays into the world of imaging 3-D soil microstructure, as well as the experienced user looking for special applications and practical solutions for μCT acquisition and image analysis.

References

Feeney, D. S., Crawford, J. W., Daniel, T., Hallett, P. D., Nunan, N., Ritz, K., Rivers, M., & Young, I. M. (2006). Three-dimensional microorganization of the soil–root–microbe system. *Microbial Ecology, 52*, 151–158.

Elliott, J. C., & Dover, S. D. (1982). X-ray microtomography. *Journal of Microscopy, 126*, 211–213.

Hainsworth, J. M., & Aylmore, L. A. G. (1983). The use of computed assisted tomography to determine spatial distribution of soil water content. *Australian Journal of Soil Research, 21*, 435–443.

Hounsfield, G. N. (1973). Computerized transverse axial scanning (tomography). 1. Description of system. *The British Journal of Radiology, 46*(552), 1016–1022. https://doi.org/10.1259/0007-1285-46-552-1016

Petrovic, A. M., Siebert, J. E., & Rieke, P. E. (1982). Soil bulk density in three dimensions by computed tomographic scanning. *Soil Science Society of America Journal, 46*, 445–450.

Chapter 2
Practicalities of X-ray CT Scanning for the Soil Sciences

Andrew Ramsey

2.1 Introduction

What better application can there be of X-ray CT than to study the heterogeneous structure of soil? What more friable, fragile structure can there be but that of soil, so sensitive to the slightest touch? Yet, accurately visualising the structure of soil is so vital to understanding the passage of nutrients and water through it and the microbes residing in it. What other technique could resolve the complex three-dimensional (3-D) features without affecting them?

X-ray CT is a promising method of examining the 3-D structure of soil since it is completely non-destructive and typically requires no special preparation of samples. An X-ray CT system can image samples from a few particles of soil (~1mm) up to a large core of soil of 20–30 cm diameter and 100 cm long. The resolution will vary, being far higher for the smaller samples and lower for the larger samples. It will typically be the diameter of the sample divided by a few thousand. The very highest resolution that can be expected is around 1 μm (although some new systems state possibilities beyond this).

X-ray CT volumes are generated from a set of projection images, so the amount of information in them is limited by the number of pixels across the detector. It is important though to make sure that the X-ray source is small enough that fine details are not blurred over more than one pixel and that the mechanics of the sample manipulation does not introduce motion blur into the reconstruction. There is a distinction between the voxel size, which is typically the effective pixel size at the position of the sample (only affected by the number of pixels across the detector and the geometric magnification), and the spatial resolution which gives the finest details separable (resolvable) in the CT volumes. This latter measure depends also upon the

A. Ramsey (✉)
Nikon Metrology Inc., Brighton, MI, USA
e-mail: Andrew.Ramsey@nikon.com

S. Jon Mooney et al. (eds.), *X-ray Imaging of the Soil Porous Architecture*,
https://doi.org/10.1007/978-3-031-12176-0_2

sharpness of the projection images and the stability and accuracy of the sample manipulation stage as well as knowledge of the position and alignment of the X-ray source, stage and detector.

X-ray CT produces a full 3-D map of the internal structure of a specimen by measuring the X-ray linear attenuation coefficient at each 3-D point in a small-pitch volumetric grid. The pitch of the grid can go as low as 1 μm but is defined by the size of the sample divided by the number of pixels across the detector, being typically in the 10s of μm range, depending on the sample size (larger for larger samples). The volumetric grid can be thought of as the 3-D analogue of a 2-D pixel grid in a digital image and in fact the volume grid elements are known as "voxels". The X-ray linear attenuation at each voxel is calculated from the millions of total attenuation measurements, one in each pixel of every projection radiograph collected at hundreds or thousands of projection angles. Since the linear attenuation is proportional to the electron density in the sample, then the CT results show a good indication of material density changes, as well as being affected by chemical changes, allowing different materials to be discriminated.

2.2 Manufacturers of μCT Systems

With so many manufacturers and designs of X-ray CT systems available (some example X-ray CT system cabinets are shown in Fig. 2.1), how can you choose the right one for your application? Below is a (non-exhaustive) list of current X-ray CT system manufacturers, though it is inevitable that new ones will appear within a year or two:

- *Bruker* (originally *SkyScan* of Belgium), based in Mass., USA: mainly suited to small samples. https://www.bruker.com/en/products-and-solutions/microscopes/3d-x-ray-microscopes.html.
- *Diondo*—based in Hattingen, Germany (formerly owned by Yxlon). www.diondo.com.
- *Nikon Metrology*–originally X-Tek Systems (UK), bought by Nikon Corporation of Japan, HQ in Belgium, manufacturing a range of cabinet sizes and now collaborating with US-based *Avonix Imaging Inc.* of Minnesota, USA for larger enclosures. www.nikonmetrology.com.
- *North Star Imaging (NSI)*—in Minnesota, USA, owned by ITW of Chicago, manufacturing a wide range of system sizes. www.4nsi.com.
- *Rayscan Technologies*—Germany. https://www.rayscan.eu/.
- *Rigaku*—Japan. Use both sealed and open-tube (rotating target) sources. imaging.rigaku.com.
- *Scanco Medical*—Switzerland. CT systems using sealed sources designed mainly for small life sciences samples. www.scanco.ch.
- *Shimadzu*—Japan. Both sealed and open-tube microfocus CT systems. www.shimadzu.com.

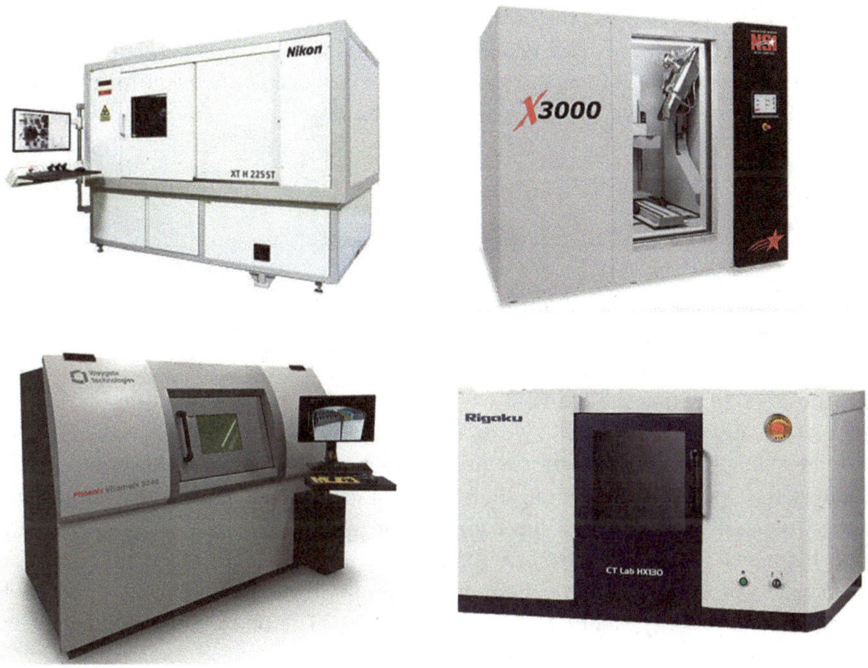

Fig. 2.1 Top-left: A Nikon Metrology XTH225ST cabinet; Top-right: An NSI X3000 cabinet. Bottom-right: A Waygate v|tome|x 225 kV system; Bottom-right: A Rigaku CT Lab HX cabinet. [Image ©Rigaku Corporation. Used by permission]

- *ThermoFisher Scientific*—makers of *HeliScan,* designed for small rock cores. www.thermofisher.com.
- *VJ Technologies*—New York, USA. www.vjt.com.
- *Waygate*—originally "*Phoenix|x-ray*", bought by GE, then Baker Hughes, based in Germany. https://www.bakerhughesds.com/.
- *Yxlon*—also in Germany (grew out of Philips X-ray). www.yxlon.com.
- *Zeiss* (both the Metrotom and Xradia product ranges)—Germany: Zeiss's measurement reputation married with Xradia's nanofocus technology. https://www.zeiss.com/metrology/products/systems/computed-tomography.html.
- *ProCon*—Germany. https://procon-x-ray.com.

Note that some of these are measurement companies relatively new to X-ray technologies; others have decades of X-ray imaging experience. Rather than discussing the range of systems from each of these companies, we note the characteristics of systems that make them suitable for inspecting soil samples. Is it better to buy a system with a high-resolution detector or a high-resolution source, or both? How does an open-tube X-ray source compare with a closed-tube source? How accurate does the sample manipulator need to be? Is helical scanning better than circular scanning?

All CT systems have an X-ray source, a sample manipulator and an X-ray detector. Industrial systems rotate the sample while keeping the source and detector static (the converse to medical system, as the need to keep a patient still rather than being rotated through 360° is not there). For helical scans, the sample is translated vertically during the rotation to form a helical path.

2.3 X-ray Sources

There is one characteristic of an X-ray source that will greatly affect the quality of the CT results and the running costs of the system. X-ray tubes come in two types: open and closed. Closed tubes are evacuated and sealed once for their lifetime. This has the advantage that the filament never needs to be replaced and often lasts a few years (typically between 3 and 7 years depending on the dose output). Once it blows though, a new X-ray tube insert is required which greatly adds to the running costs of the system. Open tubes maintain their vacuum using a constantly running vacuum pump (typically a high-speed turbo-molecular pump backed by a backing pump). The filaments in these tubes last typically a couple of months and then need replacing, a procedure which usually takes less than an hour can be easily done by operators and costs typically a few tens of dollars. The open tubes themselves can last decades.

If the electron beam, in an X-ray tube, is focused onto the target, the tube is said to be a microfocus tube, with a spot size of typically less than 10 μm. If it is not focused, the spot size may be much larger and high-resolution images can only be obtained by acquiring low-magnification images of samples placed close to a high-resolution detector. These are known as minifocus tubes. Most, but not all, closed tubes are minifocus tubes.

X-ray tubes are also characterised by their penetrating power, determined by their maximum voltage (measured in kiloVolts, or kV) and by their maximum power in Watts, which determines the number of X-ray photon they can produce. X-rays up to 225 keV can easily pass-through soil samples up to 100 mm (4 inches) in diameter. These sources often have the smallest spot sizes and are mostly microfocus.

It is worth asking the manufacturer about maintaining an open tube; some tube materials require monthly deep cleaning, while others need only an occasional wipe with an alcohol-soaked cloth. It is also worth asking about the cycle time of sources: can the source, for example, work 24/7 thus allowing for batch scanning of more samples overnight. It is advisable, regardless of system purchased, to take out maintenance cover, with local expertise (through training by the manufacturer) also being present. Well-maintained systems are often in use more than a decade after purchase.

2.4 Detectors

Most X-ray detectors work by converting the X-rays into visible light using a fluorescent material layer (known as a scintillator) in front of a large array of photodiodes, which convert the visible light into electronic charge that can be read by the digitiser. Some detectors directly capture X-ray photons. While such detectors are generally more sensitive than standard detectors, they are often much more expensive (Fig. 2.2).

Following the merger of PerkinElmer and Varian a couple of years ago, Varex Imaging Inc. has become the world's largest manufacturer of X-ray sensitive flat panel detectors (Fig. 2.2). A few other companies make competitive detectors, such as iRay, Hamamatsu and Detection Technologies Inc., but these have yet to make their way into mainstream CT systems. Waygate (being ex-GE themselves) uses a detector created by GE which is claimed to be both high-resolution and high efficiency (in terms of converting X-ray photons to electrical signals).

2.5 Obtaining High-Resolution Images

Increasing the image resolution can lead to a whole new level of detail being visible, as seen in Fig. 2.3. It is often debated as to whether it is quicker to obtain high-resolution CT data using a high-resolution detector or a high-resolution X-ray source. A detector is deemed to be high-resolution if its pixels are smaller than 150 μm. It is worth noting that the efficiency of capture of X-rays by detector pixels is proportional to the area of the pixel and so to the square of the quoted pixel size. Furthermore, smaller pixels require thinner scintillators, to prevent the spread of visible light over several pixels, and so many more X-rays pass straight through the

Fig. 2.2 Left: Varex 4343 (2880 × 2880 150 μm pixels); Right: Varex 2520DX (1900 × 1600 127 μm pixels). ©Varex Imaging. Used by permission

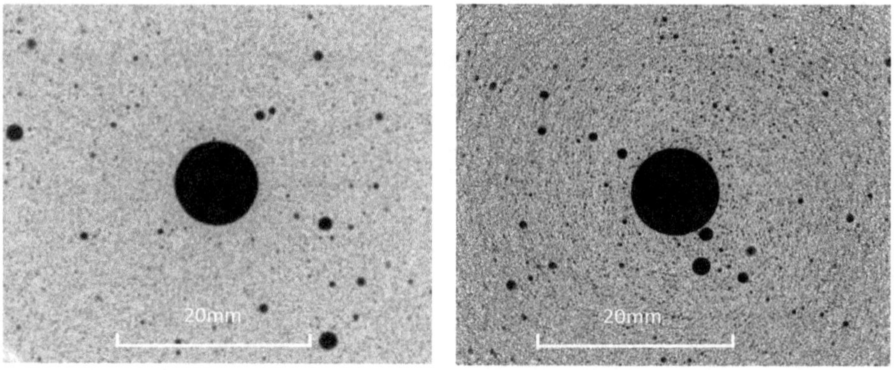

Fig. 2.3 Increasing the image resolution can lead to a whole new level of detail being visible, as seen in these images of a grinding wheel using (left) an image intensifier and (right) a Varex 1620 flat panel. [Images courtesy of Nikon Metrology UK Ltd.]

scintillator without being detected. This typically makes the efficiency of capture of X-rays by a detector inversely proportional to the cube of the quoted pixel size.

An X-ray source is considered microfocus if its spot size is less than around 100 μm. At higher magnifications the size of the X-ray spot becomes the limit on the spatial resolution that can be obtained. Since microfocus X-ray sources must limit their power at small spot sizes, or expand their spot size to prevent target damage, the scan times are often assumed to be longer. But the power of a small X-ray spot is proportional only to the spot size itself, and so doubling the resolution needs only twice the scan time instead of the eight times required when the detector resolution is doubled.

Some manufacturers have techniques for increasing the detector resolution by moving the detector within the enclosure (such as NSI's *matriX* or Nikon's *PanelShift*), or by making sub-pixel movements and interlacing image pixels (such as NSI's *subpiX* or Nikon's *PixelPush*) (see Fig. 2.4).

2.6 Image Quality

The quality of an image is often measured by its spatial resolution (the ability to resolve separate but neighbouring features) and its signal-to-noise ratio (SNR). The higher the spatial resolution, the smaller features can be seen in the images. However, it is important that the signal-to-noise ratio is kept high to prevent small features disappearing into the background noise. Background noise, which can be seen as speckle on the projection images, is due to a different number of X-rays being in each consecutive image due to the random nature of X-ray production in the target. The SNR of X-ray images, like those produced by any random process, is proportional to the square root of the signal, which in turn is proportional to the number of X-rays imaged. This latter is proportional to the X-ray beam current

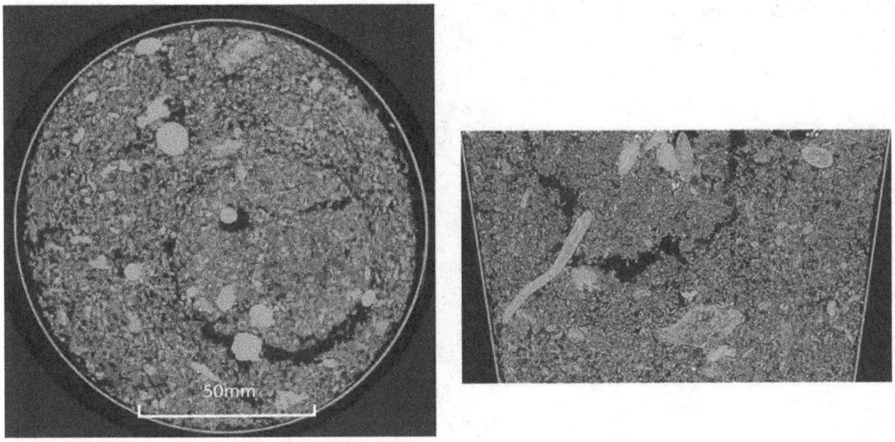

Fig. 2.4 Horizontal (left) and vertical (right) CT slices through a recently repotted plant pot showing roots (brighter) and air spaces in the soil. [Images © Nikon Metrology UK Ltd.]

(usually measured in microAmps (μA) or milliAmps (mA) – 1000 μA = 1 mA) and the efficiency of the detector as well as the scan time. To double the SNR, it is necessary to quadruple the signal. For small samples in which the beam current cannot be increased to avoid broadening the focal spot too much and losing spatial resolution, the only way to increase the signal is to lengthen the scan time. For individual images this is usually not a problem, but when a CT scan already takes tens of minutes, this can lead to unfeasibly long scan times. There is anecdotal evidence of the ends of plant roots being blurred due to long scans, i.e. they grew longer during the scan!

2.7 Sample Manipulator

The accurate manipulation of the sample during a CT scan is crucial to obtaining high-resolution CT volumes. Inaccuracy in the sample movement will blur the features in the high-resolution radiographs. For example, a precession of 100 μradians will cause a movement of 100 μm at a metre which is 10 μm 100 mm above the turntable bearings. This will prevent CT images having better resolution than, say, 20 μm at that level; which is worse higher up. Putting your sample on a pillar to raise it in front of a high source will similarly degrade your spatial resolution. Turntable runout, the slop in bearings, will cause similar blurring but this blurring will remain constant throughout the height of the sample.

A system in which the detector can be moved towards the source can save valuable scan time. If the sample is not so small that it needs the maximum system resolution, then it can be moved towards the source and the detector brought closer. Since the X-ray flux is proportional to the square of the source-to-detector distance,

Fig. 2.5 A comparison of circular CT (left) with helical CT (right) of a stack of DVDs showing how helical CT has better vertical spatial resolution throughout the CT volume away from the central slice. This is only true when the manipulator is well-aligned

bringing the detector only 30% closer to the source will double the brightness of the images and thus halve the scan time needed to get the same signal-to-noise ratio. This can lead to larger cone-beam angles and therefore greater cone-beam artefacts in non-helical scans.

Helical CT, in which the sample is moved vertically during the sample rotation to create a helical path requires more stringent manipulator alignment than purely circular CT in which the sample is simply rotated. The rotation axis must be aligned with the vertical movement axis and must be straight. Done well, helical CT though can remove cone-beam artefacts from constantly looking up at the top of the sample and down at the bottom. These artefacts, while clearly visible at the top and bottom of the CT volume are in fact present throughout except in the central slice and will degrade the vertical spatial resolution (see Fig. 2.5).

It is worth noting that a helical scan can lead to a higher-resolution CT volume than a single circular scan, especially of a tall object, since the sample can be magnified until its width almost fills the image rather than its taller height. Of course, several circular sub-scans could be performed but these will need to be stitched together using the regions where the cone-beam artefacts are greatest.

In a helical CT scan, the sample must be moved from completely below the detector to completely above it. The cabinet height often limits the height of samples which can be scanned using the technique. One method of obviating this is to crop the detector vertically for these scans which allows taller samples to be scanned, albeit more slowly due to the more rotations needed. A fixed turntable and movable source and detector combination, as provided in some CT systems, allows the helical scanning of taller samples.

2.8 Configurations

2.8.1 Cabinet or Enclosure?

In practice, most soil samples are not too large (<30 cm cube) and so will comfortably fit into a one-piece cabinet. Most will not require X-ray sources above 225 kV which helps keep the cost of an X-ray CT system down, which is often a major factor in choosing a system.

2.8.2 CT Scanning Methods

There are a few different methods of capturing CT data:

- Circular scans—these use a simple geometry since they only need a single rotation in one position.
- Helical scans—allow taller samples to be scanned in one volume; the manipulator needs to move a long way vertically, or the source and detector move instead.
- 2-D fan-beam scanning—this is a very slow method used for highly-scattering samples; it is not usually needed for soil as scatter is not such a problem.
- Dual energy scans—allow for chemical discrimination by comparing the results of scans using different X-ray energies.
- 4-D CT—time series 3-D CT scans; or continuous scanning to characterise dynamic processes e.g. infiltration.

2.8.3 Software

The software provided with a CT system is usually the operator's only interaction with the system and can make or break the choice of system. There are several software features which are considered highly desirable:

- General appearance of the user interface: How cluttered is the interface? How many features can a user interact with? How easy it is to scan several samples with similar settings? How difficult is it to set up a new sample?
- What options are available for different types of scans? Most systems provide circular cone-beam scans, reconstructed using the standard FDK algorithm (Feldkamp et al., 1984). However, helical CT allows for high-resolution scans of tall samples and the removal of cone-beam and ring artefacts (Katsevich et al., 2004).
- The ability to batch scan allows several samples to be scanned, without operator intervention, say overnight.
- Programmable software allows for custom scan methods and third-party hardware (like robotic sample handlers). With carefully designed sample holders, the sample manipulator itself, in conjunction with a sample shelf inside the cabinet,

can act as a cost-effective though slightly slower sample loader. The programmability can range from a few simple macros to full open architecture so that software engineers can write programs to control not only the X-ray CT system but of course third-party hardware like robots, or third-party software such as databases.

- 4-D CT—both time-lapse CT scans and continuous capture in which the spatial resolution can be played off against the extra time resolution for those events which happen quickly, or extra spatial resolution can be obtained during those periods of slow change (Parmesh, 2018).

2.9 Overcoming CT Artefacts

CT artefacts are unwanted features in the data not relating to real features in the sample and come in many forms. Ring artefacts are bright and/or dark rings around the sample rotation axis that often connect features which the operator desires to segment, or separate, such as particles of soil. Rings are caused by non-linear behaviour of individual pixels in a detector that have not been identified as "bad pixels" (i.e. a dead transistor in a detector) in and interpolated over. Since they do not move as the sample is rotated, they form rings in the CT volume. A single bad pixel, differing greatly in intensity from its neighbours, can produce three adjacent rings due to the filter in the filtered back-projection reconstruction algorithm [3]. A jump in sensitivity of adjacent pixels can cause two rings (for the same reason). It is rare to get a single ring.

Ring artefacts can be supressed during scanning by, for example, moving the sample or detector sideways by random amounts and subsequent correcting by shifting the image sideways by a fractional pixel amount. They can also be removed afterwards by post-processing algorithms acting on the CT volumes.

Beam hardening artefacts occur when particles in the soil filter an otherwise unfiltered X-ray beam and cause parts of the volume to be imaged with only higher energy X-rays instead of the full beam spectrum. They can complicate the grey-value thresholding of soil particles because the threshold needs to vary across the sample, being lower in the centre. It is worth noting that the voltage selected in the software is the electron beam acceleration voltage and not the energy of most of the X-rays, which is much lower and of a wide range of energies. Dense particles will filter the lower energies out of the beam leaving only more highly penetrating X-rays, making those parts of the sample appear less absorbing of X-rays and of a lower density. Placing a filter, for example, in front of the X-ray source removes those low energy X-rays from the beam and improves the CT images (Fig. 2.6), at the expense of a longer scan time. The images from the unfiltered beam are brighter but only because they contain many low energy X-rays which cannot penetrate the sample. These low energy X-rays are best removed from the equation. It is worth remembering that the mean energy of the X-ray beam is controlled much more by the thickness of filter in front of the source than by the acceleration energy of the electrons selected.

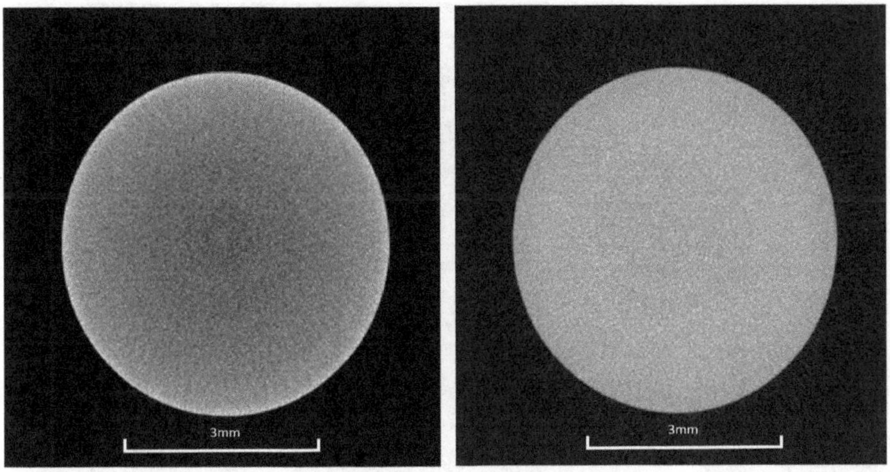

Fig. 2.6 Left: Beam hardening in a steel rod lowering the apparent density of the centre of the sample; Right: Beam hardening removed by filtering the X-ray beam and/or by software correction. [Image courtesy of Nikon Metrology UK Ltd.]

Beam hardening artefacts can also be suppressed using corrections during reconstruction. These work best when there is only one material in the sample, so often a combination of source filtering and software correction is used (see Chap. 4).

2.10 Evaluating a Potential X-Ray CT System

The best way to evaluate a potential X-ray CT system is to have the manufacturer scan some of your samples. Preferably use the same sample for all supplies under consideration for comparison purposes. For soils, a resin impregnated sample is often a good choice as it is structurally stable, presents the same challenges for all systems, and once you define the scanning characteristics you want (fast scan vs. slow, etc.), will provide the best comparison possible.

Samples to be scanned should range from the largest sample you will need to scan, down to individual soil aggregate, only few millimetres across to give a breadth of the possibilities from the instrument (Fig. 2.7). The sharpness of the images should be evaluated considering the smallest features you expect to see/segment in samples of a certain size, bearing in mind that the resolution is generally higher for smaller samples as they can be imaged at higher magnifications. Zoom in until individual voxels can be seen. The ability to scan small regions of larger samples can be very useful but is not offered by all systems. Look out for sharpening filters being used to artificially increase the resolution: such filters also increase the noise and therefore do not increase the ability to distinguish small features of interest from background noise. Noise suppression filters may also be used to reduce the

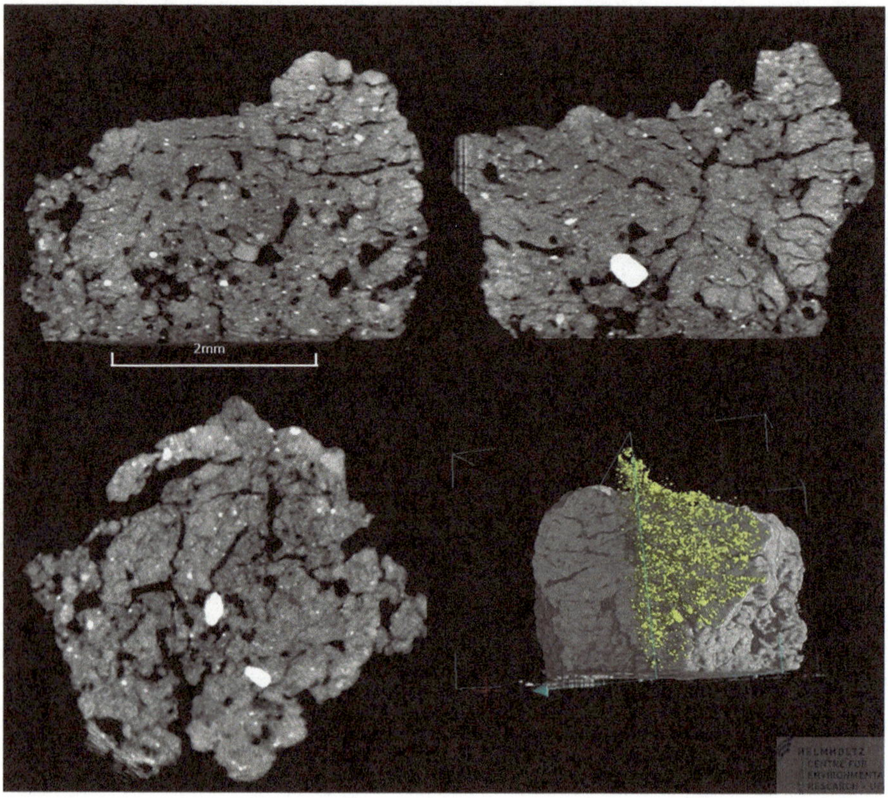

Fig. 2.7 CT volume of a small soil particle, only 3 mm across, demonstrating the spatial resolution of the CT system at high geometric magnifications with segmented pores shown in yellow (top right). [Image courtesy of the Helmholtz Centre for Environmental Research, Halle, Germany]

background noise, but they generally also reduce spatial resolution. The level of background noise can also limit the ability to distinguish small density changes. A test phantom with subtly different material densities and/or chemical composition can be used to test this discrimination. Always ask how long scans took to be collected and compare this will your expectations of sample throughput: what image quality can be achieved in an acceptable length of scan?

A heavy image contrast applied to the CT volume can be used to test how easily soil particles, pores, etc. may be distinguished from one another. Beam hardening artefacts will prevent all parts of the sample being processed with the same threshold and ring artefacts will connect separate particles. Such a heavy contrast will enhance any background noise in the data too. Always bear in mind your ultimate aim, generally for soils this will be to segment soil aggregates and particles (i.e. solid material) and measure the positions and sizes of the pores. The analysis software offered with a CT system may not be that which is best suited to these aims but check that the format the data is presented in can be used in superior software packages.

References

Katsevich, A., Basu, S., & Hsieh, J. (2004). Exact filtered backprojection reconstruction for dynamic pitch helical cone beam computed tomography. *Physics in Medicine & Biology, 49*, 14.

Parmesh, G. (2018). New software protocols for enabling laboratory based temporal CT. *Review of Scientific Instruments, 89*(9). https://doi.org/10.1063/1.5044393

Feldkamp, L. A., Davis, L. C., & Kress, J. W. (1984). Practical cone-beam algorithm. *Journal of the Optical Society of America A, 1*, 612–619.

Chapter 3
Soil Sampling and Preparation for X-ray Imaging

Fabio Terribile, Giuliano Langella, Florindo Antonio Mileti, Luciana Minieri, and Simona Vingiani

3.1 Introduction

Soil sampling is the first activity of most research performed in soil science. It is self-evident that badly designed soil sampling (e.g., location, size, soil structure disturbance) or subsampling can lead to erroneous results spoiling any subsequent analytical techniques. This is even more crucial considering the fragility of the soil pores network, which is a key property on which soil microtomography focuses. Thus, both soil sampling and soil preparation have to be considered key fundamental activities preliminary for any study in soil science.

This chapter is divided into four sections which provide an overview of: (a) the main steps to be followed for optimum sampling, (b) the definition of representative elementary volumes, (c) criteria for geospatial sampling, and (d) preparation of undisturbed samples before X-ray CT analysis.

Sampling is a critical part of all micromorphological studies and, thus, also of soil microtomography. Any mistake made in the sampling phase could affect subsequent interpretations (i.e., what is seen, processed, or analysed in the 3-D domain), so causing potentially erroneous conclusions about the soil material being studied. It is important that where possible the collected samples represent the soil material as it exists either in the field or in a laboratory (lab) experiment.

For lab experiments, sampling may generally involve the entire soil volume employed for a specific trial (e.g., small soil cylinders subjected to a specific lab treatment such as measurement of hydraulic conductivity or assessment of wetting and drying cycles), thus collecting representative soil material may not be an issue.

F. Terribile (✉) · G. Langella · F. A. Mileti · L. Minieri · S. Vingiani
Department of Agricultural Sciences and CRISP, University of Naples Federico II, Portici (NA), Italy
e-mail: fabio.terribile@unina.it; giuliano.langella@unina.it; florindoantonio.mileti@unina.it; luciana.minieri@unina.it; simona.vingiani@unina.it

S. Jon Mooney et al. (eds.), *X-ray Imaging of the Soil Porous Architecture*,
https://doi.org/10.1007/978-3-031-12176-0_3

This is the case when subsampling of a large soil column (e.g., for a given specific laboratory assessment, such as infiltration) and then a subsample is collected from this. This subsampling involves addressing the sample size issue (see section below) and the avoidance of sampling artefacts as they could occur at the outer margin of the sample.

In the case of field sampling (such as from experimental field trials), collecting representative samples is much more difficult than in laboratory experiments because soils in the field have an intrinsic heterogeneity and spatial variability.

Thus, special precautions must be taken to ensure collection of representative materials. The following questions must be carefully addressed before sample collection is undertaken: (i) What is the aim of the sampling? (ii) where to sample? (iii) what size dimensions should the sample have? (iv) what orientation of samples is needed? (v) how many samples are needed? (vi) when is the best time to sample? (vii) how to sample? and (viii) how to document, transport, and store samples?

3.2 The Main Steps for Successful Sampling

3.2.1 What is the Aim of the Sampling

The objective of the investigation is a consideration for soil sampling orientation in all microtomography studies. Samples usually consist of 3-D blocks, generally comprised of undisturbed soil aggregates. It is important to stress that there is no single technique that can be applied to all investigations. Therefore, the most suitable sampling technique must be selected to meet the requirements of a specific study and, in some cases, new sampling techniques are necessary.

Based on the aim of the study, sample volume, number of samples, and the choice of what to sample need to be defined. It must be decided in advance whether the target of the planned analyses is quantitative or descriptive, in fact X-ray tomography (which indeed produces quantitative X-ray attenuation data) could also be simply employed to quickly identify processes (e.g., in contaminated sites the mixing of heavy metal sludge into soils). Descriptive approaches require fewer sample replicates than quantitative studies to achieve their goal.

3.2.2 Where to Sample

Selection of both location and selection of the portion of the soil profile to be sampled is strictly connected with the objectives of the research. For the sake of clarity, here we separate the two sampling systems defined by the research objectives: (i) geospatial sampling design (e.g., how to sample topsoils for bulk density analysis considering the spatial variability such as that occurring in two areas (a & b)

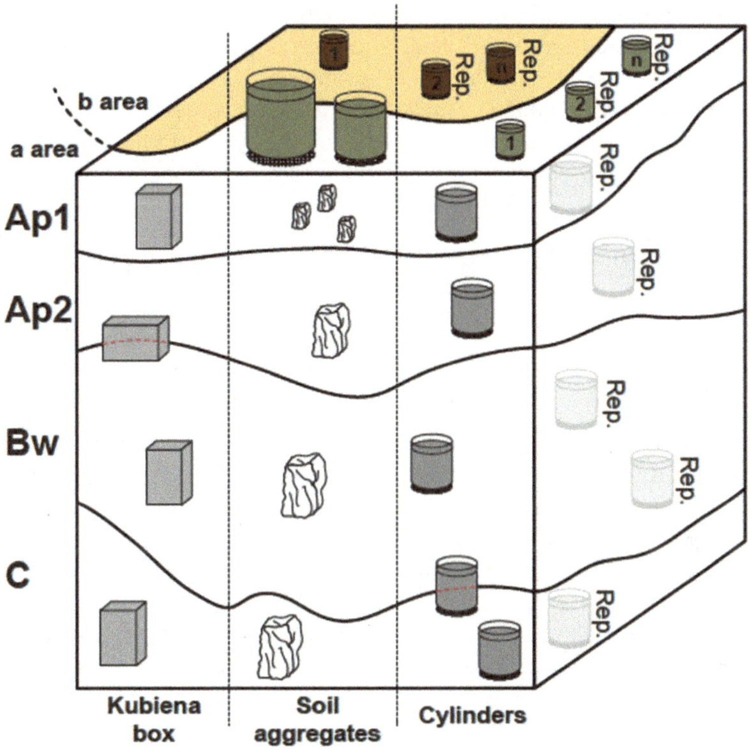

Fig. 3.1 General view about "where to sample" with respect to soil horizons (Ap, Bw, C). Abbr. Rep. = replicate

of Fig. 3.1 and (ii) how to sample a soil profile (e.g., sampling following soil horizons in the soil profile in Fig. 3.1).

Geospatial Sampling Design Over recent decades, geospatial sampling design has received a great deal of research interest (e.g., pedometrics, agronomy) and also in the case of sampling for the assessment of soil structure (e.g., Leopizzi et al., 2018) and other microtomographic (e.g., Carducci et al., 2017) analysis. Additional information will be provided in the following section on geospatial sampling.

Profile Sampling Design When a profile sampling design is applied, the soil should first be divided into soil horizons and the profile description must be as detailed and complete as possible. Samples are then taken from each horizon. If soil horizons are not well developed and expressed, samples may be spaced at regular depth intervals, while horizon boundaries and special or unique features might be sampled separately. Examples of recommended sample locations for various investigations are shown in Fig. 3.1.

For both geospatial and profile sampling design, it is good practice and thus strongly recommended that researchers always collect bulk samples for any extra laboratory analyses (e.g., particle size, clay mineralogy) that may be required to better understand tomographic results. Indeed, soil profile descriptions and lab analysis are extremely useful in order to support final evaluations of 3-D analysis.

3.2.3 Determining Sample Size

Here we define the "sample size" as the actual physical dimensions (e.g., cm^3) of the soil sample and not the number of individual samples (most often named "sample size" in statistics applied to soil studies). The required size of the sample depends on:

- the size of the specific feature of interest (e.g., pores, soil organic carbon, minerals) and their distribution. This is covered in more detail in the following section on "Representative Elementary Volume". Nevertheless, we stress that examining planar pores that occur between large prismatic peds/aggregates will require larger samples than examining pores within a crumb microstructure would.
- the sample stage of the available CT scanner is often a limiting factor. The sample size that can be accommodated might range from a few mm^3 for high-resolution acquisition systems to over 1 m^3 for industrial grade systems (e.g., GE V Tomex L).
- the required trade-off between 3-D sample size and preferred spatial resolution (e.g., du Plessis et al., 2017) which represents a powerful limiting factor in all CT studies.
- the size of the required image stack and associated image processing needs, and the need for digital subsampling for statistical analysis (e.g., digital pseudo replicate samples).

As additional observation, we wish to emphasize that larger 3-D samples (cm's rather than mm's) are often preferred to smaller ones because they support the observation of a wider range of features and how these features are interrelated. In some cases, large samples may be collected in the field and subdivided in the laboratory later in order to fit/be accommodated with the CT system being used. This is especially advisable when large aggregates are present and can be easily separated into smaller peds/aggregates.

3.2.4 How to Orientate the Sample

Although a soil block can be rotated in any direction using 3-D image processing software, it is self-evident that the orientation of the sample may be critical in some investigations, especially when sampling with cylinders is necessary (e.g., water saturation producing horizontal Fe segregation at a specific depth) or when considering specific depths within a soil profile. Therefore, it is usually important that the orientation of a sample is known. Generally, samples are collected either in the horizontal or vertical plane, while inclined samples may be required for specific purposes (e.g., studying pore–matrix interaction along a slickenside which is a sliding surface produced after wetting and drying cycles in a Vertisol). Vertically orientated samples are the most common in soil sampling procedures employed in soil microtomographic studies; they are especially useful to examine topsoil and specific soil horizons of soil profiles, bedding planes of sediments or vertical pore networks such as those created by roots or earthworms.

3.2.5 Number of Samples

The number of samples to be collected is a well-known issue in soil studies as it is very important and affects the reliability of the results obtained. Here we can consider two rather broad categories: (i) the required number of replicates in an experimental setup (e.g., soil structure studies after physical simulation of processes such as wetting and drying cycles) and (ii) the number of samples needed for a geospatial sampling design (e.g., soil structure studies of different land use mapping units). The latter of these two sampling designs will be considered in the following section on geospatial sampling. Substantial scientific literature already exists for both these considerations (e.g., Diel et al., 2019; Pöhlitz et al., 2019), but here we shall limit our investigation to a brief general overview of the subject.

With respect to the number of replicates to be acquired for a given experimental setup, the preliminary estimation of the number of replicates before sampling is necessary to ensure a given level of precision in the results. Indeed, the quantity of replicates is defined by considering the variability of the feature of interest in the soil. Therefore, it is recommended to perform preliminary tests before planning any specific experiment; these tests must enable the final decision about the required number of replicates for that specific study to be taken (e.g., Gargiulo, 2008; Gargiulo et al., 2015). However, such preliminary tests are generally not reported in scientific literature. Indeed, a short overview of the experiments involving analysis of pore networks by X-ray CT highlights the following variability in the number of replicates (although it is likely that scientists perform "look-see" scans in advance and do not report this in the literature):

- 2 replicates: (e.g., Valdez et al., 2019)
- 3 replicates: (e.g., Scotson et al., 2021; Menon et al., 2020; Ferreira et al., 2018; Gargiulo et al., 2016; Müller et al., 2018; Zhang et al., 2018)
- 4 replicates: (e.g., Singh et al., 2020)
- 5 replicates: (e.g., Diel et al., 2019; Pöhlitz et al., 2019).

However, it is fair to note that the replication used in soil CT studies has increased over time, most likely due to increased access and availability of the instruments.

There are many studies focused on soil processes (sometimes performed as physical simulations in laboratories) in which it is not evident whether replicates were present or not (e.g., Arai et al., 2019; Pires et al., 2020; Liu et al., 2021), although this is not recommended even for pilot studies. In general terms, replicate samples should be collected whenever possible because a single sample is unlikely to be representative of the variation in soil structure. Replicate samples must be collected in all studies requiring a statistical analysis of data.

From a pragmatic standpoint, the number of replicate samples is frequently connected to the total number of 3-D acquisitions that a researcher can perform given access to X-ray CT instrumentation. This number can often be limited. In this respect, Gregory et al. (2009) highlighted the difficulties in dealing with a large number of samples often required in soil science. They suggest that by CT it is usually only possible to examine less than 10 samples in 1 day in normal conditions, with a consequent reduction of the potential for replicated trials. Although the speed and availability of CT equipment has increased considerably with time, that number is still a typical number of daily samples imaged in many laboratories around the world. Since imaging has certainly become faster, higher quality images still take time and this is often prioritized rather than scanning more samples, because lower quality images usually lead to longer processing time. There is a wide literature focused on the variability of soil porosity (e.g., Nunan et al., 2006) and correlated properties (e.g., bulk density), while few studies have been conducted to explore the variability of other micromorphological features (as highlighted by Vanden Bygaart & Protz, 1999).

3.2.6 When to Sample

Many soil features, such as pore space, can change dynamically over the course of a year. Indeed, depending on the seasonal soil water status, large planar pores (e.g., cracks) open or close, soluble minerals crystallize or dissolve and organic substances can decompose. Accordingly, potential seasonal changes in the features of interest must be considered when the most appropriate time for sampling is selected. Another potential problem that can impact on sampling is the extent of soil dryness, since dry fine textured soils such as clays are difficult to dig and then artefacts, such as cracks, may form as the sampling container is forced into the soil. In such a situation, sampling should be postponed until the soil becomes wet enough to avoid the occurrence of cracks during sampling, although in some cases introducing a small amount of water to the surface in the field can facilitate sampling. Evaluating when a soil is at an optimal moisture content for sampling is subjective, requires some testing in the field and it is fundamental to achieve good sampling especially in clayey soils. Quantitative evaluation of the optimal moisture content to avoid the above disturbance would usually require a preliminary triaxial test (a standard test in soil mechanics/engineering studies but rarely performed in this context by soil scientists).

3.2.7 How to Sample

The majority of CT applications require the use of undisturbed soil samples. The most common sampling methods are those using soil cores, metal or cardboard boxes or single aggregates (see Fig. 3.1 right, left and central part of the soil profile). Cores are used in the vast majority of sampling methods employed in experimental setups particularly when topsoil has to be sampled but single aggregates can also be sampled. In addition, other methodologies are used because of the specific consistency of the soil. Methods for the sampling of friable soils, cemented and cohesive materials, loose materials are discussed below.

Friable Soils Most of these soils can be sampled by using plastic (e.g., PVC) cylinders with a cutting-edge side. These cylinders can be made at very low cost, can be manufactured easily to many different sizes (e.g., by cutting plastic PVC tubes), and allow reduced sampling disturbance in comparison with the traditional micromorphological boxes (e.g., Kubiëna tins). Cylinders are also the most common type of sample holder in soil hydrology and soil physics studies.

However, truly undisturbed soil and sediment sampling do not exist. Using X-ray CT, Carr et al. (2020) demonstrated artefacts in the pore system in all core samples due to pushing, cutting and hammering actions during sampling procedures. Any combination of hammer, rotation, percussive, and continuous push led to deformations, although the advanced trimming methods (Hvorslev, 1949) resulted in the least disturbance. This latter method refers to the slow insertion of a cylinder core vertically into the soil with simultaneous removal of the surrounding soil to minimize mechanical stress (Kemp, 1985; Carr, 2004), as should be done with a Kubiëna box (FitzPatrick, 1984; Stoops, 2009). Indeed, this demonstrates the value of the standard sampling approach, based on excavation of surrounding material as is widely used in soil micromorphological studies (FitzPatrick, 1993). Samples should ideally have a length and width that are suited to the X-ray CT scanner sample stage. The depth of the sample should be at least one-half its width to prevent its bending when filled (FitzPatrick, 1984; Murphy, 1986).

Another method, which is especially suitable for structured soils, is the sampling of soil aggregates (Fig. 3.1). This method (e.g., Peth et al., 2014) ensures that undisturbed samples are used when the element of research interest falls within the single aggregate (e.g., intra-aggregate pore space such as in Peth et al., 2008). After their collection, soil aggregates should be wrapped in protective material such as paper or plastic masking tape to minimize any disturbance damage.

Samples should be labelled with the i) orientation, ii) depth, iii) replicate number, and iv) profile number. Some researchers label the boxes on the lid first and photograph them in place before removing them from the soil, so a complete record of sample location and orientation is made (Murphy, 1986). Prior to sealing samples for transport, a few drops of fungicide or formalin can be added to each sample to prevent soil fauna burrowing within the sample during storage. All samples should then be wrapped in aluminium foil or placed in plastic bags to prevent them from either drying out or absorbing water while stored in the laboratory which can be a problem for air-dried materials, even from arid regions.

Cemented and Cohesive Materials Undisturbed specimens of cemented materials (e.g., Bkm horizons) can be obtained without sample containers by cutting or carving pieces out of the soil. However, these should then be placed in boxes to prevent breakage or separation of the structural units. All such samples, as in the case of soil aggregates, should be wrapped in protective material such as foil or plastic and labelled as described previously.

Loose Materials Some materials may be too loosely structured to be sampled without disturbance during removal from the soil. Particularly sensitive samples include sands, gravelly soils, self-mulching soil, some organic soils and recently tilled soils. These loose materials can be partially hardened in place prior to their being removed from the soil. Murphy (1986) gives step-by-step instructions for such procedures. Basically, the sample is first surrounded by a frame while it is still in the soil. The sample is then covered and partially impregnated with a cementing agent such as epoxy resin, cellulose acetate, plaster of Paris, sodium silicate, polyester resin, or a similar material. After the cementing material has hardened, the sample can be removed, labelled, and wrapped as previously described. This allows the sample to be dried and, if required, to be impregnated in the laboratory prior to scanning. Epoxy resin and plaster of Paris can be used to impregnate materials containing numerous large pores such as tilled Ap horizons (FitzPatrick 1984).

3.2.8 How to Document, Transport, and Store Samples

Prior to removal from a sampling site, all samples should be checked to make sure they are appropriately labelled and bagged. The profile description should be reviewed for any omissions. Site data should be recorded including standard GPS coordinates, landscape morphology, land use, etc., then photos (with additional sketches) of the profile or site can be made, and where possible they must show sample location (e.g., reference to specific horizons).

Before transport, samples should be placed in boxes and packed with suitable material to prevent them from coming into contact with each other and being damaged. Vibration of the samples should also be minimized using packing foam to reduce the risk of disturbance especially for friable samples. After arriving in the laboratory, samples should be stored in a cool, dry place but not frozen (as this may cause cracks). A temperature of about 4 °C or lower may be needed to retard the growth of plants and organisms. This is also especially important if reducing the disturbance of organic materials is required.

3.3 Representative Elementary Volume

The concept of representative elementary volume (REV) was introduced by Bear (1972) to describe flow in porous media. The approach deals with the definition of the minimum size of a sample necessary to present its characteristics of interest. In other words, the size at which the measured parameter (or property) becomes independent of the sample size. Its use in the quantification of soil structure, as REA (Representative Elementary Area), started with Vanden Bygaart and Protz (1999). The analysis of REV is commonly carried out by selecting consecutive soil volumes around a central point in the sample image (see Fig. 3.2). This is indeed feasible in the case of X-ray CT where adjacent constructions within the same 3-D image, but centred at different points, can be used.

The representative size is then defined as the one corresponding to the transition from the detailed scale where the "microscopic effects domain" prevails (red signal in Fig. 3.3) to the more stable "porous media domain" (green line in Fig. 3.3) (Baveye et al., 2002; Borges & Pires, 2012). Thus, the REV size established for a

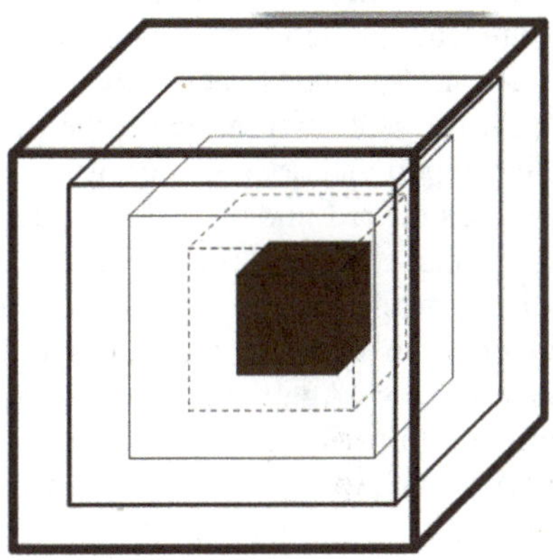

Fig. 3.2 Visual Representative Elementary Volume (REV) concept. A set of consecutive volumes where to measure a soil parameter to establish REV

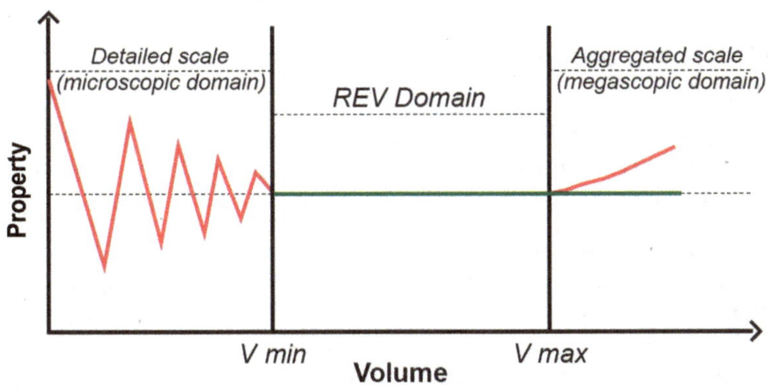

Fig. 3.3 REV conceptualization (inspired after Bear 1972)

specific property (Costanza-Robinson et al., 2011) corresponds to the smallest employed soil volume (the point named "V min" with the smallest volume on the green line in Fig. 3.3) to obtain representative measurements of the selected soil property.

Of course, REV varies according to the specific property under investigation, but also on the basis of the type of soil material. Indeed, even limiting our analysis just to the soil pores, it is evident that each porous media has its own characteristics (e.g., continuous pore space, channels, discrete pore within aggregates); therefore, the REV of a pore space parameter calculated for a specific soil may differ for the same parameter when a different soil type is analysed.

Limited research is available on the use of a REV in soils (Baveye et al., 2002; Vanden Bygaart & Protz, 1999) and, due to mild fluctuations, it is sometimes difficult to understand when it has reached the green line region (Fig. 3.3). In 1999, Vanden Bygaart and Protz (1999) highlighted a lack of suitable techniques for the study of representative size of undisturbed soil samples. But more recently, Borges and Pires (2012) analysed, respectively, the bulk density REV (50-100 cm^3) and soil porosity distribution in a representative element area (REA) (8.8 cm^2) for Brazilian soils, while Baveye et al. (2002), by analysing volumetric water content, volumetric air content, gravimetric water content, and dry bulk density of Ck horizons, demonstrated that the sizes of sampling volumes up to $60 \times 60 \times 30$ mm influence the measured values of soil parameters significantly. They report that dry bulk density did not reach the green REV region and may move towards the red line of the aggregate scale as reported in Fig. 3.3.

In this scenario, it is not feasible to have "magic" REV numbers for a specific soil feature, while it is most appropriate to suggest some preliminary tests to researchers in order to evaluate the required REA or REV for the specific feature under investigation. Again, these preliminary tests are very highly recommended, but not difficult to perform. For instance, in Fig. 3.4, an example is reported in which the REV of an Ap horizon was defined before the experiment was performed. In this case, it is evident that 50 cm^3 referred to the REV to be employed.

3.4 Geospatial Sampling

Geospatial sampling can be an important concern when samples for X-ray CT analysis are collected from the field. In general terms, we can treat this item in the following cases (moving from very specific to general cases):

(i) A degraded contaminated site that requires remediation. In many cases contaminated soils show also signs of soil physical degradation. This may be related to soil compression and reworking by heavy machines. In these cases, it is important to combine the standard geochemical soil characterization with analysis of soil porosity. In fact, a detailed analysis of the porous system (e.g., pore size distribution and shape of pores) may help in addressing the most

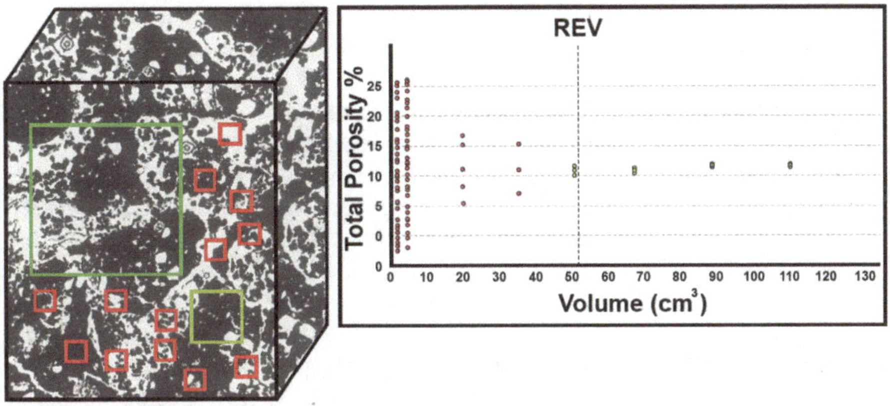

Fig. 3.4 Experimental setting to establish REV after X-ray microtomography. Here increasing soil volumes (x axis) are analysed and plotted against their corresponding total porosity. The vertical broken line identifies the experimental obtained REV (around 50 cm^3)

sustainable approach to soil remediation for each specific case. More details are reported below.

(ii) Experimental field trials. This is the standard case when X-ray CT contributes to other agronomic measures (e.g., manure application, tillage practices, etc.), including cases when measurements of soil-pore parameters are influenced by crop rotations and cover crops (e.g., Singh et al., 2020).

(iii) Standard landscape analysis. This is the case when soil survey may benefit from X-ray CT analysis to evaluate specific issues such as soil compaction.

(i) contaminated site showing also signs of physical degradation,

In contaminated sites, soil spatial variability is a key concern since it can combine to create a twofold spatial variability due to the two main sources of the contamination, natural and anthropogenic. In further detail, the spatial distribution of anthropogenic soil contamination is almost always unknown and ex-ante information is generally lacking or completely missing. In these cases, the use of proximal soil sensing devices can assist. In fact, the most widespread proximal sensing devices such as Electro Magnetic Induction (EMI) devices or Electrical Resistivity Tomography (ERT) devices are strongly affected by the soil bulk density (in addition to other soil properties such as water content, salt content, etc.). Therefore, the incoming geophysical mapping results (generally obtained by a quick scan of the site) can be profitably employed as a powerful means to better address soil sampling for the analysis of the soil porous system.

Two interesting cases, that can be profitably applied to X-ray CT, are reported by Langella et al. (2018) and Vingiani et al. (2022). These authors using different geophysical EMI and ERT sensors (Profiler, DUAL-EM and ARP) and γ-ray spectrometry recognized spatial structures and zones with different levels of anomalies, in agricultural and industrial sites. High consistency was found among the maps obtained by the different sensors. More specifically data was merged to report all the information on a regular grid at very high-resolution (50 cm at the field scale). This fusion of the data collected from the different sensors was transformed to the same raster-based geometry. This procedure produced a map of the areas in which to collect soil samples and account for the spatial variability of the target variable and highlight possible hot spots. The EMI, ERT, and γ-rays sensors approach was validated by XRF measurements of soil elemental composition by the use of a field portable technique.

In contaminated sites, soil sampling may require policy compliance: this refers to those cases where sampling must follow specific regulations and some specific sampling rules are prescribed or strongly suggested. This, for instance, is the case for soil contamination in some countries where a specific number of samples/ha are required (e.g., Italian D.M. 471/99) on the basis of the global spatial extent of the site (e.g., at least 5 sampling points for areas <1 ha).

(ii) *Experimental field trials.* This is the standard approach employed when X-ray CT contributes to an agronomic assessment. A comprehensive description is given in Gilbert (1987), de Gruijter, (2002) and Pennock (2004). Although many types of sampling designs exist, only two main types (random and systematic) are commonly used: i) Simple Random and Stratified Random Sampling, where in Stratified Sampling points refer to predefined strata and a simple random sample chosen from each stratum and ii) Systematic Sampling which refers to either transects or most commonly to grids. Where constant spacing is applied, the major caution has to be that the soil property to be sampled must not be arranged in an orderly manner which might correspond to the spacing of the grid. Bonfante et al. (2017) highlight that the implicit assumption of soil homogeneity of the agronomic design experiments may lead to erroneous results especially when subsoil condition matter. In this case, special additional care is required in identifying the location of the different soil types before performing appropriate sampling.

(iii) *Standard landscape analysis.* This can be performed by:

a. Sampling using deterministic/judgement rules. This is the classical approach applied for pedology, geomorphology, geology, etc. surveys. In the case of soils, the approach relies on the evidence that soils (S) are determined by specific soil forming factors (Jenny, 1941) following the classic Jenny eq. S = f (climate, organism, relief, parent material, time), thus data on these factors may lead to soil knowledge. In view of the factual soil complexity in real landscapes, this approach has exhibited evident limitation. Therefore, in the last three decades extensive research has developed an entirely new branch known as Digital Soil Mapping. Despite the above, currently, judgement-based rule approaches are very important. In this case, researchers use their judgement to locate sampling points in landscape positions where the soil representative of a specific land is most likely found. For instance, this type of sampling is widely used by private-sector environmental consultants and the specific objective may range from an initial evaluation of the extent of the problem to be addressed (e.g., contamination) to the final stage of problem solving through a mapping exercise. Laslett (1997) stated that consultants who undertake these surveys almost always employ judgement sampling and place their sampling points where their experience and prior knowledge of the site history suggest the contamination might be located.

b. Sampling using Geostatistical, Spectral, and Wavelet Analysis. These types of sampling have become a "must" in sound based scientific sampling over the last few decades. Details are reported in Brus and de Gruijter (1997), de Gruijter (2002), Webster and Oliver (1990), McBratney et al. (2002), Mulla and McBratney (2000). All these approaches—which fall under the broad umbrella of Digital Soil Mapping approaches—address the spatial dependence in soil properties between locations. Thus, the location of each sample point in space is critical information.

3.5 Sample Preparation

3.5.1 Preliminary Laboratory Analysis

Different types (e.g., size and shape) of samples can arrive in the laboratory for X-ray imaging such as cylinders, Kubiëna boxes, aggregates, and bulk samples (Fig. 3.1). In all cases, the start point of any investigation should be a visual assessment. This can be done by direct stereo-microscopic analysis of the soil aggregates or, when this is not feasible (e.g., cylinders and Kubiëna boxes), by analysing the corresponding bulk samples. In this case, the soil has to be observed directly with oblique incident light to identify microstructure, macropores, Fe-Mn features (such as concretions, segregation, pyrite), and $CaCO_3$ features; indeed, all these features may influence X-ray absorbance and can usually give some important insights into the sample. At this point, samples (Fig. 3.5) would usually either/both go directly for X-ray scanning without any treatment or/and undergo specific treatments.

No treatment: It is worth noting that one important advantage of X-ray CT is the opportunity to acquire 3-D soil blocks directly without any prior preparation (including impregnation) which accelerates analysis of the soil-pore system.

Treatment: even considering the advantages of no treatment, there are two major cases where pre-treatments are required:

(i) *Staining specific features to separate different soil features better during (and after) 3-D scan acquisition.* Even though the soil-pore system analysis is often the main focus of X-ray CT imaging, there is a considerable interest in analysing the pore system in combination with other soil features such as water and organic matter. Iltis et al. (2011) employed potassium iodide (KI) as a contrast-enhancing agent enabling a better segmentation of structure

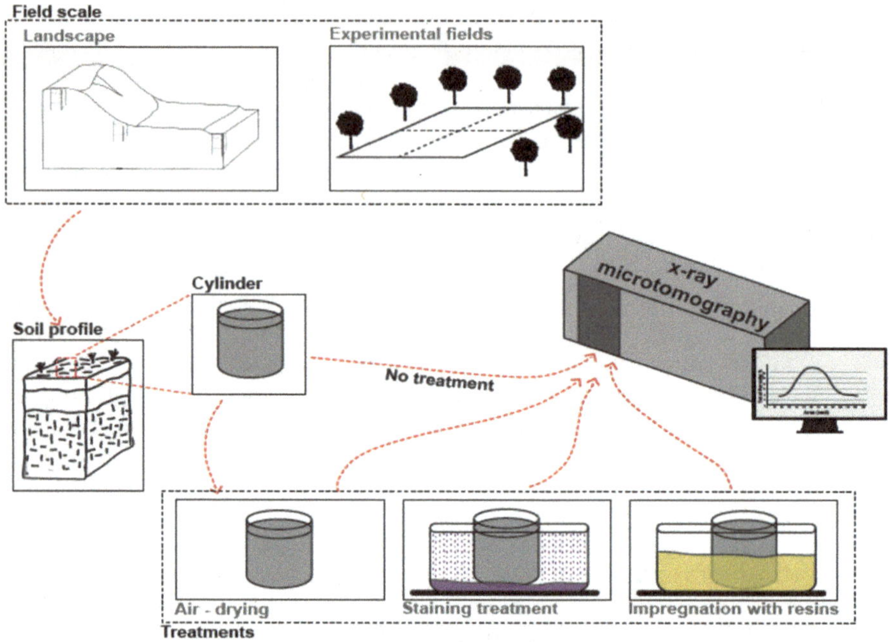

Fig. 3.5 Synoptic figure showing (i) location of sampling (either free survey in the landscape or experimental fields), soil profile sampling, different soil samples treatments before X-ray CT analysis

and porous media. Furthermore, under specific conditions, many microtomography studies also showed the ability to separate soil, air, and water (an overview is given in Tracy et al. (2015) and Mao et al. (2016)) while the situation is more complex for other soil features such as minerals (Guntoro et al., 2019) and soil organic matter (SOM) (Mao et al., 2016). The main problem lies with the evidence that, although the physical and chemical properties of the SOM are very different from those of the bulk soil (Kasteel et al., 2013), it can be very difficult to differentiate between these properties in the soil complexity due to the similarity of their X-ray attenuation coefficients (Quinton et al., 2009). The huge interest in separating the SOM phase from the bulk soil has pushed soil scientists to invest a great effort to achieve such separation by using either or both pre-treatments and image processing techniques (see Chap. 10). X-ray CT can be used, under specific conditions, to visualize large organic fragments (e.g., Particulate Organic Matter—POM) and plant roots (Kravchenko et al., 2015; Van Loo et al., 2014), but the visualization of organic matter that is well dispersed in the soil matrix (e.g., nonparticulate SOM) is very difficult (e.g., Mueller et al., 2013) because of the well-known difficulties in segmenting an image where SOM and the rest of the soils have similar ranges of varying linear attenuation coefficients. An advantageous approach to simplify image segmentation is to treat the sample with heavy elements. Indeed, soil organic matter has a marked affinity for heavy elements (with $Z > 30$)—due to their charge, ionic radius, and ionic potential and, in turn, heavy elements have an X-ray attenuation that increases with their atomic number and, thus, improves imaging contrast (Van Loo et al., 2014). This is covered in greater detail in Chap. 10.

The most well-known heavy elements for soil "staining" (here we use the word staining even though it originally referred to optical microscopy) are the following:

- Silver (Ag): (Van Loo et al., 2014) which is the most well-known staining system.
- Lead (Pb): Kettridge and Binley (2008) successfully used lead (II) nitrate solution to increase the linear attenuation of peat by flushing it over a sample before X-ray scanning. However, we should also emphasize the highly carcinogenic nature of lead (II) nitrate.
- Osmium (Os) tetroxide (Rawlins et al., 2016). In 2014, Peth et al. (2014) proposed a new method of in situ SOM visualization which could be implemented on intact soil samples up to a few centimetres in size. This method is based on the ability of osmium tetroxide, OsO4, to react with organic substances, in particular, lipids. An air-dry soil is subjected to OsO4 vapours, which, upon diffusion into the soil, bind with organic materials and stain them. Mao et al. (2016) found that the method of Os (OsO4 vapours) staining (analysed via X-ray Dual-Energy Tomography) was effective in staining organic materials of root origin and the organics associated with fine soil particles, but not biochar. Arai et al. (2019) demonstrated effective SOM localization within large macroaggregate based on synchrotron X-ray micro-computed tomography (μCT) coupled with a vapour-phase, osmium (Os)-staining pre-treatment. Unfortunately, osmium (Os) tetroxide is also toxic (Van Loo et al., 2014) and special care when handling is required.
- Iodine (I) (Boyde et al., 2014). This approach is largely used in combination with hydrological testing (Scotson et al., 2021). See Chap. 7 for more details.
- Eosin (Br based) (Lammel et al., 2019).

Several other heavy elements such as Cu, Co, Fe, Mn, Mo, and Zn may be of interest in binding SOM, but, since they already occur naturally in soil, their use must be evaluated in each specific soil case. Lammel et al., (2019) tested many different contrast enhancing agents by using synchrotron radiation microcomputerized tomography (SR-μCT) and demonstrated that I_2 was the most efficient method as it was able to improve the image contrast, so providing a powerful tool to determine the spatial location of SOM.

(ii) *Water removal:* In this context the aim is to achieve standardization of the 3-D acquisition and an improved soil-pore segmentation. Moist and wet samples can affect the quality of X-ray acquisition images because air, water, and soil phases have different linear X-ray attenuation coefficients and, therefore, soils with heterogeneous humidity may produce images that are more difficult to segment due to these variations. Water removal can be carried out by (a) air- or oven- drying of samples, and (b) alternative dehydration methods from soil micromorphology (e.g., acetone dehydration). The method used depends on the purpose of the survey and on the characteristics of the soil sample. Air-drying is a rather simple method which may take from 5 to 8 weeks (depending on the soil properties) and may still require the use of an oven. The most popular method is to oven-dry samples at 40 °C for just a few days. However, in soils characterized by the presence of expandable minerals (e.g., smectites) and/or by high organic matter content, the treatment could modify the soil structure. In such a case, standard soil micromorphology approaches, including dehydration by saturating the sample with an acetone-water mixture, become advantageous. This involves the stepwise substitution of ever-increasing concentrations of acetone (Murphy, 1986; FitzPatrick, 1993), in order to remove the water with the help of zeolite.

(iii) *Water removal and resin impregnation:* This treatment permits to submit the same sample to both X-ray CT scanning and thin section analysis. Once the water has been removed, the sample can be impregnated with different mixtures of unsaturated polyester synthetic resins or epoxies depending on how the sample was dried. For further details, see FitzPatrick (1984), Jongerius and Heitzberger (1975), and Murphy (1986). This approach is rather unusual when dealing with microtomography, but it can be very profitable both as a support for 3-D scan interpretation and analysis. Indeed, in this case, after the analysis of a 3-D scan, which provides grey levels on the basis of the X-ray attenuation coefficient, the researcher, through a standard optical microscopy analysis or through chemical microanalysis (e.g., EDS-SEM) of a soil thin section, can access the power of a much larger spectral content (e.g., visible, UV, X-ray chemical mapping) than provided by the 3-D CT scan. This combination of approaches can be very rewarding as demonstrated by Hapca et al. (2011 and 2015) and Bendle et al. (2015).

3.6 Conclusions

In this chapter, the crucial value of soil sampling and sample preparation in X-ray CT studies has been addressed. Indeed, badly designed soil sampling or poor sample preparation can lead to erroneous results spoiling any subsequent analytical techniques. Actually, the more complex and advanced applied X-ray techniques (e.g., synchrotron based) may result in very small samples which are even more affected by the above problems. Therefore, also the concepts of representative elementary area (REA) and representative elementary volume (REV) are vital, due to the importance of knowing the size at which the measured parameters (or properties) becomes independent from the sample size, and then when representative measurements are really obtained.

However, nowadays in which there is an increasing possibility to access a wide range of equipment, we also highlight the potential to integrate X-ray microtomography analysis with standard 2-D analysis (including standard optical microscopy via thin sections) and visual soil sample analysis in order to obtain a deeper knowledge on any specific soil sample, to then achieve stronger basis for upscaling the results.

References

Arai, M., Uramoto, G. I., Asano, M., Uematsu, K., Uesugi, K., Takeuchi, A., Morono, Y., & Wagai, R. (2019). An improved method to identify osmium-stained organic matter within soil aggregate structure by electron microscopy and synchrotron X-ray micro-computed tomography. *Soil and Tillage Research, 191*, 275–281. https://doi.org/10.1016/j.still.2019.04.010

Baveye, P., Rogasik, H., Wendroth, O., Onasch, I., & Crawford, J. W. (2002). Effect of sam- pling volume on the measurement of soil physical properties: Simulation with x-ray tomography data. *Measurement Science and Technology, 13*, 775–784.

Bear, J. (1972). *Dynamics of fluids in porous media*. Dover Publications.

Bendle, J. M., Palmer, A. P., & Carr, S. J. (2015). A comparison of X-ray CT and thin section analysis of Lateglacial glaciolacustrine varves from Glen Roy, Scotland. *Quaternary Science Review, 114*, 61–77. https://doi.org/10.1016/j.quascirev.2015.02.008

Bonfante, A., Sellami, M. H., Abi Saab, M. T., Albrizio, R., Basile, A., Fahed, S., Giorio, P., Langella, G., Monaco, E., & Bouma, J. (2017). The role of soils in the analysis of potential agricultural production: A case study in Lebanon. *Agricultural Systems, 156*, 67–75. https://doi.org/10.1016/j.agsy.2017.05.018

Borges, J. A. R., & Pires, L. F. (2012). Representative elementary area (REA) in soil bulk density measurements through gamma ray computed tomography. *Soil and Tillage Research, 123*, 43–49. https://doi.org/10.1016/j.still.2012.03.008

Boyde, A., Mccorkell, F. A., Taylor, G. K., Bomphrey, R. J., & Doube, M. (2014). Iodine vapor staining for atomic number contrast in backscattered electron and X-ray imaging: Iodine vapor staining for atomic number contrast. *Microscopy Research and Technique, 77*, 1044–1051. https://doi.org/10.1002/jemt.22435

Brus, D. J., & de Gruijter, J. J. (1997). Random sampling or geostatistical modelling? Choosing between design-based and model-based sampling strategies for soil (with discussion). *Geoderma, 80*, 1–44.

Carducci, C. E., Zinn, Y. L., Rossoni, D. F., Heck, R. J., & Oliveira, G. C. (2017). Visual analysis and X-ray computed tomography for assessing the spatial variability of soil structure in a cultivated Oxisol. *Soil and Tillage Research, 173*, 15–23.

Carr, S. J. (2004). Micro-scale features and structures. In D. J. A. Evans & D. I. Benn (Eds.), *A practical guide to the study of glacial sediments* (pp. 115–144). Arnold.

Carr, S. J., Diggens, L. M., & Spencer, K. L. (2020). There is no such thing as 'undisturbed' soil and sediment sampling: sampler-induced deformation of salt marsh sediments revealed by 3D X-ray computed tomography. *Journal of Soils and Sediments, 20*, 2960–2976.

Costanza-Robinson, M. S., Estabrook, B. D., & Fouhey, D. F. (2011). Representative elementary volume estimation for porosity, moisture saturation, and air-water interfacial areas in unsaturated porous media: Data quality implications. *Water Resources Research, 47*, 1–12. https://doi.org/10.1029/2010WR009655

de Gruijter, J. J. (2002). Sampling. In J. H. Dane & G. C. Topp (Eds.), *Methods of soil analysis, part 4—Physical methods* (pp. 45–79). Soil Science Society of America.

Diel, J., Vogel, H. J., & Schlüter, S. (2019). Impact of wetting and drying cycles on soil structure dynamics. *Geoderma, 345*, 63–71. https://doi.org/10.1016/j.geoderma.2019.03.018

du Plessis, A., Broeckhoven, C., Guelpa, A., & le Roux, G. S. (2017). Laboratory x-ray micro-computed tomography: A user guideline for biological samples. *Gigascience, 6*, 1–11.

Ferreira, T. R., Pires, L. F., Wildenschild, D., Heck, R. J., & Antonino, A. C. D. (2018). X-ray microtomography analysis of lime application effects on soil porous system. *Geoderma, 324*, 119–130. https://doi.org/10.1016/j.geoderma.2018.03.015

FitzPatrick, E. A. (1984). *Micromorphology of soils* (p. 433). Chapman and Hall.

Fitzpatrick, E. A. (1993). *Soil microscopy and micromorphology*. Wiley.

Gargiulo, L. (2008). *Indagini innovative sulla strutturazione del suolo. Graduation thesis in Applied Pedology*. University Federico II of Naples.

Gargiulo, L., Mele, G., & Terribile, F. (2015). The role of rock fragments in crack and soil structure development: A laboratory experiment with a vertisol. *European Journal of Soil Science, 66*. https://doi.org/10.1111/ejss.12263

Gargiulo, L., Mele, G., & Terribile, F. (2016). Effect of rock fragments on soil porosity: A laboratory experiment with two physically degraded soils. *European Journal of Soil Science, 67*. https://doi.org/10.1111/ejss.12370

Gilbert, R. O. (1987). *Statistical methods for environmental pollution monitoring* (p. 320). Van Nostrand Reinhold.

Gregory, P. J., Bengough, A. G., Grinev, D., Schmidt, S., Thomas, W. T. B., Wojciechowski, T., et al. (2009). Root phenomics of crops: Opportunities and challenges. *Functional Plant Biology, 36*, 922–929.

Guntoro, P. I., Ghorbani, Y., Koch, P. H., & Rosenkranz, J. (2019). X-ray microcomputed tomography (X-ray CT) for mineral characterization: A review of data analysis methods. *Minerals, 9*, 20–26. https://doi.org/10.3390/min9030183

Hapca, S., Wan, Z. X., Ottena, W., Wilson, C., & Baveye, P. C. (2011). Automated statistical method to align 2-D chemical maps with 3-D X-ray computed micro-tomographic images of soils. *Geoderma, 164*, Issues 3–4.

Hapca, S., Baveye, P. C., Wilson, C., Murray, L. R., & Otten, W. (2015). Three-dimensional mapping of soil chemical characteristics at micrometric scale by combining 2-D SEM-EDX data and 3-D X-ray CT images. *PLoS One, 15*, 2015. https://doi.org/10.1371/journal.pone.0137205

Hvorslev, M. J. (1949). *Subsurface exploration and sampling of soils for civil engineering purposes*. Waterways Experiment Station.

Iltis, G. C., Armstrong, R. T., Jansik, D. P., Wood, B. D., & Wildenschild, D. (2011). Imaging biofilm architecture within porous media using synchrotron-based X-ray computed microtomography. *Water Resources Research, 47*, 1–5.

Jenny, H. (1941). *Factors of soil formation, a system of quantitative Pedology*. McGraw-Hill.

Jongerius, A., & Heintzberger, G. (1975). Methods in soil micromorphology: a technique for the preparation of large thin sections. (Soil survey papers/Soil Survey Institute; No. 10). Stichting voor Bodemkartering Wageningen. https://edepot.wur.nl/304516

Kasteel, R., Schnitzler, F., Berns, A. E., Vanderborght, J., & Vereecken, H. (2013). Visualization of transport path- ways for organic compounds in undisturbed soil monoliths. *Geoderma, 195–196*, 70–78.

Kemp, R. A. (1985). *Soil micromorphology and the quaternary* (Vol. 2). QRA technical guide.

Kettridge, N., & Binley, A. (2008). X-ray computed tomography of peat soils: Measuring gas content and peat structure. *Hydrological Processes, 22*, 4827–4837.

Kravchenko, A. N., Negassa, W. C., Guber, A. K., & Rivers, M. L. (2015). Protection of soil carbon within macro-aggregates depends on intra- aggregate pore characteristics. *Scientific Reports, 5*. https://doi.org/10.1038/srep16261

Lammel, D. R., Arlt, T., Manke, I., & Rillig, M. C. (2019). Testing contrast agents to improve micro computerized tomography (μCT) for spatial location of organic matter and biological material in soil. *Frontiers in Environmental Science, 7*, 1–10. https://doi.org/10.3389/fenvs.2019.00153

Langella, G., Agrillo, A., Basile, A., De Mascellis, R., Manna, P., Moretti, P., Mileti, F. A., Terribile, F., & Vingiani, S. (2018). Geography of soil contamination for characterization and precision remediation of potentially contaminated sites. *Italian Journal of Agronomy, 13*, 6–15.

Laslett, G. M. (1997). Discussion of the paper by D. J. Brus and J. J. de Gruijter. *Geoderma, 80*, 45–59.

Leopizzi, S., Gondret, K., & Boivin, P. (2018). Spatial variability and sampling requirements of the visual evaluation of soil structure in cropped fields. *Geoderma, 314*, 58–62.

Liu, B., Ma, R., & Fan, H. (2021). Evaluation of the impact of freeze-thaw cycles on pore structure characteristics of black soil using X-ray computed tomography. *Soil and Tillage Research, 206*, 104810. https://doi.org/10.1016/j.still.2020.104810

Mao, H., Kumi, F., Li, Q., & Han, L. (2016). Combining X-ray computed tomography with relevant techniques for analyzing soil–root dynamics–an overview. *Acta Agriculturae Scandinavica, Section B–Soil & Plant Science, 66*, 1–19. https://doi.org/10.1080/09064710.2015.1067711

McBratney, A. B., Anderson, A. N., Lark, R. M., & Odeh, I. O. (2002). Newer application techniques. In J. H. Dane & G. C. Topp (Eds.), *Methods of soil analysis, part 4–Physical methods* (pp. 159–200). Soil Science Society of America.

Menon, M., Mawodza, T., Rabbani, A., Blaud, A., Lair, G. J., Babaei, M., Kercheva, M., Rousseva, S., & Banwart, S. (2020). Pore system characteristics of soil aggregates and their relevance to aggregate stability. *Geoderma, 366*, 114259. https://doi.org/10.1016/j.geoderma.2020.114259

Mueller, C. W., Weber, P. K., Kilburn, M. R., Hoeschen, C., Kleber, M., & Pett-Ridge, J. (2013). Advances in the analysis of biogeochemical interfaces: NanoSIMS to investigate soil microenvironments. In *Advances in agronomy* (Vol. 121, pp. 1–46). Elsevier.

Mulla, D. J., & McBratney, A. B. (2000). Soil spatial variability. In M. E. Sumner (Ed.), *Handbook of soil science* (pp. A321–A352). CRC Press.

Müller, K., Katuwal, S., Young, I., McLeod, M., Moldrup, P., de Jonge, L. W., & Clothier, B. (2018). Characterising and linking X-ray CT derived macroporosity parameters to infiltration in soils with contrasting structures. *Geoderma, 313*, 82–91. https://doi.org/10.1016/j.geoderma.2017.10.020

Murphy, C. P. (1986). *Thin section preparation of soils and sediments*. AB Academic.

Nunan, N., Ritz, K., Rivers, M., Feeney, D. S., & Young, I. M. (2006). Investigating microbial micro-habitat structure using X-ray computed tomography. *Geoderma, 133*, 398–407.

Pennock, D. J. (2004). Designing field studies in soil science. *Canadian Journal of Soil Science, 84*(1), 1–10.

Peth, S., Chenu, C., Leblond, N., Mordhorst, A., Garnier, P., Nunan, N., Pot, V., Ogurreck, M., & Beckmann, F. (2014). Localization of soil organic matter in soil aggregates using synchrotron-based X-ray microtomography. *Soil Biology and Biochemistry, 78*, 189–194.

Peth, S., Horn, R., & Smucker, A. J. M. (2008). Three-dimensional quantification of intra-aggregate pore-space features using synchrotron-radiation-based microtomography. *Soil Science Society of America Journal, 72*, 897–907.

Pires, L. F., Auler, A. C., Roque, W. L., & Mooney, S. J. (2020). X-ray microtomography analysis of soil pore structure dynamics under wetting and drying cycles. *Geoderma, 362*, 114103. https://doi.org/10.1016/j.geoderma.2019.114103

Pöhlitz, J., Rücknagel, J., Schlüter, S., Vogel, H. J., & Christen, O. (2019). Computed tomography as an extension of classical methods in the analysis of soil compaction, exemplified on samples from two tillage treatments and at two moisture tensions. *Geoderma, 346*, 52–62. https://doi.org/10.1016/j.geoderma.2019.03.023

Quinton, W. L., Elliot, T., Price, J. S., Rezanezhad, F., & Heck, R. (2009). Measuring physical and hydraulic properties of peat from X-ray tomography. *Geoderma, 153*, 269–277.

Rawlins, B. G., Wragg, J., Reinhard, C., Atwood, R. C., Houston, A., Lark, R. M., et al. (2016). Three-dimensional soil organic matter distribution, accessibility and microbial respiration in macroaggregates using osmium staining and synchrotron X-ray computed tomography. *The Soil, 2*, 659–671. https://doi.org/10.5194/soil-2-659-2016

Scotson, C. P., Duncan, S. J., Williams, K. A., Ruiz, S. A., & Roose, T. (2021). X-ray computed tomography imaging of solute movement through ridged and flat plant systems. *European Journal of Soil Science, 72*, 198–214. https://doi.org/10.1111/ejss.12985

Singh, J., Singh, N., & Kumar, S. (2020). X-ray computed tomography–measured soil pore parameters as influenced by crop rotations and cover crops. *Soil Science Society of America Journal, 84*, 1267–1279. https://doi.org/10.1002/saj2.20105

Stoops, G. (2009). Evaluation of Kubiëna's contribution to micropedology. At the occasion of the seventieth anniversary of his book "Micropedology". *Eurasian Soil Science, 42*, 693–698.

Valdez, A. S., Bosch-Serra, À. D., Yagüe, M. R., Poch, R. M., & Puigpinós, E. (2019). Earthworm community and soil microstructure changes with long-term organic fertilization. *Archives of Agronomy and Soil Science, 66*, 957–970. https://doi.org/10.1080/03650340.2019.1648792

Tracy, S. R., Daly, K. R., Sturrock, C. J., Crout, N. M. J., Mooney, S. J., & Roose, T. (2015). Three-dimensional quantification of soil hydraulic properties using X-ray computed tomography and image-based modelling. *Water Resource Research*, 1006–1022. https://doi.org/10.1002/2014WR016020

Van Loo, D., Bouckaert, L., Leroux, O., Pauwels, E., Dierick, M., Van Hoorebeke, L., et al. (2014). Contrast agents for soil investigation with X-ray computed tomography. *Geoderma, 213*, 485–491. https://doi.org/10.1016/j.geoderma.2013.08.036

Vanden Bygaart, A. J., & Protz, R. (1999). The representative elementary area (REA) in studies of quantitative soil micromorphology. *Geoderma, 89*, 333–346. https://doi.org/10.1016/S0016-7061(98)00089-5

Vingiani, S., Agrillo, A., De Mascellis, R., Langella, G., Manna, P., Mileti, F. A., & Terribile, F. (2022). Multi-sensor approach combined with pedological investigations to understand site-specific variability of soil properties and potentially toxic elements (PTEs) content of an industrial contaminated area. *Applied Sciences, 12*, 3993. https://doi.org/10.3390/app12083993

Webster, R., & Oliver, M. A. (1990). *Statistical methods in soil and land resource survey* (p. 316). Oxford University Press.

Zhang, Z., Liu, K., Zhou, H., Lin, H., Li, D., & Peng, X. (2018). Correction to: Three dimensional characteristics of biopores and non-biopores in the subsoil respond differently to land use and fertilization. (Plant and Soil, (2018), 428, 1-2, (453-467), 10.1007/s11104-018-3689-3). *Plant and Soil, 430*, 441. https://doi.org/10.1007/s11104-018-3727-1

Chapter 4
Optimising the Scanning Process: Demystifying the Dark Art of Optimising Microtomography Scan Settings

Craig J. Sturrock

4.1 Introduction

X-ray Computed Tomography (X-ray CT) is now becoming an accessible technology to quantify soil structure metrics. However, optimising images from X-ray CT systems remains a challenging task. Understanding the influence of the wide range of settings that control the X-ray source and detector, in addition to the procedures relating to preparing a sample for scanning, can be overwhelming. In this chapter, the aim is to demystify these factors and provide a simple, user-friendly, guide for acquiring the best quality tomography data possible from soil samples.

The value of a defined protocol is that it can allow the user to easily follow a set of instructions to achieve a specific result. However, it is always important to keep in mind to only use or develop a technique that is appropriate for the task in hand. For example, there are constraints on maximum resolution of a scan based on the detector configuration and dimensions of the sample and distances between the X-ray source, sample and detector (a more detailed description is outlined below and in Fig. 4.1). Therefore, it is not practically possible to use X-ray CT to image a large, intact 20 cm diameter field core at a voxel resolution of one micron due to the data size of the volume. However, by use of multiscale imaging, it is possible to collect images of the macropore structure of the entire 20 cm diameter core at a coarser resolution and then subsample the core to obtain smaller volumes to image at successively higher levels of resolution (e.g. small cores to aggregates). Finally, always pose the question, what is the significant advantage of using CT for any given scientific question?

C. J. Sturrock (✉)
Division of Agricultural and Environmental Sciences, School of Biosciences, University of Nottingham, Nottingham, UK
e-mail: craig.sturrock@nottingham.ac.uk

S. Jon Mooney et al. (eds.), *X-ray Imaging of the Soil Porous Architecture*,
https://doi.org/10.1007/978-3-031-12176-0_4

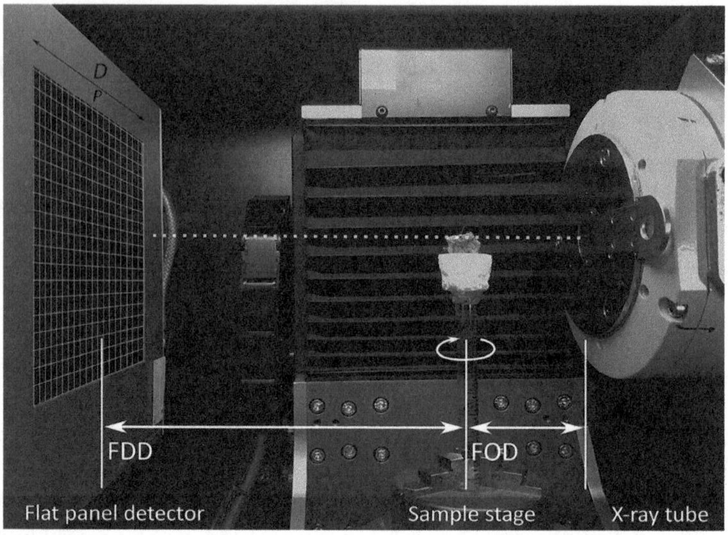

Fig. 4.1 Image of inside the cabinet of a Phoenix Nanotom 180NF CT Scanner (Waygate Technologies). The scanner consists of a transmission target X-ray tube, rotating sample stage (soil aggregate on stage) and a flat-panel detector. The relationship between the focal object distance (FOD) and the focal detector distance (FDD) influences the magnification of the sample on the detector. The smaller the FOD and greater the FDD results in greater magnification

X-ray CT is a non-invasive, non-destructive, imaging technique developed for the visualisation and quantification of the interior structure of an object in three dimensions (3-D). Its key principle is based on the differential attenuation of X-rays as they interact with the constituent materials of a sample. For example, the mineral grains and stones in a soil sample have a higher X-ray attenuation compared to lower density materials such as plant roots or organic matter. Vaz et al. (2011) discussed the current trend in types of scanners used in the soil sciences. Due to their more general availability, much of the pioneering application of CT to soil and plant research used medical CT systems (e.g. Petrovic et al., 1982; Hainsworth & Aylmore, 1983; Crestana et al., 1985; Tollner et al., 1987). Phogat and Aylmore (1989) developed a bespoke γ-ray computed tomography system, primarily with the aim to improve the quantitative abilities of CT to simultaneously image soil water content and bulk density (Aylmore, 1993). Later, Gregory et al. (2003) developed a high-resolution CT system for imaging plant roots in soil. Although this system was termed 'high-resolution' at the time of development (100-micron resolution), current tomography systems can now achieve submicron resolution highlighting the somewhat subjective nature of such terminology. During the last 15 years, the industrial sector has driven significant advances in both the speed and the image quality of CT technology resulting in the development of cone beam X-ray CT systems.

The typical cone beam CT scanner consists of three main parts, an X-ray tube, a sample manipulator stage and an image capture device such as flat-panel detector or

CCD based image intensifier (see Fig. 4.1). The exact specification of these three parts will vary with equipment supplier and model.

X-ray tube: X-rays are generated by an X-ray tube, which is essentially a vacuum tube that converts electrical input power into heat and a cone shaped beam of X-ray radiation. A tungsten filament is electrically heated inside the tube (cathode), which emits electrons that are accelerated towards a metal target (anode) at the end of the tube by application of a high voltage between the cathode and the anode. As the electrons enter the electric field of the target material atoms, they are rapidly slowed down resulting in a loss of their kinetic energy, generating heat and electromagnetic radiation. This type of radiation is termed *Bremsstrahlung* radiation (braking radiation) and is continuous as it consists of X-rays of a range of energies. Additionally, if an accelerated electron directly strikes a bound electron from the inner shell of the target atom, causing it to be ejected, X-rays are produced as an electron from the outer shell falls into the inner shell due to its reduction in energy. These X-rays are termed 'characteristic' as their energy spectra are characteristic of the material of the target. X-rays produced from tubes are therefore polychromatic as they display a range of wavelength/energies depending on the maximum acceleration voltage of the electron beam. In comparison, in synchrotron light sources X-rays are generated by electromagnetic deflection of an extremely high-energy electron beam. The X-ray beam produced is typically highly monochromatic, coherent and has a far higher photon flux than that of a benchtop microfocus source. This is important to recognise when optimising the X-ray settings to optimise the image quality for different types of material that will be discussed later.

Nano-focus systems (e.g. capable of an electron beam focal spot <1 μm) tend to have transmission X-ray tubes as they offer a greater level of control of the focusing capabilities of the electron beam used to generate the X-rays (Fig. 4.2a). These tubes are positioned horizontally in the scanner and the electron beam is transmitted through the target at the end of the gun. However, such tubes are power limited, and as such, are not as well suited for larger samples where higher X-ray energies and flux are required to fully penetrate the sample. Insufficient X-ray penetration results in dark streaks in the reconstructed data, especially around dense highly attenuating objects in the sample. This phenomenon is termed 'photon starvation' (Barrett & Keat, 2004). It should be noted that focal spot size has important consequences for the sharpness of the resultant images which will be discussed later in the chapter. Microfocus (e.g. focal spot typically >3 μm) CT systems typically have a reflection X-ray tube (Fig. 4.2b) (also known as directional target tubes), which is positioned at an approx. 45° angle inside the cabinet and the electron beam is reflected off the surface of the X-ray target material. Target materials are generally tungsten, but some manufacturers offer options to change the target material to allow the generation of X-rays of lower or higher energy spectra which allows a greater contrast in specific types of material. For example, molybdenum or copper produces lower energy spectra so they can give better results for imaging of lower density materials. For larger soil samples, such as 15–20 cm diameter x 100 cm length field cores, a large 'walk-in' cabinet system with even higher power (up to 450 kV) micro- or mini-focus (focal spot >100 μm) X-ray tubes are required.

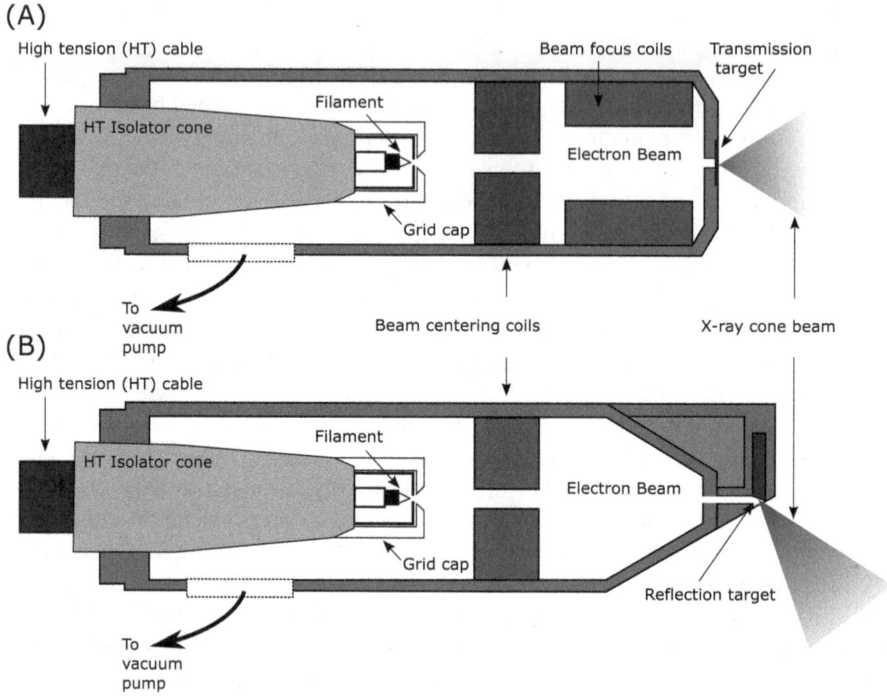

Fig. 4.2 Schematic representations of cross sections through (**a**) transmission and (**b**) reflection target X-ray tubes

Detector: The digital panel or CCD camera converts the X-rays that have penetrated through and around the sample into a digital image. Although there are numerous types of X-ray detectors, current state-of-the-art CT industrial systems tend use flat-panel detectors (Harrer & Kastner, 2011). Flat-panel detectors consist of a glass sheet coated with a layer of amorphous silicon and imprinted with a 2-D array of thin film transistors (TFT). Signal generation can be detected by either indirect or direct methods. Indirect panels use a phosphor scintillation screen (commonly gadolinium oxysulfide or caesium iodide based) to create visible light when struck by an X-ray photon which is subsequently recorded as an electrical signal by a photodiode (Seibert, 2006). Direct detectors consist of semiconductor materials sandwiched between two electrodes and convert X-rays directly into a positive and negatively charged ion pairs. Advantages of flat-panel detectors over previous image intensifier detectors are the lack of geometric distortion, flat and uniform images, thin profiles and high dynamic range. Recent advances in detector design have seen increased sensitivity, improved signal-noise and quantum efficiency, higher dynamic range and the number and size of picture elements (pixels) (e.g. the Dynamic 41|100 detector panel from Waygate Technologies (www.bakerhughesds.com/) or XRD1620xNCS panel from Perkin Elmer (https://www.perkinelmer.com/).

Irrespective of the configuration of the CT system, the process of conducting a scan is generally the same. The process starts by placing the sample of interest inside the scanner which is then exposed to X-rays generated from the X-ray tube. Numerous 2-D radiograph images (often called projections) are acquired by the detector as the sample is rotated around the y-axis, typically over 360°. The X-rays are absorbed and/or scattered by the sample materials causing attenuation of the beam. The degree of X-ray attenuation is proportional to the atomic number of the material, the photon energy of the beam and the thickness of the material. Heeraman et al. (1997) provide a detailed mathematical description of the physics behind this process. The degree of X-ray attenuation can be numerically calculated from the resultant 2-D radiographic image acquired by the detector. Differences in X-ray attenuation of the sample manifest as variations in contrast to the radiograph image, with higher density regions attenuating more of the X-rays and appearing as darker regions on the image (see Fig. 4.5a). The 2-D images are then processed to generate a 3-D volumetric dataset of X-ray attenuation of the sample. This process, called reconstruction, and for commercially supplied benchtop systems is generally undertaken via 'simple to use' proprietary software supplied with scanner. In this case, only a limited number of options in the software are available to the operator to complete this task. The reconstructed images can then be viewed and rendered to give a 3-D model of the material. Visualisation software then permits numerous analysis options from creating sections through the 3-D image to permit microstructural quantification of the sample. For soils, common measurements, all based on segmentation of constituent materials depending on their differences in density, are of soil pore network (volume, number, size, shape, internal porosity, surface area, etc.) (Mooney, 2002; Schlüter et al., 2011, 2014; De Gryze et al., 2006), plant root development (Bao et al., 2014; Flavel et al., 2012; Fry et al., 2018; Gregory et al., 2003; Hargreaves et al., 2009; Tracy et al., 2012), impact of earthworms (Amosse et al., 2015; Bottinelli et al., 2017) and insect larvae on soil structure (Johnson et al., 2007; Booth et al., 2020), characterisation of chemical tracers (Grayling et al., 2018) and, of recent interest, the distribution of microplastics in soils (Totzke et al., 2021).

4.2 Preparing the Sample

How to prepare a soil sample for scanning depends on its size, shape, state (dry or hydrated) and the research question that is being addressed. One of the major advantages of using CT is that minimal sample preparation is required with often no need for time-consuming chemical fixing or resin impregnation such as that described in Chap. 3. The only critical condition for preparing the sample is that it remains static during the duration of the scan. The majority studies using CT to study soils work with various sizes of either soil aggregates/clods (1– 200 mm) or soil cores (5–2500 mm diameter). Core studies may be on field structured soils or sieved and repacked to specific conditions (moisture content, bulk density, etc.). The later have been extensively used for the investigation of plant root growth for root phenotyping applications and studies of the rhizosphere (Nunan et al., 2006; Moradi et al., 2011; Mooney et al., 2012; Flavel et al., 2012; Ahmed et al., 2016; Helliwell et al., 2017; Rabbi et al., 2018; Koebernick et al., 2019). The most important factor in preparing a sample for scanning is that it is securely fixed to a sample holder and will not move during the scan. To ensure optimal quality data, the first and the last radiograph image should be the same position. Samples that are subject to movement will have blurred edges and 'cup-like' ghostly artefacts in the reconstructed images (see Fig. 4.3e). When working with small dry samples such as soil aggregates, they can be either fixed to glass, plastic or wood rods using a suitable adhesive such as cyanoacrylate (superglue), epoxy resin or for larger samples, hot melt glue. Alternatively, in some cases the aggregates can be suspended between low-density polystyrene. Disposable plastic pipettes can be used to stack numerous samples in this manner. This method has the advantage that the scanner can be programmed to automatically scan each sample in a batch mode and minimising user interaction to increase sample throughput. If working with wet soil samples, they need to be scanned inside a sealed container, wrapped in cling film or covered in wax, to limit evaporation which will result in shrinkage of the sample especially if the scan times are long (e.g. > 1 hr). Depending on the size of the sample the above framework can be implemented by using different diameter pipettes or plastic tubes (Fig. 4.4).

As the popularity, ease of access and relatively low cost of additive manufacturing also known as 3-D printers become more readily available to many laboratories, the ability to produce bespoke sample holders offers opportunities for improved precision to optimise sample preparation strategies. The work of Keyes et al. (2017) is a good example of the use of 3-D printing to produce a specialised microcosm to enable isolation of individual plant roots and soil for high-resolution scanning at a synchrotron light source. An additional benefit of 3-D printing approaches is the 3-D models can be easily shared between labs that may be working with similar sample types. For example, at the Hounsfield Facility, University of Nottingham a 3-D printed sample holder was designed for soil columns that can be used when undertaking plant growth trials. Sections of high-density polyethylene drainage pipe can be cut to the desired length and then nylon mesh taped to the bottom end (Mairhofer et al., 2017). The column, when filled with soil and a plant, is then placed inside the sample holder so there is greater consistency between samples and facilitates easier automation of multi -scan and -reconstruction operations. A selection of 3-D models for sample holders used at the Hounsfield Facility are available for download here: https://github.com/UoNMakerSpace/Xray-CT-Sample-Holders

For investigations of the soil surface (e.g. Garbout et al. (2018)), it is advised to tilt the sample to reduce cone beam artefacts from the flat surface. This may not be required if you have access to a scanner capable of helical scanning functionality. In-situ testing rigs to investigate mechanical stress on soils are now commercially available. Combined with rapid scanner acquisition times and correlative image analysis approaches temporal changes in soil samples can be performed (for

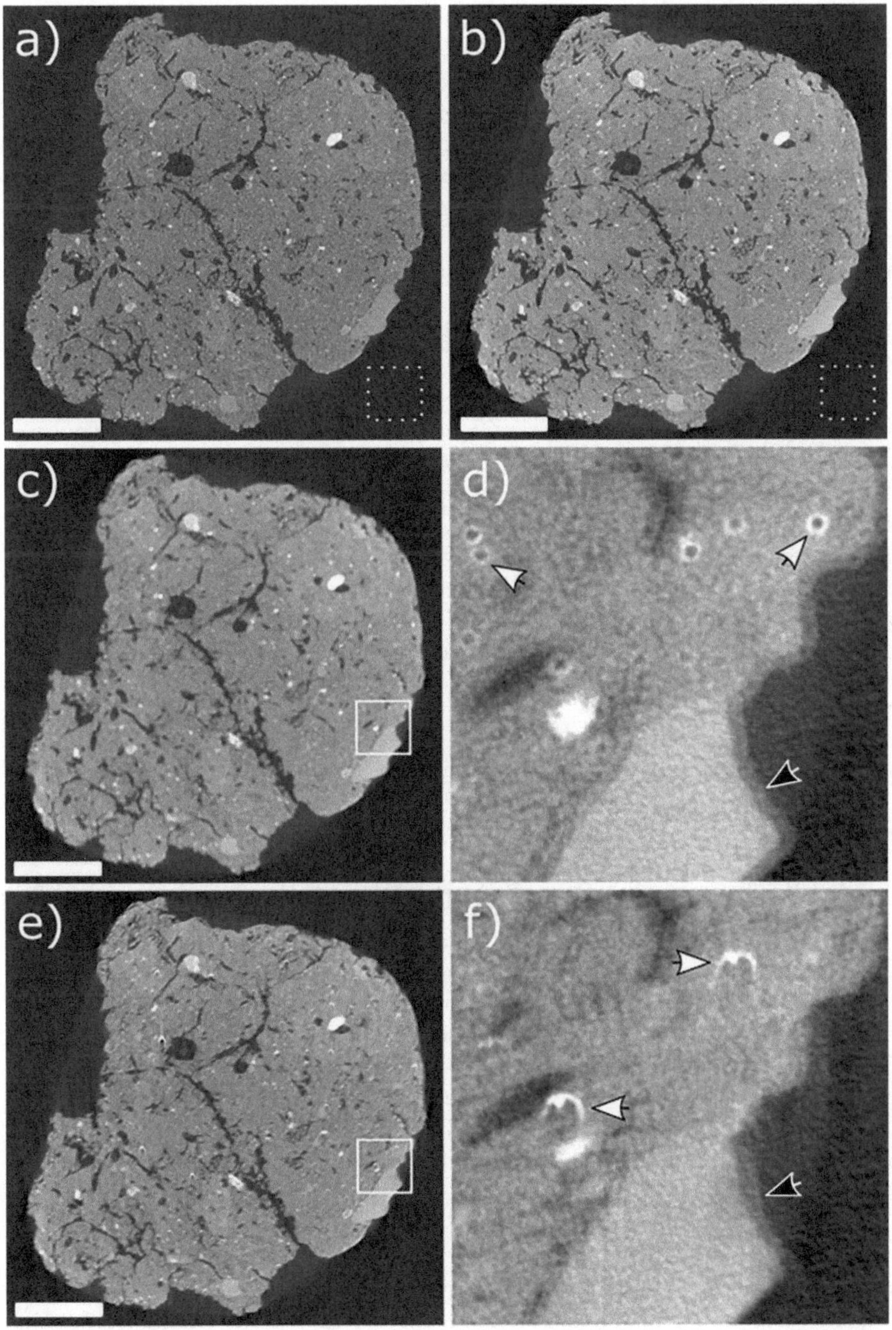

Fig. 4.3 X-ray CT XY slice images of a soil aggregate to illustrate common artefacts collected using a Phoenix Nanotom NF180 scanner. (**a**) The X-ray energy is too low (50 kV) resulting in a noisy image poor quality image with low contrast. Standard deviation of the inset dotted square grey level (16-bit depth) = 6230. (**b**) Optimal X-ray energy settings showing a good quality low noise and no obvious artefacts. Standard deviation of inset dotted square = 2490. (**c**) Example of an image

examples see Hall et al., 2010, Wang et al., 2019, Cheng et al., 2020, Bull et al., 2020). See also Chap. 8.

Inclusion of a reference object on the sample can be useful when using multiple scans to image a single sample. Using a high-density reference material placed on the sample while it is being scanned allows greater accuracy when mapping histogram outputs between individual sections of the scans. A 0.05 mm thick × 1 mm width strip of copper taped down the length of the column can be used to achieve this (See Figure 4.4b). This method is particularly useful when scanning field cores which may display considerable heterogeneity in structural composition. The presence of a high-density mineral object such as a stone in one part of the multi-scan can drastically shift the reconstructed data histogram which can cause problems for downstream analysis.

One final point to remember is to acclimatise the sample before scanning to the ambient environmental conditions inside the scanner cabinet. Scanning a field soil core directly from storage in a cold room will tend to expand during the scan and deteriorate the quality of the data. The opposite may be the case in summer months if the laboratory is at a higher temperature than the interior of the scanner, especially if the scanner is fitted with air conditioning. An additional benefit of an air-conditioned cabinet is the mitigation of any internal temperature increases that can occur during very long scans resulting from heat output of the X-ray tube and detector panel. For precision high-resolution scanning, particularly for soil aggregates, it is recommended to allow samples to acclimatise inside the scanner before starting the scan.

4.3 Optimising the Image

4.3.1 Positioning the Sample in the Scanner

Once the sample is prepared, it can be placed on the rotation stage of the scanner and suitably positioned between the X-ray source and the detector depending on the nature of the investigation. The relationship between the distance of the sample (Focal Object Distance: FOD) from the X-ray source and the source and the detector (Focal Detector Distance: FDD) determines the magnification of the scan. Decreasing the FOD or increasing the FDD results in increased magnification of the sample but also reduces the field of view (see Fig. 4.1). Additionally, detector signal to noise

---

Fig. 4.3 (continued) with a rotation offset error at reconstruction. Image is blurred and has characteristic halo artefacts. (d) Magnified region (white square in (**c**)) highlighting halo artefacts (white arrow) and double edge on features in image (black arrow). (**e**) Example of motion artefact due to lateral sample moment during the scan. The image is blurred and has characteristic cup artefacts. (**f**) Magnified inset of (**e**), showing cup artefacts (white arrows) and double edge on features in the image. Scale bar = 3 mm

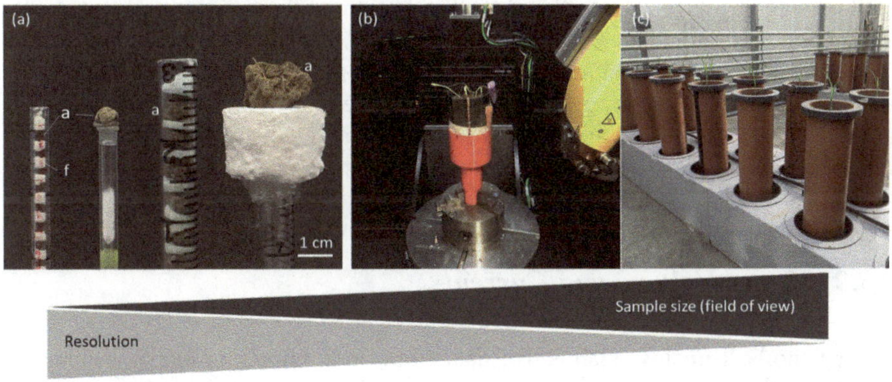

Fig. 4.4 Examples of different sample preparation methods depending on size. (**a**) Small soil aggregates either glued or suspended in low-density polystyrene foam. (**b**) A medium scale field soil core in a PVC column (15 cm length × 7 cm Ø) in a v|tome|x M 240 kV CT system (**c**) Large 100 soil cores (100 cm length × 20 cm Ø) made from commercial drainage pipe in the glasshouse of the Hounsfield Facility, University of Nottingham. Highest magnification and voxel resolution is achieved with small samples

ratio is inversely proportional to the FDD due to the reduction in intensity the farther from the X-ray source. The spatial voxel resolution of a scan is determined by detector size (width) and the sample diameter and can be described as

$$V = P \cdot \frac{d}{D}$$

where V is the voxel resolution, P is the pixel size, d is the sample diameter and D is the detector width (Fig. 4.1). For example, the voxel resolution for a 25 mm diameter sample in a CT system with a 100 mm detector and 0.05 mm pixel size would be 0.0125 mm.

For the best quality of data collection, the sample should be positioned so it does not leave the sides of the detector at any of the projection angle during the scan as the information relating to the edges of the sample that leaves the field of view will not be present in the final reconstructed data. However, these parts of the sample will still be present in projections of the front and back views of the sample and can lead to artefacts in the data. Commercial reconstruction software now includes processing options to reduce these artefacts (Fig. 4.3d-f). For the optimal scan resolution of a given sample, it should be moved closer or further away from the X-ray tube so that the projection image fills approximately 90% of the detector width. If the sample is longer than it is wide, then it can be moved further from the source but note this will decrease the resolution. However, multiple acquisitions (e.g. at top, middle and bottom) can also be used to scan the sample in parts to maintain the maximal resolution of the scan. The data can then be digitally stitched together either automatically via the reconstruction software or manually by the user in a suitable visualisation software. As a rule of thumb, the maximal resolution equals the sample width divided by the number of pixels on the detector of the projection. For example, a soil core of 50 mm diameter, positioned to fill 90% of a 2000 pixel width detector (1800 px) will have a resolution of 27.7 microns. Some scanner configurations allow for the detector to be moved closer to the source. This can be advantageous to increase to detector signal to noise ratio as the X-ray intensity is higher. However, this is at the expense of the maximum resolution as the magnification is lower.

4.3.2 X-ray Voltage and Current

Getting the best quality quantitative data from your sample primarily centres on optimising the CT scan settings to obtain good quality radiographs with low image noise. Ensuring your sample is penetrated by X-rays at all angles of rotation whilst not over saturating your background is one of the golden rules of CT. Optimising the X-ray tube voltage and current are the first steps to obtain the best quality projection image. The tube voltage controls the penetrating power of the generated X-rays. As explained earlier, X-rays are generated from the tube by crashing electrons onto or through its metal target. Increasing the tube voltage produces an electron beam with more energy resulting in the creation of X-rays with a higher energy spectra and thus have greater penetrating power. Increasing the tube current increases the quantity electrons in the beam so although more X-rays are generated per unit time their energy spectra are the same.

Before starting the optimisation process, it is important to ensure that the flat-field correction images are disabled in the acquisition software. Depending on the manufacturer of the CT system software, these may be known as different names such as black and white or offset and gain calibration images. As a starting point, it is sensible to choose a moderate X-ray tube voltage and current setting (e.g. Table 4.1). With the X-rays 'on', you can observe the greyscale histogram window of the live projection image (Fig. 4.5). When optimising the voltage, fully rotate the sample to ensure good penetration is achieved through all projection angles. This is particularly important when scanning field structured soils with high numbers of high-density objects such as stones where it can be possible to have some projection angles where multiple stones are present in one path. Good penetration is obtained when the X-ray intensity measured on the detector through the longest path and/or highest density part of sample is significantly higher than the detector response when the X-rays are off. Theoretically, the minimum recommended penetration value is approx. 10–15% of the maximum detector response. For example, for a 14-bit detector panel with 16,384 grey values and a black level of 1600, then the darkest part of your projection images should not be less than $1600 + 0.10 \times (14384–1600) = 2878$. The background (air) should not be >85% of the detector's maximum grey value to avoid pixel saturation. Even a single saturated pixel can be the source of ring artefacts in reconstructed data due to it continually outputting its maximal value. Ideally separate peaks in the histogram should be resolved between materials of different densities (e.g. soil and air).

Table 4.1 Example starting settings for a range of soil samples. Settings can be adjusted to the exact requirement of your own scans sample size, material composition and scanner type/configuration

Soil type	Class	Sample Diameter (mm)	Scanner Model	Target type	Max Tube Voltage (kV)	Tube Voltage (kV)	Tube Current (μA)	Detector timing (ms)	Focus Mode	Filter Material
Sandy	Aggregate	2	GE Nanotom	Transmission	160	85	140	750	M1	–
Clay	Aggregate	2	GE Nanotom	Transmission	160	85	140	750	M1	–
Sandy	Aggregate	10	GE Nanotom	Transmission	160	80	100	750	M0	–
Clay	Aggregate	10	GE Nanotom	Transmission	160	80	100	750	M0	–
Sandy	Sieved soil	50	GE v|tomel|x M	Directional	240	120	100	120	–	–
Clay	Sieved soil	50	GE v|tomel|x M	Directional	240	120	100	120	–	–
Sandy	Sieved soil	100	GE v|tomel|x M	Directional	240	180	150	180	–	0.5 mm cu
Clay	Sieved soil	100	GE v|tomel|x M	Directional	240	180	150	180	–	0.5 mm cu
Sandy	Field soil	50	GE v|tomel|x M	Directional	240	140	100	140	–	–
Clay	Field soil	50	GE v|tomel|x M	Directional	240	140	100	140	–	–
Sandy	Field soil	100	GE v|tomel|x M	Directional	240	180	150	180	–	0.5 mm cu
Clay	Field soil	100	GE v|tomel|x M	Directional	240	180	150	180	–	0.5 mm cu

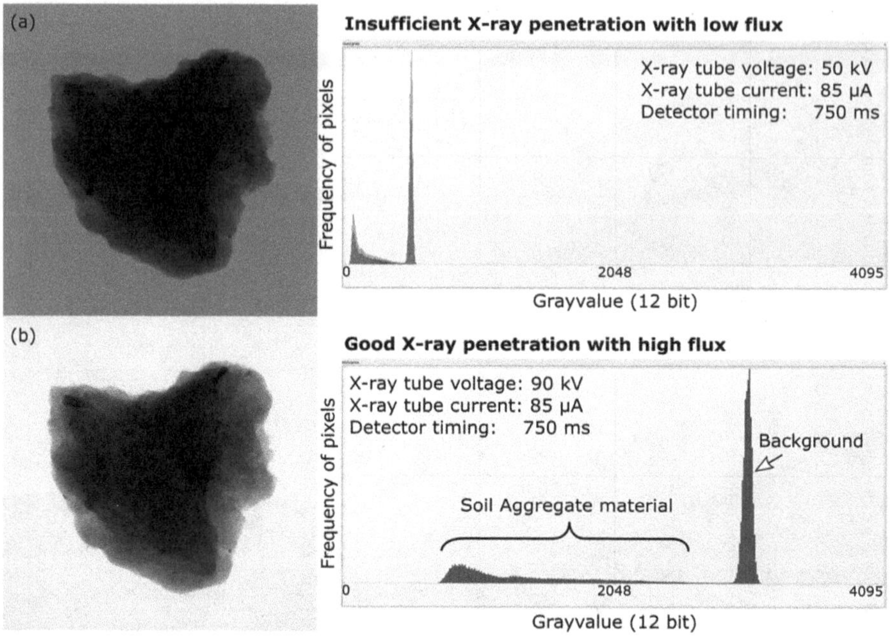

Fig. 4.5 The importance of ensuring your sample has good X-ray penetration. (**a**) A radiograph of a 20 mm soil aggregate collected using insufficient X-ray energy (50 kV). The histogram shows the narrow range of grey values in the image with low X-ray flux. (**b**) A radiograph of the same aggregate but with a higher X-ray energy (90 kV) which has higher contrast and flux. This is evidenced by the broad spread of grey values in the histogram and clearly defined peaks from the different materials in the sample. Results of CT scans using these settings are shown in Figure 4.3a-b

As higher tube voltages generate X-rays with higher energies and therefore higher penetrating power, the contrast of the lower density materials in the sample from the background noise is reduced. It is therefore important to take care and avoid losing these details in scan images. However, this can be challenging for the typical multi-phase heterogeneous nature of soils that commonly consist of a multitude of components with a wide range of material densities (e.g. air and water filled pores, organic matter in various stages of decomposition, micro- and macro- biota, plant roots, clay, silt and sand). In such cases, it may be appropriate to apply a metal filter to the X-ray tube to reduce the lower energy X-rays in the beam.

By increasing the tube current, the number of X-rays passing through the sample is increased. However, if the tube voltage is not high enough to penetrate the densest part/path length of your sample, increasing the current will not improve the resultant image. If the X-ray beam does not have enough energy to penetrate the sample, then having more of them will not make any difference. Conversely, for a sample where the penetration is sufficient, increasing the current is often a better choice, rather than increasing the voltage further, thus minimising the risk of reducing the contrast of low-density materials. A further consideration for tube settings relates to their influence on the electron beam focal spot size on the tube target. The focal spot size is a function of the tube voltage and current often measured as the target power. Higher voltages and current therefore increase the focal spot size and result in a larger penumbra effect of the projected image. Care should be taken not to exceed the target power (W) above the spatial resolution or this will reduce the sharpness of the image.

The X-ray beam may require to be filtered to achieve good contrast of a sample without saturating the detector. Thin sheets (approx. 0.1–2 mm) of aluminium, copper or tin are commonly used to attenuate part of the X-ray spectra from the X-ray source at a particular energy. The filter removes the lower energy X-rays in the spectra and hardens the beam which can help reduce beam hardening artefacts. Filters are also useful to avoid saturation of the detector if a higher X-ray energy is needed to achieve a good penetration of a denser sample.

4.3.3 Detector Settings

The exposure timing and sensitivity of detector can also be used to optimise the contrast of projection images but have important consequences for both scan time and image quality. The exposure timing can be thought of as shutter speed on a camera. The longer the shutter is open, the greater the amount of light reaches the film and creates a brighter image. As more X-rays are collected with a longer exposure time, image noise levels can be lower although overall scan time will inherently increase. Typical detector timings in benchtop CT systems range from approximately one hundred milliseconds to several seconds. The detector sensitivity is similar to a digital camera ISO value. Higher values make the detector more sensitive to X-rays and again will increase the image brightness, although the noise levels in the image are also amplified. Longer detector timings can be useful for higher density samples where a high degree of X-ray beam filtering has been required to obtain good penetration of the sample. Whereas higher sensitivity settings can be useful when imaging low-density dynamic samples at high-resolution when you need to keep scan times short and are limited on X-ray target power. One of the most common questions asked when enquiring about scanning a soil sample using X-ray CT is 'how long will it take?' Scans can be made in less than 5 min, but the data will have a much higher level of noise compared to ones that are acquired at times >2 h. However, depending on the nature of your investigation this may be acceptable. Flavel et al. (2014) used scans of 4 min to successfully quantify the response of wheat root length density, branch density and angle to phosphate application in small diameter (30 mm) columns containing soil. If you are trying to identify and track the movement of insect larvae in soil and quantify relatively large burrow structures in the soil, then limiting the X-ray dose to the insects and avoiding motion artefacts is of greater importance than the noise level in the data. Bont et al. (2017) and Booth et al. (2020) provide examples for such a strategy where rapid 6–7 min scans and noise reduction image processing techniques to successfully capture insect larvae position and the structure of their burrow networks.

4.4 Acquisition Modes

The images captured during a CT scan can be made continuously (often called 'fast scans') where the sample is in constant motion during its rotation, or stepwise, where it stops at each rotational increment (step scans/long scans). In continuous scanning mode, only a single radiograph image is collected for each angular projection, and as such, the image noise is higher compared to a step scan where several images can be integrated to average out noise and improve image quality. Depending on the application, both methods have their strengths and weaknesses. As continuous scans are faster than step scans, they are useful where the sample is dynamic and at risk of motion artefacts e.g. following the movement of a contrast agent in soil. Furthermore, if the image processing protocol can handle a certain level of image noise, it is possible to greatly increase throughput using continuous scan modes.

However, step scans are generally the standard mode of operation and offer improved image quality due to the option to integrate a series of images to reduce noise. A skip value defines the number of images to discard between each step so that the sample has time to rotate to the next position and stop before the next series of images are integrated. If using very short detector timings, the skip value will need to be higher to allow adequate time for the sample to turn otherwise a potentially blurred image due to sample motion will be added to the projection image integration. A suitably long skip time is also required to reduce the ghost image of the previous angular projection as these can also contribute to reduce quality of the final scan. Care must also be taken if scanning very small samples on thin glass or plastic rods as the stop-start motion of the stage can induce a vibration the sample. In such a case, a continuous scan may provide a better result or alternatively a longer scan with an increased detector skip could be preferable.

The number of projection images per scan is determined by the width of the sample on the detector panel. When the sample occupies up to 95% of the detector width, then a greater number of projection images will be required to avoid interpolation between edge voxels of the sample. Ideally, the number of projection images should ensure that each incremental position moves on by no more than one voxel. In practice, this is achieved by multiplying the number of pixels of your sample across the width of the detector by $\pi/2$ (approx. 1.57). For example, scanning a soil core that occupies 90% of a 2000 pixel width detector (1800px) would require a total of 2826 projections for optimal image quality. If it is appropriate to use a lower image resolution, it may be possible to reduce acquisition time as less projection would be required.

4.5 Conclusions

The aim of this chapter was to provide a guide to a beginner/novice user of X-ray CT with an introduction to the technique and demystify the various settings available on a benchtop CT system. The work presented here is not exhaustive but should give a basic grounding to build experience.

In summary, to achieve the best quality scans for most standard scans of soil aggregates and cores the following actions are useful to adhere to:

1. The sample must remain static during the scan.
2. The sample should stay within the horizontal field of view of the detector.
3. The sample must be penetrated by X-rays at all angles of rotation.
4. The sample or background surrounding the sample must not be over penetrated or "washed" out at any angle of rotation.
5. Investing the time in obtaining good quality radiographs ensures good quality reconstructed data which ultimately aids image processing.

References

Ahmed, S., Klassen, T. N., Keyes, S., Daly, M., Jones, D. L., Mavrogordato, M., Sinclair, I., & Roose, T. (2016). Imaging the interaction of roots and phosphate fertiliser granules using 4-D X-ray tomography. *Plant and Soil, 401*, 125–134.

Amosse, J., Turberg, P., Kohler-Milleret, R., Gobat, J. M., & Le Bayon, R. C. (2015). Effects of endogeic earthworms on the soil organic matter dynamics and the soil structure in urban and alluvial soil materials. *Geoderma, 243*, 50–57.

Aylmore, L. A. G. (1993). Use of computer-assisted tomography in studying water-movement around plant-roots. *Advances in Agronomy, 49*(49), 1–54.

Bao, Y., Aggarwal, P., Robbins, N. E., Sturrock, C. J., Thompson, M. C., Tan, H. Q., Tham, C., Duan, L., Rodriguez, P. L., Vernoux, T., Mooney, S. J., Bennett, M. J., & Dinneny, J. R. (2014). Plant roots use a patterning mechanism to position lateral root branches toward available water. *Proceedings of the National Academy of Sciences of the United States of America, 111*, 9319–9324.

Barrett, J. F., & Keat, N. (2004). Artifacts in CT: Recognition and avoidance. *Radiographics, 24*, 1679–1691.

Bont, Z., Arce, C., Huber, M., Huang, W., Mestrot, A., Sturrock, C. J., & Erb, M. (2017). A herbivore tag-and-trace system reveals contact-and density-dependent repellence of a root toxin. *Journal of Chemical Ecology, 43*, 295–306.

Booth, S., Kurtz, B., De Heer, M. I., Mooney, S. J., & Sturrock, C. J. (2020). Tracking wireworm burrowing behaviour in soil over time using 3-D X-ray computed tomography. *Pest Management Science, 76*, 2653 2662.

Bottinelli, N., Zhou, H., Capowiez, Y., Zhang, Z. B., Qiu, J., Jouquet, P., & Peng, X. H. (2017). Earthworm burrowing activity of two non-Lumbricidae earthworm species incubated in soils with contrasting organic carbon content (vertisol vs. Ultisol). *Biology and Fertility of Soils, 53*, 951–955.

Bull, D. J., Smethurst, J. A., Sinclair, I., Pierron, F., Roose, T., Powrie, W. & Bengough, A. G. 2020. Mechanisms of root reinforcement in soils: An experimental methodology using four-dimensional X-ray computed tomography and digital volume correlation. *Proceedings of the Royal Society a-Mathematical Physical and Engineering Sciences,* 476.

Cheng, Z., Wang, J. F., Coop, M. R., & Ye, G. L. (2020). A miniature triaxial apparatus for investigating the micromechanics of granular soils with in situ X-ray micro-tomography scanning. *Frontiers of Structural and Civil Engineering, 14*, 357–373.

Crestana, S., Mascarenhas, S., & Pozzimucelli, R. S. (1985). Static and dynamic 3-dimensional studies of water in soil using computed tomographic scanning. *Soil Science, 140*, 326–332.

De Gryze, S., Jassogne, L., Six, J., Bossuyt, H., Wevers, M., & Merckx, R. (2006). Pore structure changes during decomposition of fresh residue: X-ray tomography analyses. *Geoderma, 134*, 82–96.

Flavel, R. J., Guppy, C. N., Tighe, M., Watt, M., Mcneill, A., & Young, I. M. (2012). Non-destructive quantification of cereal roots in soil using high-resolution X-ray tomography. *Journal of Experimental Botany, 63*, 2503–2511.

Flavel, R. J., Guppy, C. N., Tighe, M. K., Watt, M., & Young, I. M. (2014). Quantifying the response of wheat (Triticum aestivum L) root system architecture to phosphorus in an Oxisol. *Plant and Soil, 385*, 303–310.

Fry, E. L., Evans, A. L., Sturrock, C. J., Bullock, J. M., & Bardgett, R. D. (2018). Root architecture governs plasticity in response to drought. *Plant and Soil, 433*, 189–200.

Garbout, A., Sturrock, C. J., Armenise, E., Ahn, S., Simmons, R. W., Doerr, S., Ritz, K., & Mooney, S. J. (2018). Top cap: A tool to quantify soil surface topology and subsurface structure. *Vadose Zone Journal, 17*.

Grayling, K. M., Young, S. D., Roberts, C. J., De Heer, M. I., Shirley, I. M., Sturrock, C. J., & Mooney, S. J. (2018). The application of X-ray micro computed tomography imaging for tracing particle movement in soil. *Geoderma, 321*, 8–14.

Gregory, P. J., Hutchison, D. J., Read, D. B., Jenneson, P. M., Gilboy, W. B., & Morton, E. J. (2003). Non-invasive imaging of roots with high-resolution X-ray micro-tomography. *Plant and Soil, 255*, 351–359.

Hainsworth, J. M., & Aylmore, L. A. G. (1983). The use of computer-assisted tomography to determine spatial-distribution of soil-water content. *Australian Journal of Soil Research, 21*, 435–443.

Hall, S. A., Bornert, M., Desrues, J., Pannier, Y., Lenoir, N., Viggiani, G., & Besuelle, P. (2010). Discrete and continuum analysis of localised deformation in sand using X-ray mu CT and volumetric digital image correlation. *Geotechnique, 60*, 315–322.

Hargreaves, C. E., Gregory, P. J., & Bengough, A. G. (2009). Measuring root traits in barley (Hordeum vulgare ssp vulgare and ssp spontaneum) seedlings using gel chambers, soil sacs and X-ray microtomography. *Plant and Soil, 316*, 285–297.

Harrer, B., & Kastner, J. (2011). In F. A. Lasagni & A. F. Lasagni (Eds.), *X-ray microtomography: Characterisation of structures and defect analysis fabrication and characterization in the micro-Nano range*. Springer.

Heeraman, D. A., Hopmans, J. W., & Clausnitzer, V. (1997). Three dimensional imaging of plant roots in situ with x-ray computed tomography. *Plant and Soil, 189*, 167–179.

Helliwell, J. R., Sturrock, C. J., Mairhofer, S., Craigon, J., Ashton, R. W., Miller, A. J., Whalley, W. R., & Mooney, S. J. (2017). The emergent rhizosphere: Imaging the development of the porous architecture at the root-soil interface. *Scientific Reports, 7*.

Johnson, S. N., Crawford, J. W., Gregory, P. J., Grinev, D. V., Mankin, R. W., Masters, G. J., Murray, P. J., Wall, D. H., & Zhang, X. X. (2007). Non-invasive techniques for investigating and modelling root-feeding insects in managed and natural systems. *Agricultural and Forest Entomology, 9*, 39–46.

Keyes, S. D., Zygalakis, K. C., & Roose, T. (2017). An explicit structural model of root hair and soil interactions parameterised by synchrotron X-ray computed tomography. *Bulletin of Mathematical Biology, 79*, 2785–2813.

Koebernick, N., Daly, K. R., Keyes, S. D., Bengough, A. G., Brown, L. K., Cooper, L. J., George, T. S., Hallett, P. D., Naveed, M., Raffan, A., & Roose, T. (2019). Imaging microstructure of the barley rhizosphere: Particle packing and root hair influences. *New Phytologist, 221*, 1878–1889.

Mairhofer, S., Pridmore, T., Johnson, J., Wells, D. M., Bennett, M. J., Mooney, S. J., & Sturrock, C. J. (2017). X-ray computed tomography of crop plant root systems grown in soil. *Current Protocols in Plant Biology, 2*, 270–286.

Mooney, S. J., Pridmore, T., Helliwell, J., & Bennett, M. (2012). Developing X-ray computed tomography to non-invasively image 3-D root systems architecture in soil. *Plant and Soil, 352*, 1–22.

Mooney, S. J. (2002). Three-dimensional visualization and quantification of soil macroporosity and water flow patterns using computed tomography. *Soil Use and Management, 18*, 142–151.

Moradi, A. B., Carminati, A., Vetterlein, D., Vontobel, P., Lehmann, E., Weller, U., Hopmans, J. W., Vogel, H. J., & Oswald, S. E. (2011). Three-dimensional visualization and quantification of water content in the rhizosphere. *New Phytologist, 192*, 653–663.

Nunan, N., Ritz, K., Rivers, M., Feeney, D. S., & Young, I. M. (2006). Investigating microbial micro-habitat structure using X-ray computed tomography. *Geoderma, 133*, 398–407.

Petrovic, A. M., Siebert, J. E., & Rieke, P. E. (1982). Soil bulk-density analysis in 3 dimensions by computed tomographic scanning. *Soil Science Society of America Journal, 46*, 445–450.

Phogat, V. K., & Aylmore, L. A. G. (1989). Evaluation of soil structure by using computer-assisted tomography. *Australian Journal of Soil Research, 27*, 313–323.

Rabbi, S. M. F., Tighe, M. K., Flavel, R. J., Kaiser, B. N., Guppy, C. N., Zhang, X. X., & Young, I. M. (2018). Plant roots redesign the rhizosphere to alter the three-dimensional physical architecture and water dynamics. *New Phytologist, 219*, 542–550.

Schlüter, S., Sheppard, A., Brown, K., & Wildenschild, D. (2014). Image processing of multiphase images obtained via X- ray microtomography: A review. *Water Resources Research, 50*, 3615–3639.

Schlüter, S., Weller, U., & Vogel, H. J. (2011). Soil-structure development including seasonal dynamics in a long-term fertilization experiment. *Journal of Plant Nutrition and Soil Science, 174*, 395–403.

Seibert, J. A. (2006). Flat-panel detectors: How much better are they? *Pediatric Radiology, 36*, 173–181.

Tollner, E. W., Verma, B. P., & Cheshire, J. M. (1987). Observing soil-tool interactions and soil organisms using X-ray Computer-Tomograph Y. *Transactions of the Asae, 30*, 1605–1610.

Totzke, C., Oswald, S. E., Hilger, A., & Kardjilov, N. (2021). Non-invasive detection and localization of microplastic particles in a sandy sediment by complementary neutron and X-ray tomography. *Journal of Soils and Sediments, 21*, 1476–1487.

Tracy, S. R., Black, C. R., Roberts, J. A., Sturrock, C., Mairhofer, S., Craigon, J., & Mooney, S. J. (2012). Quantifying the impact of soil compaction on root system architecture in tomato (Solanum lycopersicum) by X-ray micro-computed tomography. *Annals of Botany, 110*, 511–519.

Vaz, C. M. P., De Maria, I. C., Lasso, P. O., & Tuller, M. (2011). Evaluation of an advanced benchtop micro-computed tomography system for quantifying porosities and pore-size distributions of two Brazilian Oxisols. *Soil Science Society of America Journal, 75*, 832–841.

Wang, Y., Zhang, D., & Hu, Y. Z. (2019). X-ray computed tomography characterization of soil and rock mixture under cyclic triaxial testing: The effects of confining pressure on meso-structural changes. *Environmental Earth Sciences, 78*.

Chapter 5
X-ray Computed Tomography Image Processing & Segmentation: A Case Study Applying Machine Learning and Deep Learning-Based Strategies

Talita R. Ferreira, Fábio A. M. Cássaro, Hu Zhou, and Luiz F. Pires

5.1 Image Processing: Cropping and Filtering

Depending on the type and the size of the sample, different decisions on the selection of a region of interest (ROI), i.e. cropping of X-ray CT images for subsequent image processing will be necessary. For instance, large samples collected in plastic tubes might present some structural damage in their peripheries depending on how the sample has been collected and it may be desirable to avoid such regions in any subsequent analysis (see Chap. 3 on soil sampling). In this case, extracting (or cropping) a regular-shaped ROI (cubic or cylindrical) in the adjacencies of the centre of the reconstructed image, which corresponds to the core of the sample, is frequently carried out (Galdos et al., 2020; Borges et al., 2019; Ferreira et al., 2018; Pires et al., 2017; Backeberg et al., 2017; Vogel, 1997). In contrast, if the intention is to evaluate the effects of the sampling procedure on the soil structure itself, it is often more appropriate to choose a ROI that encompasses the sample's external surroundings. Some software (e.g. Avizo) even offer options for selecting irregular shaped ROI, based on a free-hand drawing of the desired region (Ferreira et al., 2019). When the sample is irregular, such as natural soil aggregates, it is more appropriate to consider the entire volume for the image processing since closed and open intra-

T. R. Ferreira
Brazilian Synchrotron Light Laboratory (LNLS), Brazilian Center for Research in Energy and Materials (CNPEM), Campinas, São Paulo, Brazil
e-mail: talita.ferreira@lnls.br

F. A. M. Cássaro · L. F. Pires (✉)
Department of Physics, State University of Ponta Grossa (UEPG), Ponta Grossa, Paraná, Brazil
e-mail: fcassaro@uepg.br; lfpires@uepg.br

H. Zhou
Department of Soil and Water Sciences, China Agricultural University, Beijing, China
e-mail: zhouhu@cau.edu.cn

S. Jon Mooney et al. (eds.), *X-ray Imaging of the Soil Porous Architecture*,
https://doi.org/10.1007/978-3-031-12176-0_5

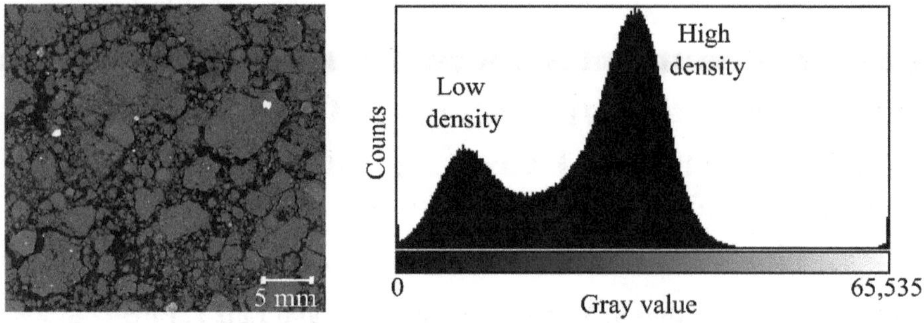

Fig. 5.1 Histogram of a 16-bit colour depth section (2-D) obtained from a soil X-ray CT image. The minimum value (0) represents low-density materials while the maximum (65,535), 2^n-1, represents high-density materials (soil minerals) (not in Hounsfield units)

aggregate pores can be concomitantly investigated. In this analysis, the aggregate must be carefully separated from its background instead of simply extracting a regular or irregular ROI (Bacq-Labreuil et al., 2019; Khan et al., 2012; Wang et al., 2012).

As previously described, after scanning and reconstruction, a set of 2-D digital images, each represented by a matrix of elementary data (pixels), is obtained. When stacked, the 2-D images slices form a 3-D image volume, which consists of elementary units called voxels. Usually, for visualization and subsequent analysis of the X-ray CT images, materials of diverse attenuation coefficients are linked to distinct shades of grey that characterizes the radiodensity of each voxel. X-ray CT images are normally generated with 256 (2^8) or 65,536 (2^{16}) grey values, which represent images with greyscale depths of 8- and 16-bits, respectively (Galdos et al., 2019; Marcheschi, 2008; Gillespy 3rd & Rowberg, 1994). Figure 5.1 presents a greyscale histogram in a typical soil X-ray CT image (16-bit). The frequency of appearance of a given grey value and consequently the presence of peaks in the histogram represent differences in the attenuation properties of the material: lower and higher grey values in the greyscale represent the photon attenuation by low and high-density materials in the soil, respectively. The presence of two noticeable peaks in the histogram is characteristic of a two-phase porous medium.

The correct delimitation of the peaks permits the identification of the phases of interest (segmentation) in the X-ray CT image. In the case of a 'dried' soil sample, the left and the right peak are associated with the soil pore space and the soil solid matrix, respectively (Fig. 5.1). However, it is common to find histograms from soil images presenting overlapping peaks, which requires more effort, and on some occasions considerably so, to define the peaks and identify the sample phases in the image. Among the factors that affect the possibility of delimiting the peaks in the histogram of an X-ray CT image is the signal-to-noise ratio attained during the acquisition of the image due to non-ideal photon statistics and the partial volume effect caused by the averaging of the X-ray intensity on the corresponding detector pixel that has to image a sharply contrasting boundary of the scanned object (Buzug,

2008). The use of specific filters, after image reconstruction, is recommended for denoising the image while preserving phase edges. Filters aimed at smoothing, sharpening, edge detection, and contrast enhancement generally improve the separation of peaks in the greyscale histogram and facilitate subsequent segmentation. In particular, edge detection filters can have significant impacts on the identification of phases and consequently on the obtention of reliable quantitative results. Thus, having the segmentation step in mind, the choice on the best filtering technique should consider its capability of producing a high-quality image without losing relevant original details.

Denoising algorithms can be classified as filtering methods in the spatial domain (operations are applied directly on the image matrix), transform domain (operations are applied after the image matrix is transformed into the frequency domain), or other domains (based, for instance, on statistical models schemes and random fields) (Goyal et al., 2020). Spatial domain methods for image denoising are classical and often applied when considering X-ray CT images. These methods exploit similarities between pixels or patches of an image through local filters, in which the denoising of a pixel is restricted by spatial distance, or non-local filters, which take advantage of the correlation amongst the entire range of pixels in an image (Goyal et al., 2020). The image filtering (in the spatial domain) is based on a spatial convolution operation between the image itself, represented by an M × N (in the simplest 2-D case) dimension matrix, and a pre-defined K × K matrix, known as the kernel or mask, which results in an image with the same original dimension (M × N). In other words, the filtering process consists of multiplying pixel values from an image by a kernel matrix that represents the filter (Marcheschi, 2008). In the filtering computation, the kernel is virtually moved pixel by pixel (voxel by voxel, in 3-D) over the input image. Commonly applied filters for X-ray CT image enhancement are: low-pass, Gaussian, median, non-local means, unsharp mask, among others (Tuller et al., 2013; Marcheschi, 2008).

The low-pass filter, sometimes called box filter, involves replacing each pixel of the image by the average of the neighbourhood pixels specified by the kernel (Keselbrener et al., 1992). Only low spatial frequencies are maintained in the digital images after the filter application. The elimination of high frequencies, associated with image noise, has the disadvantage of also eliminating well-defined edges, consequently blurring the resulted image (Marcheschi, 2008). The Gaussian filter considers a discrete 2-D Gaussian function to define the kernel (Pathmanabhan & Dinesh, 2007), so that the kernel coefficients are inversely proportional to the central value of the matrix (K × K). In this way, the pixels located in the centre have a higher weighting, i.e. greater importance, than those at the image border. One of the drawbacks of this method is there is no preservation of the image edges, producing a blurred and unclear filtered image (Marcheschi, 2008). The median filter represents a nonlinear method that, instead of considering weights as the kernel coefficients, computes the median of the pixel values in the specified kernel neighbourhood (Heinzl et al., 2018). For being characterized by preserving the edges between the different phases found in the digital images (Müter et al., 2012; Sun & Neuvo, 1994), the median filter is particularly useful and one of the most popular filters for processing of soil CT images (Oliveira et al., 2021; Zhou et al., 2020; Jarvis et al., 2017; Burr-Hersey et al., 2017).

The non-local means (NLM) filter was first introduced by Buades et al. (2005) and seeks to exploit similarity amongst pixels in a non-local manner. It is a linear filter that, instead of using small-sized kernels, potentially utilizes the entire image as a search window (Schlüter et al., 2014). The presence of similar features or patterns in the image is explored by the NLM filter (Goyal et al., 2020). Buades et al. (2011) highlighted that the most similar pixels to a given pixel have no reason to be spatially close, which is the assumption for the previously described local filters; instead, similar pixel neighbourhoods can occur anywhere in the image. This justifies the necessity to scan a large portion of the image in search of the pixels that are similar to the pixel under denoising. Thus, the neighbourhood of a pixel i can be defined as any set of pixels j in the image such that a window around j resembles a window around i (this can be performed in both 2-D and 3-D approaches). Therefore, the pixels in that 2-D or 3-D neighbourhoods are averaged to determine the new intensity value at i (the denoised pixel) (Buades et al., 2004). The NLM filter is known to be very efficient at reducing image noise without significantly losing information at phase edges and, therefore, has been the choice of many authors to filter X-ray CT images of porous media such as rocks (Garfi et al., 2020) and soils (Gao et al., 2019; Pöhlitz et al., 2019; Ferreira et al., 2018, 2019). However, a sharpening filter, for instance, the unsharp masking and sharpening high-boost filter, is often necessary after the image denoising to accentuate the high frequencies present in the image, making the discontinuous structures (e.g., contours) sharper (Pires et al., 2020; Distante & Distante, 2020; Tuller et al., 2013).

5.2 Image Segmentation

After filtering, the next step in image processing is usually segmentation, that is, the identification or separation of the phases of interest (e.g. solid matrix, pores, water, organic matter). This is the most important step for obtaining reliable quantitative information. A digital greyscale image is characterized by pixels each containing intensity value information. As mentioned, the grey values are distributed among pixels between black (0) and white (65,535) (assuming a 16-bit image) (Fig. 5.1). Depending on the objectives and the characteristics of the scanned material, different segmentation methods can be applied using either manual, semi-automatic, or full-automatic procedures (Yen et al., 1995). The manual procedure is limited by the time required for the segmentation of the samples. This is mainly due to the necessity to carefully check the edges that separate the phases of interest (e.g. pores from the solid matrix when the soil porous system is analysed to ensure an accurate segmentation). Semi-automatic procedures have the advantage that specific functions can be chosen to define the regions occupied by specific phases. These functions are used to select pixels having similar grey values and classifying them as pores or solids. This procedure speeds up the segmentation procedure, but typically requires powerful computer processing for the analysis of 3-D images. For studies on porous media, such as soils and rocks, global and locally adaptive segmentation methods are the most frequently used (Schlüter et al., 2014; Houston et al., 2013; Tuller et al., 2013; Iassonov et al., 2009; Wang, 2008; Jan, 2006), although machine learning-based segmentation has also been applied recently (Han et al., 2019; Chauhan et al., 2016).

5.2.1 Global Segmentation

Global segmentation uses a single threshold value for all pixels in the images. The value 1 (white) is assigned to pixels whose grey values are higher than the pre-defined threshold value, while 0 (black) is assigned to the remaining pixels (those whose grey values are equal or lower than the threshold value) (Iassonov et al., 2009). Methods to determine the threshold value are based on the greyscale histogram of the digital image in question (Fig. 5.1). Several widely used methods of global segmentation are based on maximum between-class variances, minimum error, maximum entropy, or Fuzzy C-Means (FCM) (Schlüter et al., 2014).

Otsu's (1979) method is one of the most popular global segmentation methods used in soil sciences (Oliveira et al., 2020; Pires et al., 2019; Pöhlitz et al., 2019; Leue et al., 2019). The Otsu method was created to find a value that provides better separation between the background and foreground (i.e. object or phase of interest) by minimizing their weighted within-class variance. The method works better for images with good contrast between background and foreground, images with bimodal histograms, and uniform lighting conditions (Iassonov et al., 2009). The minimum error thresholding method assumes the existence of two Gaussian functions related to the background and the foreground, and the optimum threshold is determined by optimizing a criterion function related to the Bayes risk (Jiulun & Winxin, 1997; Kittler & Illingworth, 1986). The maximum entropy method is based on the calculation of the entropy of the object and background (Gull & Skilling, 1984). The FCM is related to a combination of the classic k-means, which is a hard clustering technique in which each pattern is allocated to a single cluster, and the fuzzy set theory (Landini et al., 2008). The FCM is an iterative clustering method that involves minimizing the objective function (weighted squared root error function) (Schlüter et al., 2014). The minimization of the c-means functional, which is represented by a nonlinear optimization problem, is solved by different methods such as genetic algorithms, iterative minimization, etc. The objective function is minimized to a fuzzy membership (U) and a set of cluster centroids. The fuzzy membership is recomputed for each iterative step until an unchangeable U is minimized.

Although the threshold values can be determined automatically by the above-mentioned global segmentation methods and others, they might fail to segment an object from the background when there are no distinct peaks in the histogram because of the complexity of the soil, i.e. heterogeneity in the soil fabric. In some cases, researchers have determined the threshold value via visual inspection of the soil images; however, this is subjective and lacks consistency and reproducibility, therefore great caution is needed to ensure consistent phase allocation when choosing the threshold values for different samples.

Despite their simplicity and low computational cost, global segmentation methods do not consider how the grey values of the pixels/voxels are spatially distributed in images (Schlüter et al., 2014). The quality of the generated binary image greatly depends on the extent and possibility of separation between the peaks corresponding to the foreground and background in the histogram. Frequently, even departing from properly filtered X-ray CT images, global methods fail in providing a good quality segmentation, which justifies the need for alternative approaches.

5.2.2 Local Segmentation

Different from the global methods, locally adaptive segmentation methods are known to account for pixel neighbourhood statistics in the class assignment problem (Schlüter et al., 2014). Examples of local segmentation methods are the hysteresis, indicator kriging, and watershed methods. The hysteresis segmentation is employed in problems that involve two-class segmentation (Pini & Madonna, 2016). The procedure is carried out first segmenting the digital image with an upper threshold (high-edge) to identify object pixels for which the uncertainty of class assignment is highest. Thus, pixels brighter than the upper threshold value are considered as belonging to the object. These pixels are assigned as true foreground pixels. The second threshold (low-edge) identifies pixels that belong to a low-intensity class. Pixels darker than this lower threshold will be assigned as the background. The use of a pre-defined kernel (for example, a square) is employed for connecting the pixels defined as high- and low-edges in the segmentation procedure (Pini & Madonna, 2016). Pixels with intermediate grey values (uncertain) are assigned to the low-intensity class when their neighbouring pixels belong to the low-intensity class. Otherwise, they are considered as objects when connected to pixels of this class (Schlüter et al., 2014). Unassigned pixels are classified as high-intensity class pixels. The quality of the segmentation by this method depends mainly on the choice of the low- and high-edge thresholds.

The indicator kriging method completes the segmentation via the calculation of the indicator covariance functions from the image-data (Oh & Lindquist, 1999). In the method, a lower threshold value is defined and the pixels with grey values below this cut-off belong to the background, while the second class includes those pixels with grey values larger than a defined higher threshold value, for example, the objects (Oh & Lindquist, 1999). Unassigned pixels, with grey values between the higher and lower threshold values, will be assigned to the first or second classes depending on the probability that the voxel belong to the background or foreground. The probability is estimated employing the indicator kriging method, which considers the spatial location of the undefined pixel and its neighbours (Oh & Lindquist, 1999).

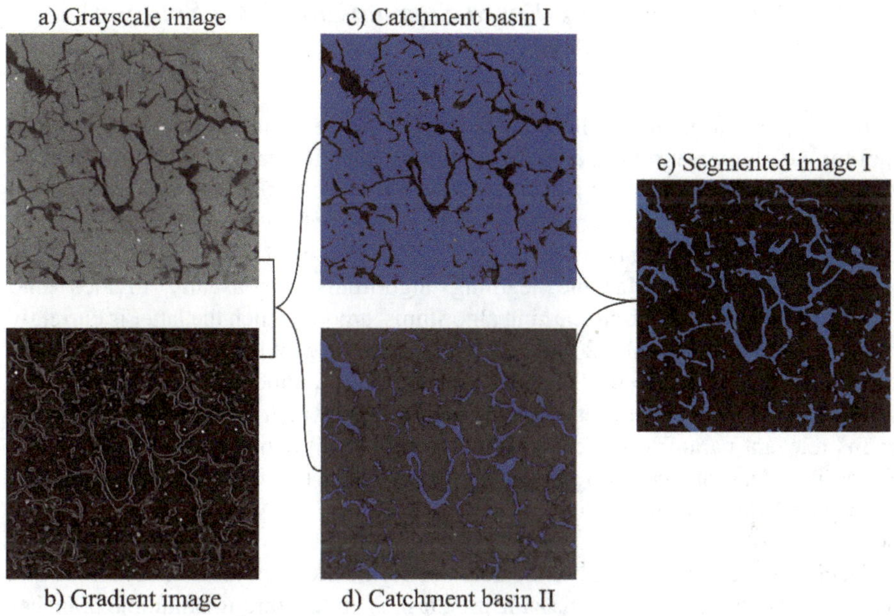

Fig. 5.2 Schematics of watershed segmentation: (**a**) Greyscale 2-D section of a soil X-ray CT image showing the pore and solid phases in darker and brighter grey tones, respectively. (**b**) Output image of the gradient operator applied on the greyscale image; dark regions represent the 'valleys' (local minimum) corresponding to the interior of pores and solids; bright lines represent the 'mountains' or 'dams' corresponding to the boundaries of these pores and solids. (**c**, **d**) Catchment basins filled with markers (seeds) representing the solids (I) and pores (II). (**e**) Result of the 'flooding' process applied in the watershed segmentation

The watershed segmentation method considers a digital image as a topographic surface, and the grey value of the pixels is interpreted as altitudes on the surface (Jan, 2006). By applying the gradient operator on the greyscale image, an output image is obtained in which the valleys (local minimum) correspond to the interior of phases (e.g. pores or solids) and are surrounded by 'mountains' that correspond to the edges of phases (boundaries between pores and solids) (Fig. 5.2) (Beucher & Meyer, 1993). In other words, the watershed transform decomposes an image into catchment basins that are filled with 'water' in a 'flooding process'. When the waterfront from different basins reaches the highest altitudes, they are prevented from merging by the dams represented in the gradient image by the phase edges (Jan, 2006). The selection of markers (seeds selected to start the immersion process) inside the catchment basins is a crucial step for this method and can be manually or automatically defined, considering specific knowledge of the objects (phases) (Fig. 5.2). The method generally succeeds, even when there is poor contrast between regions of the digital image.

5.2.3 *Machine Learning-Based Segmentation as a Solution to New X-ray CT Imaging Challenges*

In recent years, artificial intelligence (AI) has become an interesting and powerful tool for the segmentation and analysis of tomographic images (Weikert et al., 2020). Among several definitions of AI, the following by Ertel (2017) is most appropriate: 'the study of how to make computers do things at which, at the moment, people are better'. Machine learning is one of the techniques used by AI to achieve this purpose (Chowdhary, 2020). Machine learning algorithms are usually divided into unsupervised and supervised learning algorithms, among which the latter is currently more widely applied (Joshi, 2020). For a classification problem such as the segmentation of a digital image in different classes (phases), supervised algorithms need samples with labelled classes (training data) from which a mathematical model learns relevant parameters and becomes able to make predictions on new samples. Supervised learning comprises different approaches based on, for example, the concepts of decision trees, support vector machines (SVM), and artificial neural networks (Joshi, 2020; Chauhan et al., 2016).

Decision trees consist of a hierarchical decision-making process at each node of a tree-type structure, so that the classification task is guided through multiple branches of alternative decisions. Thus, creating and aggregating multiple trees of similar architecture results in an ensemble method (e.g., bagging, random forest, and boosting ensembles) that permits producing robust predictive models with greater classification performances (Joshi, 2020; Polikar, 2012). The SVM, on the other hand, was originally developed to perform binary classification based on the construction of an optimal hyperplane that imposes a maximum separation between two classes. Some other approaches are currently being considered to extend the SVM method for multi-class classification. Despite increasing the complexity of the problem, which requires tuning of some hyperparameters, it may provide high accuracy and generalization capabilities (Tzotsos & Argialas, 2008). With the recent technological advances, neural networks are emerging in the context of the learning process called *deep learning*, which is a particular area of machine learning. Deep artificial neural networks contain multiple hidden layers of neurons aimed at learning complex patterns in large volumes of data (Wani et al., 2020). It can only converge for satisfying accuracy if trained with enough data followed by consistent computational manipulation. The full potential of this technology has been enabled by the advent of graphics processor-based computation since it allows parallel training of deep networks (Joshi, 2020).

Known challenges when applying X-ray CT to soil-related investigations are (1) satisfying the repetition/replication requirements for appropriate statistical analysis, since X-ray CT measurements of dense samples (those substantially composed of high attenuation materials) are time-consuming and often costly, and (2) processing and segmenting the whole image dataset to obtain quantitative information. The former challenge will no longer be an issue for those with access to fourth-generation Synchrotron Light Source imaging as the high photon flux provided at these facilities, in the hard X-ray range, enables fast image acquisition even of materials composed of high Z (atomic number) elements (Craievich, 2020). This characteristic is essential for 4-D (time-resolved three-dimensional) experiments which consist of acquiring a sequence of 3-D images that allows understanding dynamic processes (Ferreira et al., 2022). For instance, Pak et al. (2020) recently demonstrated that this experimental approach, with a scanning time of ~6 min per 3-D image, had great importance for observations of pore-scale processes governing the multiphase flow in porous media. Comparatively, at MOGNO beamline (MicrO and NanO Tomography beamline at the Brazilian Synchrotron Light Source, SIRIUS), it is expected that for future research 3-D imagery will be possible in few seconds (1–5 s) (Costa et al., 2017; Ferreira et al., 2022). Thus, the necessity of fast processing and segmentation of very large image datasets is going to be a reality at MOGNO, and in upcoming beamlines dedicated to X-ray CT throughout the world (though it is worth noting the majority of studies using CT to explore soils are currently undertaken using benchtop systems).

In this chapter we propose the application of strategies based on Machine Learning followed by Deep Learning (Pinto et al., 2022; Spina et al., 2018; Vasconcelos et al., 2018) to segment soil 3-D images into its pore and solid phases. This will serve as a preparation to deal with large soil image datasets such as those used in time-resolved soil studies.

5.3 An X-ray CT Image Segmentation Protocol Based on Machine Learning and Deep Learning Strategies: A Case Study

5.3.1 X-ray CT Image Acquisition and Preparation for the Machine Learning Pipeline

3-D images of soil aggregates (~3 mm diameter each) generated at the X-ray microtomography beamline (IMX)—second-generation light source—at the Brazilian Synchrotron Light Laboratory, Brazilian Center for Energy Research and Materials (LNLS/CNPEM) were selected as demonstration specimens. The samples (33 in total) were scanned using a polychromatic beam, filtered by a 550 μm thick Si filter, and 1024 projections were acquired over 180° rotation of the sample. The exposure time for each projection was 2 s, leading to a total scan time of ~35 min per sample. An in-house filtered back projection-based algorithm (Miqueles et al., 2018) was applied to reconstruct the images in 16-bit, raw type, which had a maximum array of $2048 \times 2048 \times 2048$ voxels, with a voxel side length of 1.64 μm, resulting in image volumes of $3.36 \times 3.36 \times 3.36$ mm^3 each.

For the development of the current segmentation protocol, a series of steps was adopted and is illustrated in Figs. 5.3, 5.4, 5.5. The steps presented in Fig. 5.3 were performed using the Avizo software v. 2019.4. They served for preparing the images

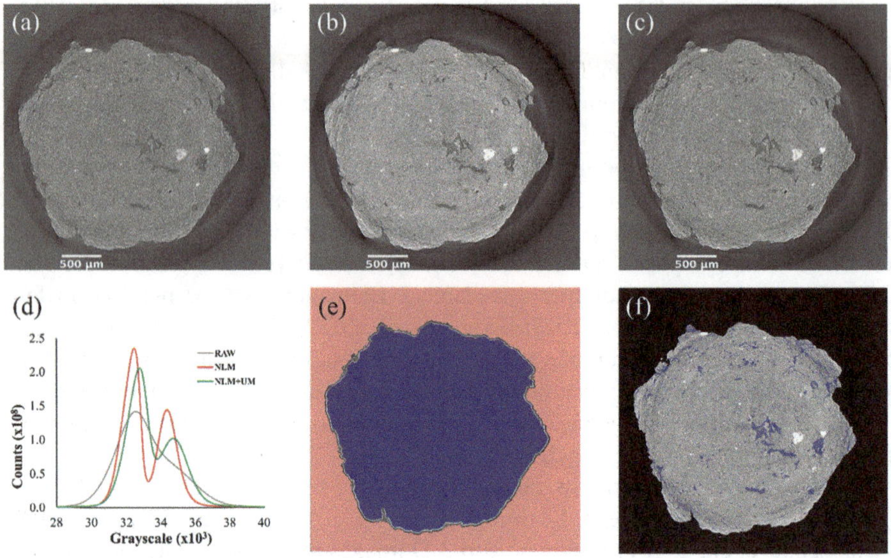

Fig. 5.3 Tomographic slice at the central region of a soil aggregate (z = 900): (**a**) raw image, (**b**) after non-local means filter (I_{NLM}), and (c) non-local means filter followed by the unsharp mask filter ($I_{NLM + UM}$). (**d**) Greyscale histograms of the raw, I_{NLM}, and $I_{NLM + UM}$ images. (**e**) Seeds for an immersion process at the aggregate (purple) and background (pink). (**f**) I_{NLM} after background masking and simple thresholding based on the valley between peaks (33,289) of I_{NLM} histogram (blue represents grey values <33,289)

for examination via the machine learning techniques (Note: this task can be performed using other software or included in the machine learning pipeline). To start, one 3-D image volume among the entire dataset was randomly selected, and a sub-volume of 2048 × 2048 × 1938 voxels (x, y, z coordinates), containing the aggregate, was extracted to avoid unnecessary computational processing in regions exclusively containing background information. Fig. 5.3a shows a slice from the raw image at the central region of the aggregate (z = 900). Intrinsically X-ray CT images are characterized by having a certain level of noise, mainly introduced by an unfavourable conversion rate of X-ray photons to visible light when scintillator-based detection systems are used (Banhart, 2008). Before any analysis, the image noise needs to be reduced to allow a more accurate segmentation, especially if the noise is too prominent, as the case presented in Fig. 5.3a. In this study, the image was denoised by a non-local means filter (I_{NLM}) (Mode: GPU Adaptive Manifold 3-D—spatial standard deviation: 1, intensity standard deviation: 0.3, search window: 10, local neighbourhood: 5) (Fig. 5.3b) and subsequently had its edges enhanced by an unsharp mask ($I_{NLM + UM}$) filter (Edge size: 4, Edge contrast: 0.7, Brightness threshold: 0) (Fig. 5.3c). This produced evident effects on the greyscale histogram associated with the images (Fig. 5.3d), imposing a separation between the peaks corresponding to the air (left peak) and the solid (right peak) phases. It is seen that the effect was more pronounced for I_{NLM} than for $I_{NLM + UM}$.

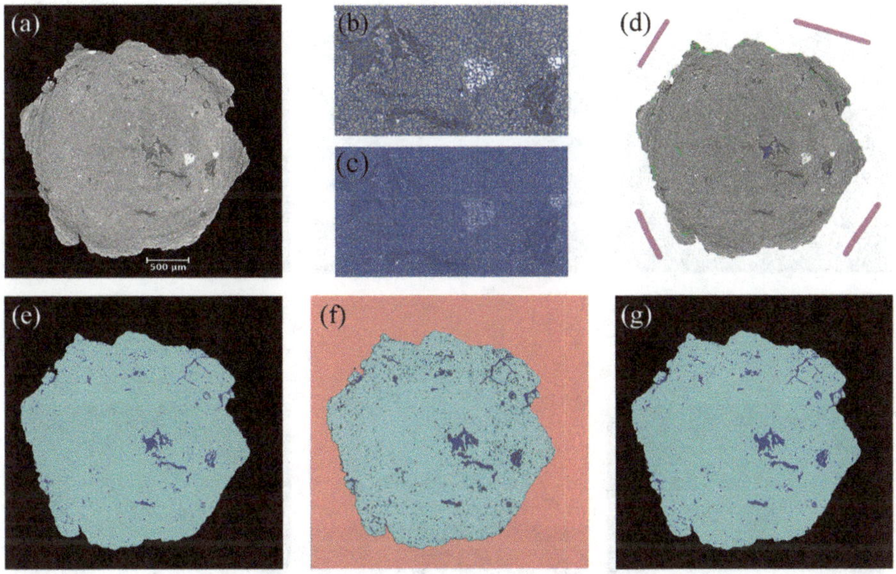

Fig. 5.4 (**a**) Tomographic slice at the central region of a soil aggregate (z = 900) after non-local means filter followed by unsharp mask filter ($I_{NLM + UM}$) and background masking (in black). Waterpixels in a zoomed fraction of (**a**) with seed spacing and compactness values, respectively, set to (**b**) 9 and 1000, and (**c**) 3 and 10. (**d**) Handcrafted annotations on (**a**) at background (pink), solid matrix (green) and pores (blue). (**e**) Classification result of (**a**) via machine learning, where pores and solid matrix are in royal and turquoise blue, respectively. (**f**) Seeds for an immersion process at the background (pink), solid matrix (turquoise blue), and pores (royal blue). (**g**) Result of watershed algorithm computation based on (**f**)

It is important to explain that restricting the X-ray CT analysis to a regular ROI (e.g. cube) inside a naturally irregular soil aggregate image, something frequently performed in studies involving soil aggregate image analysis (Camargo et al., 2022; Ma et al., 2020; Ferreira et al., 2019; Peth et al., 2008), disregards some information from the borders of the aggregate. An interesting way of exploring the entire aggregate volume is to assign a label to the background to account for its volume whenever necessary (e.g. in Zhao et al., 2020). For that, markers slightly smaller (purple-coloured, Fig. 5.3e) and larger (pink-coloured) than the aggregate boundaries were generated, representing seeds for the aggregate and background, respectively, to be subsequently considered in an immersion process. The markers were created using a combination of simple thresholding, morphological filters such as erosion/dilation, and fill-holes operations. High-gradient zones in the greyscale image were identified by an algorithm of edge detection (default Canny method in Avizo) and the unclassified zones were then filled by applying a watershed algorithm (Schlüter et al., 2014; Beucher & Lantuejoul, 1979), resulting in well-defined labels for the background and the aggregate regions. In Fig. 5.3f, the black area (zero value assigned) represents the background region after using the result from the watershed method to mask the I_{NLM} (Fig. 5.3b). Figure 5.3f shows, in blue, the result of simply

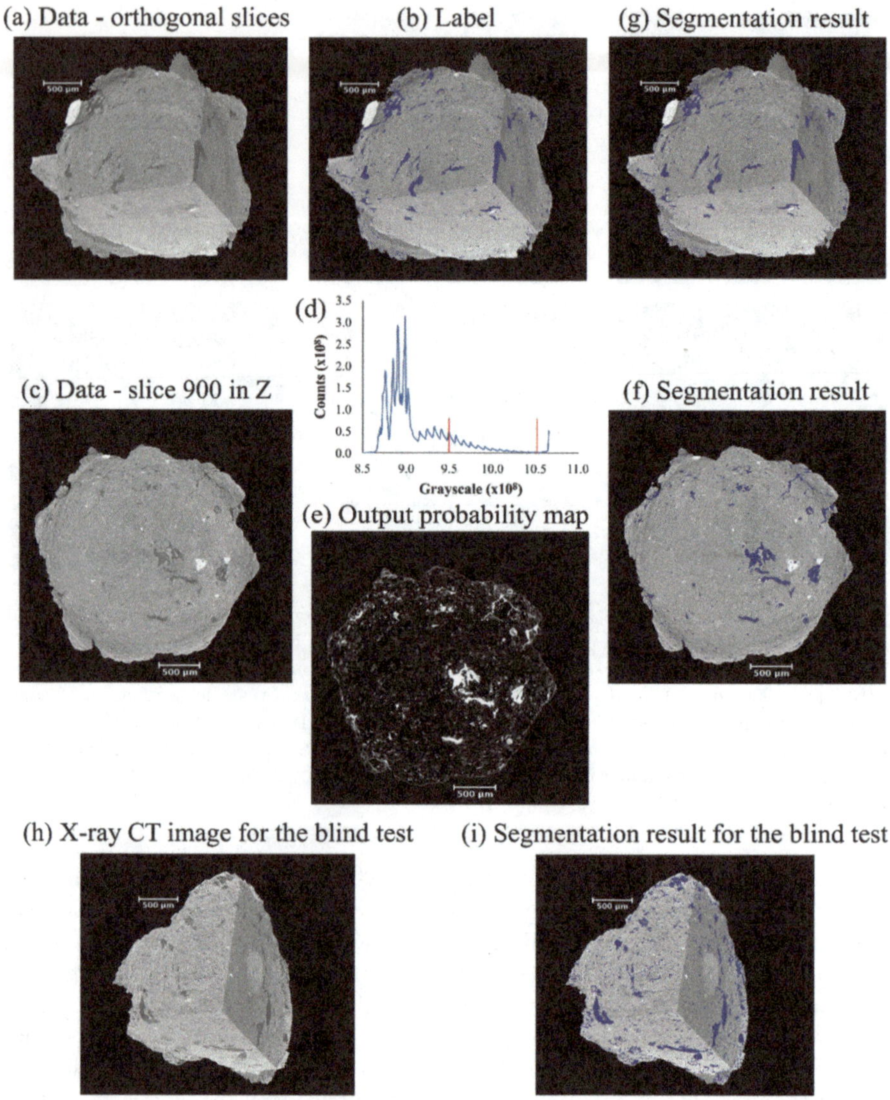

Fig. 5.5 (**a**) Orthogonal tomographic slices of a soil aggregate after non-local means filter (I_{NLM}), and background masking (in black), constituting the Data (see Table 5.1). (**b**) Same orthogonal slices showing pores in blue from the corresponding label image resulted from Fig. 5.4g. (**c**) Slice z = 900 of (**a**). (**d**) Greyscale histogram of the (**e**) inference probability distribution output for the pore class, with the threshold of 9.5×10^8. (**f**) The final classification of pores in blue with the threshold of 10.5×10^8. (**g**) Same orthogonal slices of (**a**) showing pores in blue from the final classification via deep learning. (**h**) Orthogonal tomographic slices of another soil aggregate (I_{NLM}) with the masked background (in black) for a blind inference test and (**i**) the resulting classification of pores with the threshold of 10.5×10^8

considering the grey value between the air and solid peaks (33,289, I_{NLM} curve, Fig. 5.3d) as a threshold value to segment the pore and solid phases. The outcome of this choice is, in general, regions of under and overestimation of pores mainly at the centre and close to the borders of the aggregate, respectively. Similar strategies using thresholding, morphological filters, and fill-holes operations failed in the attempt of generating accurate markers for the pore and solid phases, to apply a watershed algorithm. This happened as slight variations in the chosen threshold value eliminated important markers inside pores or created fake markers in the solid matrix. Thus, this reinforces the necessity of alternative segmentation solutions that not only make the traditional methods faster but also provide reliable results.

5.3.2 Machine Learning Pipeline

Aiming to accomplish the real-time segmentation of large 4-D X-ray CT data especially at SIRIUS synchrotron facility, Annotat3D*, a software based on Machine Learning and High-Performance Computing (HPC) techniques (Pinto et al., 2022; Spina et al., 2018; Vasconcelos et al., 2018) has been developed by the Scientific Computing Group at the Brazilian Synchrotron Light Laboratory, Brazilian Center for Energy Research and Materials (LNLS/CNPEM) (*in-house software, i.e. not currently released for wider use). Annotat3D was employed to conduct the steps illustrated in Fig. 5.4a-e. Different from the available machine learning-based segmentation algorithms, such as Waikato Environment for Knowledge Analysis (WEKA) (Arganda-Carreras et al., 2017), which works on a pixel-level classification, the Annotat3D software is based on a superpixel classification (Vasconcelos et al., 2018). A superpixel is formed by a group of neighbouring pixels with similar intensities, meaning that fewer data, composed of local contextual information, need to be processed for the final segmentation. In addition, since the method is implemented using HPC with parallel programming via multiple GPUs, the superpixel classification requires much less computational time than a pixel-level classification. In this context, a method to classify superpixels in soil images, belonging to either solid or pore space, must be optimized by setting important parameters related to: feature extraction, superpixel arrangement, and the type of classifier to be used.

Feature extraction consists of filtering the input greyscale image to enrich the information captured by the superpixels, i.e. ensuring that relevant properties of the objects of interest are considered. In other words, extracted feature maps are filtered versions of the input image. The following filter options are currently available: None (Original Image), Fast Fourier Transform (FFT) Gauss, FFT Gabor, FFT DoG, Sobel, Membrane Projections, Minimum, Maximum, Average, Variance, Median, and LBP (local binary patterns). Also, a feature selection procedure can be enabled to impose a percentage threshold criterion (1% is the default) on the importance of features for classification accuracy. In this way, those contributing with less than the importance threshold value may be disregarded, resulting in less computational processing time without a reduction in the classification quality. The current version of the Annotat3D software offers two approaches to estimate the superpixels using multi-GPU code: the Simple Linear Iterative Clustering (SLIC) algorithm (Achanta et al., 2012) and a method based on the watershed transformation (Gonçalves et al., 2019). In the latter case, the resulting superpixels are called 'waterpixels' and they may be computed in 2-D, for each of the z-slices, or in 3-D. Both approaches allow controlling the spatial distribution of the superpixels by setting the two following parameters: seed spacing (increasing this number generates a lower amount of superpixels) and compactness (increasing this number smooths the borders of superpixels, making them more regular). Additionally, two types of classifiers are currently available: the Random Forest and the Support Vector Machines (Ho, 1998; Cortes & Vapnik, 1995).

Figure 5.4a shows the same slice (z = 900) presented in Fig. 5.3, after using the result from the watershed method (Fig. 5.3e) to mask the background (zero value assigned) of $I_{NLM + UM}$ (Fig. 5.3c). Hence, the masked $I_{NLM + UM}$ was used as an input image for the machine-learning pipeline. $I_{NLM + UM}$ was chosen over I_{NLM} for this study but there would be no restriction to use the I_{NLM}, which as well demonstrated good results. For the tested protocol, all filters were enabled for feature extraction with the 'default feature selection procedure'. The watershed transform-based method was selected to compute 2-D waterpixels and the Random Forest classifier with 200 trees was selected to be trained.

It is important to emphasize that when choosing among the approaches for superpixel estimation, one must keep in mind that to avoid errors in the subsequent classification, the superpixels should fit the edges of the object of interest in the image, e.g. pores in the case of the soil. For instance, for better visualization, Fig. 5.4b shows the waterpixels formed in a zoomed fraction of the same slice with seed spacing and compactness values set to 9 and 1000, which seemed to be well adapted to the pore borders at first glance. However, this approach provided low-quality classification, especially for narrower pores (not presented). The best classification result was obtained by setting seed spacing and compactness values to 3 and 10, respectively. This produced a narrower distribution of waterpixels (Fig. 5.4c), which was tested to confirm its functionality.

Once the superpixels were generated, the next step consisted in declaring the classes (labels) for image segmentation, which were in the present case: background, soil solid matrix, and pores. As a supervised method, the Annotat3D software operator uses his/her judgement to add markers (or annotations) on a given slice of the input image in regions of background (pink), solid matrix (green), and soil pores (blue), as exemplified in Fig. 5.4d (superpixels not shown). The marked superpixels serve as a training set for the classifier. In other words, the classifier learns from the input labelled superpixels, which are mathematically correlated with the extracted feature maps and provide a full classification (segmentation) of the remaining superpixels of that particular slice or of the image as a whole. As the classifier takes longer in the latter case, the best-tested approach was to successively check the classification result for slices where markers were added (saving the cumulative markers at the end of the optimization for each slice), until the operator judges it is time to check the result for the entire image. In this process, which is optimized as more slices are marked, it is important to identify the existence of frequently misclassified regions and make sure that correct markers are placed on them, so that the classifier can be optimized.

One can assume that the necessity of few markers for a good classification makes the classifier more efficient and consequently less operator time will be required in the process. For the tested protocol, markers were added in approximately 200 out of 1938 slices when the classification quality for the whole image was considered reasonable. It is important to mention that reaching an acceptable classification for the whole image allows the operator to save the classifier method and/or the training features as different files. One of these files can be later uploaded and used for classifying any other image from the original dataset. Nonetheless, these images need to undergo the same preparation steps (i.e. filtering and background masking, Fig. 5.3). The file containing the classifier model cannot be further modified for other input images, whereas the file containing the training features allows the operator to change the superpixel estimation algorithm, the classifier type, and, if non-reliable results are achieved, add new markers to improve the classification.

Figure 5.4e shows the classification result for the previously mentioned slice ($z = 900$) which shows pores were better delimited via machine learning than by simple thresholding (Fig. 5.3f). There was less under and overestimation of pores at the centre and the outer regions of the aggregate, respectively. Nevertheless, it is still possible to identify some flaws by carefully comparing Fig. 5.4e with Fig. 5.4a. Therefore, the resultant labels from the machine learning classification were used to define seeds for a final watershed segmentation of $I_{NLM + UM}$ in Avizo. Aiming to leave an unclassified zone (grey) for the watershed algorithm computation (Fig. 5.4f), the pore (royal blue) and solid matrix (turquoise blue) labels were minimally eroded, while the background (pink) was not modified. The result of this procedure is depicted in Fig. 5.4g, which demonstrates a very good agreement of pore delimitations when compared with Fig. 5.4a.

With respect to the segmentation protocol described so far, one aspect worth noting is that the steps illustrated in Fig. 5.3, which were conducted using the Avizo software v. 2019.4, could alternatively have been included in the machine learning pipeline. One alternative would be to train the classifier from scratch, that is, without masking the background of $I_{NLM + UM}$. However, this imposes a higher level of difficulty for the optimization of the classifier, and, based on the current experience, it would be more time-consuming. Another option would be to split the machine learning pipeline, creating two different classifiers: one specialized in classifying the soil aggregate and background, which would replace the steps described in Fig. 5.3 to create the background mask (not tested here), and a second one specialized in classifying the solid matrix and pores, just like that described in Fig. 5.4.

Another relevant aspect to consider is that approximately 10 days were required to reach the result presented in Fig. 5.4g. In these 10 days, the conducted steps were: the preparation of the input image for the machine learning pipeline, the addition of markers in ~200 slices, and final watershed segmentation. The addition of markers was the most time-consuming step and was more related to the current efficiency of the machine learning algorithms than with computer processing power limitations. Presently, this operation is far from being a fast process, but the great advantage lies in the possibility of applying the trained classifier or training features originated from this laborious process for fast segmentation of the remaining images from the same dataset (this currently takes just few minutes per image), which was tested and presented a visual classification quality similar to that obtained in Fig. 5.4e (not presented here). Annotat3D software is under continuous development and has been shown to become faster and more efficient with the on-going improvements, thus requiring fewer markers per slice and fewer slices in total containing markers, which is currently what slows the process. Although there are alternative machine learning tools available that can be used for image processing, such as ImageJ (WEKA) and Dragonfly, Annotat3D is being specifically designed to handle large amounts of data in a multi-node/multi-GPU HPC environment. The software encompasses highly optimized tools to visualize, process, and segment these data in timely manner, preferably faster than other available alternatives.

5.3.3 Deep Learning Pipeline

At this point, the proposed segmentation protocol advances from the classification result obtained via machine learning and watershed methods (Fig. 5.4g) to the application of deep learning techniques to make the whole process more general, robust, and automated. More specifically, deep learning techniques aim at finding good representations of the input data as a hierarchy of features, with more abstract representations computed in terms of less abstract ones (Goodfellow et al., 2016). While traditional machine learning algorithms cannot learn features directly from the data and thus rely more on human effort (Fig. 5.4d), deep learning algorithms perform both feature extraction and classification from previously computed training label images (Bengio, 2009). The same in-house software, Annotat3D, was used to apply deep learning techniques to train a convolutional neural network to reproduce a classification considered as reference (ground truth), and also classify new similar data. Thus, it was necessary to methodically test the parameters listed in Table 5.1 to reach an optimized set of training constraints.

In this study, as a start, a workspace was created, which consists of a directory automatically structured by the Annotat3D software to save all the decisions from the deep learning pipeline. Next, a dataset was built to serve as a basis for the training step (Table 5.1). The dataset included: the masked I_{NLM} as the input data (16-bit; raw type—orthogonal planes of the image shown in Figs. 5.5a, c), the corresponding label image resulting from the machine learning and watershed methods (8-bit; raw type—orthogonal planes of the image with pores represented in blue, Fig. 5.5b), and the weight map (8-bit; raw type) based on the label image. The label image consists of values 0, 1, and 2, attributed to the background, pores, and solid matrix, respectively, and these values essentially identify the different classes present in the label image. On the other hand, the weight map was used to compel the network to focus

Table 5.1 Parameters considered in the deep learning pipeline

Dataset manager		Network manager			Inference	
Sampling	Augmentation	Network	Dataset	Settings	Inference	Settings
Data	Vertical Flip	Unet2-D	Number of images	*System:*	Network	*System:*
Label	Horizontal Flip	Unet3-D	Number of samples	Number of GPUs	Input images	Number of GPUs
Weight	Contrast	Vnet	Dimensions	*Training:*	Output folder	*Patches:*
Sampling:	Gaussian blur		Data info	Batch size		Volume padding
Number of classes	Elastic deformation		Label info	Iterations		Patch border
Sample size			Weight info	Learning rate		
Patch size				Loss function		

on the pores, as this class represented, in the analysed case, only 1% of the entire image and consequently represented a more challenging task in the performed classification. In the weight map from the investigated image, intensities of 1, 255, 100, and 20 were attributed to the background, centre of pores, borders of pores, and solid matrix, respectively. After loading the data, label, and the weight map (images with dimensions of 2048 × 2048 × 1964 voxels and a voxel side length of 1.64 μm), the sampling was arranged by assigning the following values to the number of classes, sample size, and patch size: 3, 40, and 500 × 500 × 500, respectively, resulting in a dataset of 30.7 GB. In the Annotat3D software, the sample size characterizes the number of sub-images, called patches, to be selected from the input images (data, label, and weight map) to subsequently feed the neural network training, while the patch size represents the fixed 3-D dimensions of each of these sub-images. One can also include image augmentation strategies such as vertical/horizontal flip, contrast variation, additional Gaussian blur, and elastic deformation (Table 5.1) to enrich the training data (not tested here). However, some precautions should be taken as the size of the dataset may be dramatically increased by doing this.

Convolution Neural Networks (CNNs) are the most successful artificial neural networks for extracting features from sub-images used for image classification (Peixinho, 2017). In essence, CNNs are composed of a sequence of layers that combine linear convolution and non-linear operations such as activation, pooling, and normalization. The Annotat3D software offers two different CNN architectures: the Unet (2-D and 3-D approaches) and the Vnet (only for the 3-D approach) (Milletari et al., 2016; Ronneberger et al., 2015). It was observed that Vnet provided the best classification results for the current segmentation protocol. After choosing the neural network architecture (Vnet), the dataset was loaded, and, at this point, the Annotat3D software shows a list of the dataset's characteristics (Table 5.1) that allows the verification of its correctness. Afterwards, in the settings menu, four GPUs (NVIDIA Tesla V100 32 Gb) were selected for the training step. It is known that the gradient descent-based backpropagation is a useful learning algorithm that aims to minimize a loss function calculated using the desired output (the label image) and the obtained output classification (Wani et al., 2020). The loss function can take into consideration one, all, or a fixed intermediate number of patches (the batch size) per iteration of the backpropagation algorithm (Peixinho, 2017). The loss function tends to provide lower values as the number of iterations increases. Nonetheless, one needs to find a balance among error minimization, training time, and visual accuracy between the obtained classification and the label image. In the present case, the batch size, number of iterations, learning rate, and type of loss function that provided the best results were: 1, 60,000, 10^{-5}, cross-entropy, respectively. It resulted in a training time of ~24 h. It is important to mention that, in the current implementation of the Annotat3D software, each GPU processed one batch, which means that setting a batch size equal to one, and using four GPUs, resulted in a real batch size of 4.

When the training was finished, the network was exported and loaded together with the greyscale I_{NLM} (same data used for training, Figs. 5.5a, c) in the inference menu (Table 5.1). Four GPUs were selected to perform the inference task. By default, the inference generates a 32-bit float tiff greyscale image for each class. Each image contains the probability distribution over the specified class (the brighter the regions, the higher the probability of that region to belong to the specified class: 0—background, 1—pores, or 2—solid matrix). For instance, Fig. 5.5d shows the greyscale histogram of the probability distribution output for class 1 (pores) (Fig. 5.5e), after converting it to a 32-bit unsigned raw image type, for convenience. To reach a final segmentation and consequently limit the pore regions, a threshold value was carefully chosen from the greyscale histogram. For example, the threshold value of 9.5×10^8 overestimated the pore class (Fig. 5.5e) while the threshold value of 10.5×10^8 provided an excellent classification of pores (Fig. 5.5f). Figure 5.5g shows orthogonal slices of the resulting segmentation, which showed excellent visual agreement with the label used for training of the neural network (Fig. 5.5b).

As the main purpose of the developed protocol was to classify new data, a greyscale tomographic image from another soil aggregate (Fig. 5.5h), belonging to the same experimental set described at the beginning of this section, was submitted to the trained CNN, in a blind test mode. The segmentation result for this blind test, using the same threshold value of 10.5×10^8, is shown in Fig. 5.5i. Visually the pore class was also correctly segmented for this other soil aggregate. The remaining 31 tomographic images were also segmented, taking 2 to 4 hours of inference per image, depending on the volume occupied by the soil aggregate. So, all images could be segmented with good quality within a few days. Although this trained CNN has demonstrated to be suitable for the segmentation of a specific set of images, which were all generated in approximately the same experimental conditions, this can be useful for other sets of similar data, e.g., soil images generated with other magnifications (different voxel side length) and other X-ray beam and detector characteristics. Occasionally, the segmentation result might not be as good as it was for the original dataset, but the user can add new datasets as examples to fine-tune the CNN training and optimize its performance on the inference of new data.

Another important aspect is that, for this protocol, a whole segmented image was provided to build the dataset, but this is not mandatory. In other words, smaller fractions of the image could have been tested to investigate the potential of training a CNN. For certain, it would demand less prior effort to generate a target label via machine learning and watershed techniques. Similarly, extending its applicability to other sets of images can be performed by providing small datasets based on such different images. The timeframe for training (24 hours) and inference (a few days for 31 images) has already accelerated considerably compared to traditional segmentation methods (where months might have been necessary). Nonetheless, Annotat3D software is being improved to handle an increasing number of image repetitions and new time-resolved imaging demands at Sirius Synchrotron Facility. It is worth mentioning that the current version of Annotat3D is being deployed for usage by the beamlines of Sirius to enhance the user experience and do final debugging/optimization. The work presented in this chapter was paramount for aiding in this process. It is expected that the software will become available to the community in the near future.

5.4 Conclusions

X-ray CT imaging has become an important tool for soil sciences research. Performing high-quality segmentation is a critical step for revealing details of inner sections or volumes of soil X-ray CT images. This can open up the possibility of modelling and/or characterizing morphometric characteristics of the intricate structure of the porous system. Frequently, depending on some characteristics of the X-ray CT measurement and the condition of the sample, the related greyscale histogram does not exhibit a clear distinction between peaks associated with its phases (air and solids in general). In such cases, smoothing the image noise and enhancing phase edges by filtering are necessary, but often not sufficient to allow segmentation by a traditional global threshold method. Even local segmentation methods such as the watershed transform, which usually outperform global thresholding, becomes challenging as it depends on accurate markers for the immersion process.

Recent developments have presented complementary, or perhaps alternative, properties to the global and local segmentation methods, some of which have been employing AI for the segmentation process. In this chapter, we presented the use of a new software (Annotat3D) and protocols, based on machine learning and deep learning, that permitted the investigation, with good results, of samples with poor phase distinction. Besides providing good quality segmentation for challenging images, the followed pipelines led to faster than the usual segmentation which will extend the analysis capacity and facilitate the conduction of time-resolved studies.

Acknowledgements This research used resources of the Brazilian Synchrotron Light Laboratory (LNLS), an open national facility operated by the Brazilian Center for Research in Energy and Materials (CNPEM) for the Brazilian Ministry for Science, Technology, Innovations, and Communications (MCTIC). The IMX beamline staff is acknowledged for assistance during the experiments. The Scientific Computing Group is acknowledged for assistance on the use of the in-house software (Annotat3D). TRF thanks Lucas Portes Ramos, participant student in the 29° Summer Scholarship Program held by CNPEM, for collaborating in the methodical testing of the deep learning parameters. LFP would like to acknowledge the financial support provided by the Brazilian National Council for Scientific and Technological Development (CNPq) and the Coordination for the Improvement of Higher Education Personnel (Capes) through the Grants 304925/2019-5 (Productivity in Research) and 88881.119578/2016-01 (Visiting Professor).

References

Achanta, R., Shaji, A., Smith, K., Lucchi, A., Fua, P., & Süsstrunk, S. (2012). SLIC superpixels compared to state-of-the-art superpixel methods. *IEEE Transactions on Pattern Analysis and Machine Intelligence, 34*, 2274–2281. https://doi.org/10.1109/TPAMI.2012.120

Arganda-Carreras, I., Kaynig, V., Rueden, C., Eliceiri, K. W., Schindelin, J., Cardona, A., & Seung, H. S. (2017). Trainable Weka segmentation: A machine learning tool for microscopy pixel classification. *Bioinformatics, 33*, 2424–2426.

Backeberg, N. R., Iacoviello, F., Rittner, M., Mitchell, T. M., Jones, A. P., Day, R., Wheeler, J., Shearing, P. R., Vermeesch, P., & Striolo, A. (2017). Quantifying the anisotropy and tortuosity of permeable pathways in clay-rich mudstones using models based on X-ray tomography. *Scientific Reports, 7*, 14838.

Bacq-Labreuil, A., Crawford, J., Mooney, S. J., Neal, A. L., & Ritz, K. (2019). Cover crop species have contrasting infuence upon soil structural genesis and microbial community phenotype. *Scientific Reports, 9*, 7473.

Banhart, J. (2008). *Advanced tomographic methods in materials research and Engeneering*. Oxford University Press.

Bengio, Y. (2009). Learning deep architectures for AI. Found. Trends® Mach. *Learn., 2*, 1–127. https://doi.org/10.1561/2200000006

Beucher, S., & Lantuejoul, C. (1979). Use of watersheds in contour detection. In *International Workshopon image processing: Real-time edge and motion detection/estimation* (pp. 12–21). Rennes.

Beucher, S., & Meyer, F. (1993). The morphological approach to segmentation: The watershed transformation. In E. Dougherty (Ed.), *Mathematical morphology in image processing* (pp. 433–481). Marcel Dekker.

Borges, J. A. R., Pires, L. F., Cássaro, F. A. M., Auler, A. C., Rosa, J. A., Heck, R. J., & Roque, W. L. (2019). X-ray computed tomography for assessing the effect of tillage systems on topsoil morphological attributes. *Soil & Tillage Research, 189*, 25–35. https://doi.org/10.1016/j.still.2018.12.019

Buades, A., Coll, B., & Morel, J.-M. (2011). Non-local means Denoising. *Image Processing On Line, 1*, 208–212. https://doi.org/10.5201/ipol.2011.bcm_nlm

Buades, A., Coll, B., & Morel, J. M. (2005). *A non-local algorithm for image denoising, in: Proceedings of the IEEE computer society conference on computer vision and pattern recognition* (pp. 60–65). IEEE Comput. Soc. https://doi.org/10.1109/CVPR.2005.38

Buades, A., Coll, B., Morel, J.M., (2004). *On image denoising methods*. Technical Report 2004, Centre de Mathematiques et de Leurs Applications (CMLA), 2004.

Burr-Hersey, J. E., Mooney, S. J., Bengough, A. G., Mairhofer, S., & Ritz, K. (2017). Developmental morphology of cover crop species exhibit contrasting behaviour to changes in soil bulk density, revealed by X-ray computed tomography. *PLoS One, 12*, e0181872. https://doi.org/10.1371/journal.pone.0181872

Buzug, T. M. (2008). Fundamentals of X-ray physics. In *Computed tomography: From photon statistics to modern cone-beam CT* (pp. 15–73). Springer. https://doi.org/10.1007/978-3-540-39408-2_2

Camargo, M. A., Cássaro, F. A. M., & Pires, L. F. (2022). How do geometric factors influence soil water retention? A study using computerized microtomography. *Bulletin of Engineering Geology and the Environment, 81*, 137. https://doi.org/10.1007/s10064-022-02632-z

Chauhan, S., Rühaak, W., Khan, F., Enzmann, F., Mielke, P., Kersten, M., & Sass, I. (2016). Processing of rock core microtomography images: Using seven different machine learning algorithms. *Computers & Geosciences, 86*, 120–128.

Chowdhary, K. R. (2020). *Introducing artificial intelligence, in: Fundamentals of artificial intelligence* (pp. 1–23). Springer. https://doi.org/10.1007/978-81-322-3972-7_1

Cortes, C., & Vapnik, V. (1995). Support-vector networks. *Machine Learning, 20*, 273–297. https://doi.org/10.1007/bf00994018

Costa, G. S. R., Vasconcelos, G. J. Q., O'Dowd, F. P., & Archilha, N. L. (2017). Automation solutions and prototypes for the X-ray tomography beamline of Sirius, the new brazilian synchrotron light source. In I. Costa (Ed.), *ICALEPCS 2017* (pp. 923–927).

Craievich, A. F. (2020). Synchrotron radiation in Brazil. Past, present and future. *Radiation Physics and Chemistry, 167*, 108253. https://doi.org/10.1016/j.radphyschem.2019.04.003

Distante, A., & Distante, C. (2020). *Handbook of image processing and computer vision: Volume 2: From image to pattern.* Springer International Publishing. https://doi.org/10.1007/978-3-030-42374-2

Ertel, W. (2017). *Introduction to artificial intelligence* (2nd ed.). Springer International Publishing. https://doi.org/10.1007/978-3-319-58487-4_1

Ferreira, T. R., Pires, L. F., & Reichardt, K. (2022). 4-D X-ray computed tomography in soil science: An overview and future perspectives at Mogno/Sirius. *Brazilian Journal of Physics, 52*, 33. https://doi.org/10.1007/s13538-021-01043-x

Ferreira, T. R., Pires, L. F., Wildenschild, D., Brinatti, A. M., Borges, J. A. R., Auler, A. C., & dos Reis, A. M. H. (2019). Lime application effects on soil aggregate properties: Use of the mean weight diameter and synchrotron-based X-ray μCT techniques. *Geoderma, 338*, 585–596.

Ferreira, T. R., Pires, L. F., Wildenschild, D., Heck, R. J., & Antonino, A. C. D. (2018). X-ray microtomography analysis of lime application effects on soil porous system. *Geoderma, 324*, 119–130. https://doi.org/10.1016/j.geoderma.2018.03.015

Galdos, M., Brown, E., Rosolem, C., Pires, L. F., Hallett, P., & Mooney, S. J. (2020). Brachiaria species influence nitrate transport in soil by modifying soil structure with their root system. *Scientific Reports, 10*, 5072.

Galdos, M. V., Pires, L. F., Cooper, H. V., Calonego, J. C., Rosolem, C. A., & Mooney, S. J. (2019). Assessing the long-term effects of zero-tillage on the macroporosity of Brazilian soils using X-ray computed tomography. *Geoderma, 337*, 1126–1135.

Gao, W., Schlüter, S., Blaser, S. R. G. A., Shen, J., & Vetterlein, D. (2019). A shape-based method for automatic and rapid segmentation of roots in soil from X-ray computed tomography images: Rootine. *Plant and Soil, 441*, 643–655. https://doi.org/10.1007/s11104-019-04053-6

Garfi, G., John, C. M., Berg, S., & Krevor, S. (2020). The sensitivity of estimates of multiphase fluid and solid properties of porous rocks to image processing. *Transport in Porous Media, 131*, 985–1005. https://doi.org/10.1007/s11242-019-01374-z

Gillespy, T., 3rd, & Rowberg, A. H. (1994). Dual lookup table algorithm: An enhanced method of displaying 16-bit-grey-scale images on 8-bit RGB graphic systems. *Journal of Digital Imaging, 7*, 13–17.

Goodfellow, I., Bengio, Y., & Courville, A. (2016). *Deep learning*. MIT Press.

Goyal, B., Dogra, A., Agrawal, S., Sohi, B. S., & Sharma, A. (2020). Image denoising review: From classical to state-of-the-art approaches. *Information Fusion, 55*, 220–244. https://doi.org/10.1016/j.inffus.2019.09.003

Gonçalves, H. M., Vasconcelos, G. J. Q., Rosa, P. R. R., Carvalho, M., Archilha, N. L., Spina, T. V., 2019. *cudaIFT: 180x faster image foresting transform for Waterpixel estimation using CUDA*. In: 14th International Conference on Computer Vision Theory and Applications (VISAPP), 395–404.

Gull, S. F., & Skilling, J. (1984). Maximum entropy method in image processing. *IEE Proceedings F, 131*, 646–659.

Han, Q., Zhao, Y., Liu, L., Chen, Y., & Zhao, Y. (2019). A simplified convolutional network for soil pore identification based on computed tomography imagery. *Soil Science Society of America Journal, 83*, 1309–1318. https://doi.org/10.2136/sssaj2019.04.0119

Heinzl, C., Amirkhanov, A., & Kastner, J. (2018). Processing, analysis and visualization of CT data. In S. Carmignato, W. Dewulf, & R. Leach (Eds.), *Industrial X-ray computed tomography* (pp. 99–142). Springer International Publishing.

Ho, T. K. (1998). The random subspace method for constructing decision forests. *IEEE Transactions on Pattern Analysis and Machine Intelligence, 20*, 832–844. https://doi.org/10.1109/34.709601

Houston, A. N., Schmidt, S., Tarquis, A. M., Otten, W., Baveye, P. C., & Hapca, S. M. (2013). Effect of scanning and image reconstruction settings in X-ray computed microtomography on quality and segmentation of 3-D soil images. *Geoderma, 207*(208), 154–165.

Iassonov, P., Gebrenegus, T., & Tuller, M. (2009). Segmentation of X-ray computed tomography images of porous materials: A crucial step for characterization and quantitative analysis of pore structures. *Water Resources Research, 45*, W09415.

Jan, J. (2006). *Medical image processing reconstruction and restoration: Concepts and methods*. Taylor & Francis.

Jarvis, N., Larsbo, M., & Koestel, J. (2017). Connectivity and percolation of structural pore networks in a cultivated silt loam soil quantified by X-ray tomography. *Geoderma, 287*, 71–79. https://doi.org/10.1016/j.geoderma.2016.06.026

Jiulun, F., & Winxin, X. (1997). Minimum error thresholding: A note. *Pattern Recognition Letters, 18*, 705–709.

Joshi, A. V. (2020). *Machine learning and artificial intelligence*. Springer International Publishing. https://doi.org/10.1007/978-3-030-26622-6_4

Keselbrener, L., Shimoni, Y., & Akselrod, S. (1992). Nonlinear filters applied on computerized axial tomography: Theory and phantom images. *Medical Physics, 19*, 1057–1064.

Khan, F., Enzmann, F., Kersten, M., Wiegmann, A., & Steiner, K. (2012). 3-D simulation of the permeability tensor in a soil aggregate on basis of nanotomographic imaging and LBE solver. *Journal of Soils and Sediments, 12*, 86–96. https://doi.org/10.1007/s11368-011-0435-3

Kittler, J., & Illingworth, J. (1986). Minimum error thresholding. *Pattern Recognition, 19*, 41–47.

Landini, L., Positano, V., & Santarelli, M. F. (2008). 3-D medical imaging processing. In E. Neri, D. Caramella, & C. Bartolozzi (Eds.), *Image processing in radiology*. Springer-Verlag.

Leue, M., Uteau-Puschmann, D., Peth, S., Nellesen, J., Kodešová, R., & Gerke, H. H. (2019). Separation of soil macropore types in three-dimensional X-ray computed tomography images based on pore geometry characteristics. *Vadose Zone Journal, 18*, 180170. https://doi.org/10.2136/vzj2018.09.0170

Ma, R., Jiang, Y., Liu, B., & Fan, H. (2020). Effects of pore structure characterized by synchrotron-based micro-computed tomography on aggregate stability of black soil under freeze-thaw cycles. *Soil & Tillage Research, 104855*. https://doi.org/10.1016/j.still.2020.104855

Marcheschi, P. (2008). Elaboration of the images in the spatial domain. In E. Neri, D. Caramella, & C. Bartolozzi (Eds.), *Image processing in radiology*. Springer-Verlag.

Milletari, F., Navab, N., & Ahmadi, S.-A. (2016). V-Net: Fully convolutional neural networks for volumetric medical image segmentation. In *Proceedings of the fourth international conference on 3-D vision (3-DV)* (pp. 565–571).

Miqueles, E., Koshev, N., & Helou, E. S. (2018). A Backprojection slice theorem for tomographic reconstruction. *IEEE Transactions on Image Processing, 27*, 894–906. https://doi.org/10.1109/TIP.2017.2766785

Müter, D., Pedersen, S., Sφrensen, H. O., Feidenhans'l, R., & Stipp, S. L. S. (2012). Improved segmentation of X-ray tomography data from porous rocks using a dual filtering approach. *Computers & Geosciences, 49*, 131–139.

Oh, W., & Lindquist, W. B. (1999). Image thresholding by indicator kriging. *IEEE Transactions on Pattern Analysis and Machine Intelligence, 21*, 590–602.

Oliveira, J. A. T., Cássaro, F. A. M., & Pires, L. F. (2020). The porous size distribution obtained and analyzed by free acess software. *Revista Brasileira de Ensino de Física, 42*, e20200192.

Oliveira, J. A. T., Cássaro, F. A. M., & Pires, L. F. (2021). Estimating soil porosity and pore size distribution changes due to wetting-drying cycles by morphometric image analysis. *Soil & Tillage Research, 205*, 104814.

Otsu, N. (1979). A threshold selection method from grey-level histograms. *IEEE Transactions on Systems, Man, and Cybernetics, 9*, 62–66.

Pak, T., Luz, L. F., Tosco, T., Costa, G. S. R., Rosa, P. R. R., & Archilha, N. L. (2020). Pore-scale investigation of the use of reactive nanoparticles for in situ remediation of contaminated groundwater source. *Proceedings of the National Academy of Sciences, 2019*, 18683. https://doi.org/10.1073/pnas.1918683117

Pathmanabhan, A., & Dinesh, S. (2007). The effect of Gaussian blurring on the extraction of peaks and pits from digital elevation models. *Discrete Dynamics in Nature and Society, 62137*, 1–12.

Peixinho, A.Z., 2017. *Learning image features by convolutional networks under supervised data constraint*. Ph.D. Thesis (Unicamp).

Peth, S., Horn, R., Beckmann, F., Donath, T., Fischer, J., & Smucker, A. J. M. (2008). Three-dimensional quantification of intra-aggregate pore-space features using synchrotron-radiation-based microtomography. *Soil Science Society of America Journal, 72*, 897–907.

Pini, R., & Madonna, C. (2016). Moving across scales: A quantitative assessment of X-ray CT to measure the porosity of rocks. *Journal of Porous Materials, 23*, 325–338.

Pinto, A., Borin, G., Carlos, B., Bernardi, M., Sarmento, M., Peixinho, A., Spina, T., & Miqueles, E. (2022). Annotat3D: a modern web application for interactive segmentation of volumetric images at Sirius/LNLS. *Synchrotron Radiation News*. https://doi.org/10.1080/08940886.2022.2112501

Pires, L. F., Auler, A. C., Roque, W. L., & Mooney, S. J. (2020). X-ray microtomography analysis of soil pore structure dynamics under wetting and drying cycles. *Geoderma, 362*, 114103.

Pires, L. F., Roque, W. L., Rosa, J. A., & Mooney, S. J. (2019). 3-D analysis of the soil porous architecture under long term contrasting management systems by X-ray computed tomography. *Soil & Tillage Research, 191*, 197–206. https://doi.org/10.1016/j.still.2019.02.018

Pires, L. F., Borges, J. A. R., Rosa, J. A., Cooper, M., Heck, R. J., Passoni, S., & Roque, W. L. (2017). Soil structure changes induced by tillage systems. *Soil & Tillage Research, 165*, 66–79.

Pöhlitz, J., Rücknagel, J., Schlüter, S., Vogel, H. J., & Christen, O. (2019). Computed tomography as an extension of classical methods in the analysis of soil compaction, exemplified on samples from two tillage treatments and at two moisture tensions. *Geoderma, 346*, 52–62. https://doi.org/10.1016/j.geoderma.2019.03.023

Polikar, R. (2012). Ensemble learning. In Y. Ma (Ed.), *Zhang, C* (pp. 1–34). Ensemble Machine Learning. https://doi.org/10.1007/978-1-4419-9326-7_1

Ronneberger, O., Fischer, P., & Brox, T. (2015). U-net: Convolutional networks for biomedical image segmentation. In N. Navab, J. Hornegger, W. Wells, & A. Frangi (Eds.), *Medical image computing and computer-assisted intervention–MICCAI 2015* (pp. 234–241). Springer. https://doi.org/10.1007/978-3-319-24574-4_28

Schlüter, S., Sheppard, A., Brown, K., & Wildenschild, D. (2014). Image processing of multiphase images obtained via X-ray microtomography: A review. *Water Resources Research, 50*, 3615–3639. https://doi.org/10.1002/2014WR015256

Spina, T. V., Vasconcelos, G. J. Q., Gonçalves, H. M., Libel, G. C., Pedrini, H., Carvalho, T. J., & Archilha, N. L. (2018). Towards real time segmentation of large-scale 4-D micro/Nanotomography images in the Sirius synchrotron light source. *Microscopy and Microanalysis, 24*, 92–93. https://doi.org/10.1017/s1431927618012849

Sun, T., & Neuvo, Y. (1994). Detail-preserving median based filters in image processing. *Pattern Recognition Letters, 15*, 341–347.

Tzotsos, A., & Argialas, D. (2008). Support vector machine classification for object-based image analysis. In T. Blaschke, S. Lang, & G. J. Hay (Eds.), *Object-based image analysis* (pp. 663–677). Springer. https://doi.org/10.1007/978-3-540-77058-9_36

Tuller, M., Kulkarni, R., & Fink, W. (2013). Segmentation of X-ray CT data of porous materials: A review of global and locally adaptive algorithms. In S. H. Anderson & J. W. Hopmans (Eds.), *Soil–water–root processes: Advances in tomography and imaging* (pp. 157–182). John Wiley & Sons, Ltd. https://doi.org/10.2136/sssaspecpub61.c8

Vasconcelos, G. J. Q., Antonieti, G., Libel, G. C., Rosa, P. R. R., Archilha, N. L., Carvalho, T. J., Pedrini, H., & Spina, T. V. (2018). Evaluation of segmentation methods based on classification patterns for micro-tomography applications in rock analysis. In *Conference on Graphics, Patterns and Images, 31. (SIBGRAPI)*. Sociedade Brasileira de Computação, Foz do Iguaçu, PR, Brazil.

Vogel, H. J. (1997). Morphological determination of pore connectivity as a function of pore size using serial sections. *European Journal of Soil Science, 48*, 365–377.

Wang, W. (2008). Rock particle image segmentation and systems. In P.-Y. Yin (Ed.), *Pattern recognition techniques, technology and applications*. I-Tech.

Wang, W., Kravchenko, A. N., Smucker, A. J. M., Liang, W., & Rivers, M. L. (2012). Intra-aggregate pore characteristics: X-ray computed microtomography analysis. *Soil Science Society of America Journal, 76*, 1159. https://doi.org/10.2136/sssaj2011.0281

Wani, M. A., Bhat, F. A., Afzal, S., & Khan, A. I. (2020). *Introduction to deep learning, in: Advances in deep learning* (pp. 1–11). Springer. https://doi.org/10.1007/978-981-13-6794-6_1

Weikert, T., Cyriac, J., Yang, S., Nesic, I., Parmar, V., & Stieltjes, B. (2020). A practical guide to artificial intelligence–based image analysis in radiology. *Investigative Radiology, 55*, 1–7. https://doi.org/10.1097/RLI.0000000000000600

Yen, J.-C., Chang, F. J., & Chang, S. (1995). A new criterion for automatic multilevel thresholding. *IEEE Transactions on Image Processing, 46*, 82–95.

Zhao, Y., Hu, X., & Li, X. (2020). Analysis of the intra-aggregate pore structures in three soil types using X-ray computed tomography. *Catena, 193*, 104622. https://doi.org/10.1016/j.catena.2020.104622

Zhou, H., Whalley, W. R., Hawkesford, M. J., Ashton, R. W., Atkinson, B., Atkinson, J. A., Sturrock, C. J., Bennett, M. J., & Mooney, S. J. (2020). The interaction between wheat roots and soil pores in structured field soil. *Journal of Experimental Botany*. https://doi.org/10.1093/jxb/eraa475

Chapter 6
Quantification of Soil Porous Architecture

Steffen Schlüter and Hans-Jörg Vogel

6.1 Introduction

In recent decades X-ray Computed Tomography (X-ray CT) has become a standard technique for non-invasive imaging of the three-dimensional structure of intact soil. The number of working groups worldwide that routinely apply X-ray CT for soil-related research is steadily growing. A consequence of this expanding community is a wealth of methods for the quantification of soil pore structure which are provided through different software toolboxes. There has not yet been a consensus about what a minimum set of metrics would be that is absolutely essential to quantify pore structure. The reasons for that are two-fold. First, soil structure analysis is not an end in itself. It is meant to reveal the heterogeneous small-scale architecture of soil, which is required for a mechanistic understanding of observable processes and phenomena at larger scale such as water movement, solute fluxes, plant–soil interactions, gas emissions or the maintenance of biodiversity. Likewise, the existence of hot spots in biological activity and geochemical processes is a manifestation of microscale processes regulated by soil structure. Depending on the specific focus, different characteristics of the pore structure might be most relevant, e.g. the volume fraction, continuity, size distribution, clustering, or roughness of pores. It is good practice that the set of investigated metrics is tailored accordingly. This is supported by theoretical or empirical evidence regarding the sensitivity of the investigated process to one or another structural feature or their combination. For example, if we are interested in hydraulic conductivity, then we should analyse not only the pore volume but also the pore size distribution and pore connectivity. In contrast, if we are interested in soil aeration, then the volume and spatial distribution pattern of

S. Schlüter (✉) · H.-J. Vogel
Department Soil System Science, Helmholtz Centre for Environmental Research–UFZ, Halle, Germany
e-mail: steffen.schlueter@ufz.de; Hans-Joerg.Vogel@ufz.de

S. Jon Mooney et al. (eds.), *X-ray Imaging of the Soil Porous Architecture*,
https://doi.org/10.1007/978-3-031-12176-0_6

macropores is highly relevant. Also, researchers all tend to stick to established image analysis protocols irrespective of the underlying research question either out of habit or due to limited capacity for adapting alternative protocols. This can have two negative side effects: First, the structural description may remain suboptimal with regard to the processes investigated and second, an untargeted shot-gun approach might be carried out to analyse as many metrics as possible to discover correlations, without a clear hypothesis on the causal relationship between form and function. This could be a promising approach in case the causal relationships are not sufficiently clear or to discover unexpected relationships. However, more often than not, such studies merely report the correlations that were found without further interpretation. Thus, the link between correlation and causation is frequently missing, and generalizations are impossible as the outcome could have been very different under different conditions. This chapter is an attempt to prioritize and identify some highly relevant pore structure metrics well-grounded in mechanistic process understanding. This chapter is not meant to provide a thorough review of the relationship between soil structure and soil functions. For this we refer the reader to Rabot et al. (2018).

There are more issues to consider for employing X-ray CT successfully to discover the microscale drivers for macroscale soil functioning than just identifying suitable pore structure metrics. An equally important question in the design of an X-ray CT study is at what scale should the pore structure be analysed. There is a well-known trade-off between sample size and image resolution that is constrained by the hardware (see Chaps. 2 and 4). There are two aspects that need to be balanced for identifying the appropriate scale. The soil volume needs to be large enough to not just capture the pore structure representatively (see Chap. 3) but also the targeted soil process (e.g. solute transport, soil respiration). In addition, the resolution must be fine enough to resolve the pores that are most relevant for a given process (e.g. macropores for preferential flow, meso and micropores for water retention and microbial activity). Very often both aspects cannot be fulfilled at the same time. In this chapter, we will survey hierarchical sampling approaches that reconcile both aspect and address how pore structure metrics change with scale.

Finally, X-ray CT is only capable of imaging the X-ray photon attenuation in soil (see Chaps. 2 and 4), which limits its application to mapping the physical structure of soil, i.e. local changes in bulk density, spatial distribution of large pores and perhaps the spatial distribution of air and water, if the voxel size is small enough. The detection of roots in soil, even if they are bigger than the image resolution, already poses a huge methodological challenge (see Chap. 9). Moreover, X-ray CT is virtually blind for any biochemical information. New protocols are now available to improve the detection of organic matter with X-ray CT by employing staining agents containing heavy metals that selectively bind to organic compounds to make them visible for X-rays (Loo et al., 2014; Peth et al., 2014; Zheng et al., 2020) and hence allow their integration into spatially explicit modelling approaches (see Chap. 11). Yet, this is only a proxy for local concentrations but insensitive to the quality of organic matter. Likewise, information on three-dimensional (3-D) local elemental or isotopic composition and spatial distribution and composition of soil microorganisms is lacking. Different microscopic and micro-spectroscopic techniques (fluorescence microscopy, electron microscopy, secondary ion mass spectroscopy, etc.) are required to retrieve this biochemical information on exposed surfaces. These two-dimensional (2-D) maps need to be aligned with structural information obtained via X-ray CT through image registration. This correlative imaging approach is evolving rapidly (Hapca et al., 2015; Kravchenko et al., 2019; Lucas et al., 2020a; Schlüter et al., 2019a) as it is currently the only viable option for a more holistic view on soil microenvironments. In sect. 6.4 we will give a brief introduction into some practical aspects of correlative imaging.

Finally, this chapter is also meant to facilitate easy access to pore structure analysis for readers that are new to the subject by stating free software options for each of the explained methods. Here, we restrict ourselves to Fiji/ImageJ (Schindelin et al., 2012) and associated plugins, but point out that there is other free or proprietary software that comes with comparable functionality like scikit-image, QuantIm, Matlab, Dragonfly or Avizo.

6.2 Quantification of Pore Structure

For this chapter it is assumed that the X-ray CT raw data has already been reconstructed and image datasets processed providing segmented images with soil structural features of interest (Chap. 4). Image processing steps from image enhancement to image segmentation and post-processing are surveyed in Chap. 5. In principle, segmented images could contain any number of material classes. Since this chapter is dealing with the quantification of soil pore architecture, we focus on binary images with pores as the foreground and the soil matrix as the background.

The list of metrics that can be used for the quantification of pore architecture is virtually endless. Likewise, there are multiple criteria by which they can be grouped and sorted. Here we make an attempt to identify frequently used metrics and group them from a technical point of view according to the method by which the binary image, i.e. pores vs. background, is transformed to retrieve these metrics. This survey is underpinned with two examples for vastly different pore system with similar visible porosity of 0.1-0.14 $mm^3\ mm^{-3}$ (Fig. 6.1). It comprises a ploughed (left) and a no-till (centre) topsoil sample with a silt loam soil matrix scanned at a voxel size of 60 μm (Schlüter et al., 2020a).

6.2.1 Metrics Based on Binary Image

Many metrics are directly calculated on binary images of the segmented pore space without any other previous transformations. Prominent examples are the so-called Minkowski functionals M_{0-3} (Armstrong et al., 2019; Vogel et al., 2010), which in 3-D consists of volume M_0, surface area M_1, integral of mean curvature M_2 and

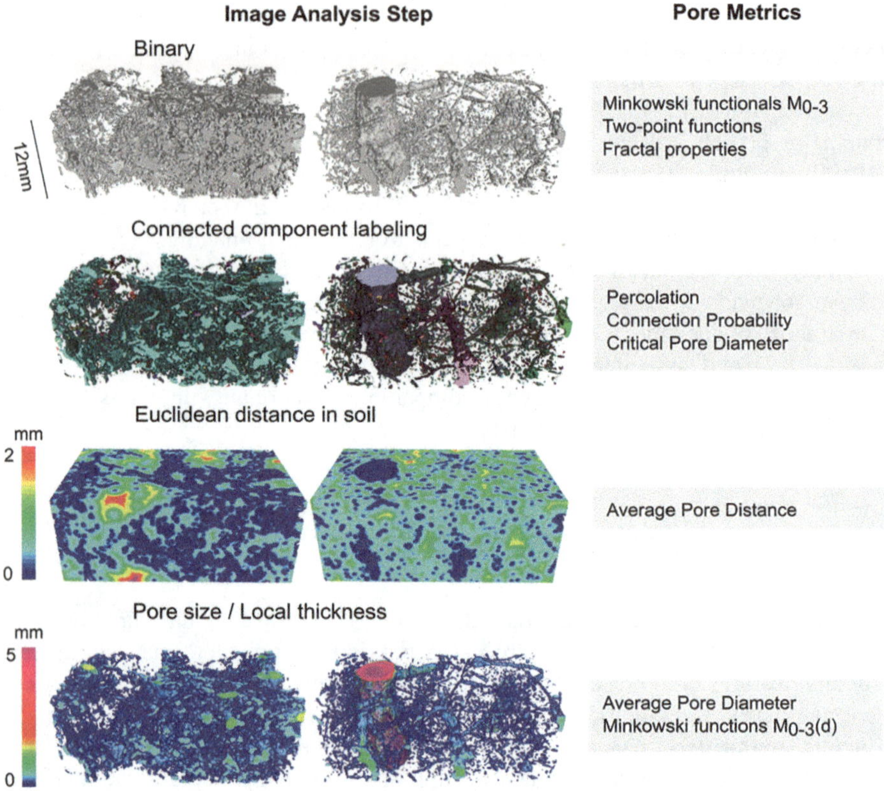

Fig. 6.1 A selection of pore metrics (listed on the right) derived by different analysis methods of the segmented pore space of a ploughed (left) an undisturbed soil (centre) with the similar visible macroporosity (>60 μm). The values for each metric are listed in Table 6.1

integral of total curvature M_3. The meaning of the first two is rather intuitive. M_2 represents the mean curvature radius of the pore–solid interface integrated over the entire pore surface. For convex structures such as spherical bodies it reflects the mean breadth or the average caliper diameter (Ohser & Mücklich, 2000). If simplified shapes can be assumed a priori, such as the cylindrical shape of roots, then M_2 relates to total root length irrespective of root diameter (Koebernick et al., 2014). M_2 can also be normalized by the volume and surface area of a structure to obtain a dimensionless structure model index with well-defined values for spheres, cylinders, plates and so on that is independent of their size (Andersson et al., 2018; Hildebrand & Rüegsegger, 1997). M_3 is a topological measure describing the connectivity of the structure (e.g. pore space) by counting the number of isolated objects, adding the number of cavities and subtracting the number of redundant connections. This measure is equivalent to the Euler characteristic (Vogel et al., 2010). Fully enclosed cavities in pores (i.e. particles floating in pore space without touching other particles) are considered unphysical features which can be ignored. The appeal of Minkowski

Table 6.1 Summary of morphological properties for the pore structures of a ploughed soil and no-till soil depicted in Fig. 6.1

metric	Range	unit	ploughed soil	no-till soil
Porosity m_0	0 – 1	$mm^3\ mm^{-3}$	0.14	0.10
Surface area density m_1	0- +∞	$mm^2\ mm^{-3}$	1.24	0.36
Mean breadth m_2	−∞ - +∞	$mm\ mm^{-3}$	0.40	0.25
Euler characteristic χ	−∞ - +∞	mm^{-3}	−0.38	0.89
Fractal dimension D	1-3	–	1.41	1.31
Connection probability Γ	0 – 1	–	0.91	0.57
Percolation	0, 1	–	1	1
Mean pore distance	0- +∞	mm	0.50	0.44
Critical pore diameter d_c	0- +∞	mm	0.36	4.56
Mean pore size	0- +∞	mm	0.41	3.66

functionals is that they can be directly and very efficiently computed on the digital grids of binary images (Ohser & Mücklich, 2000). By employing stereological principles, uncertainties and biases associated with box-shaped geometry of voxels are avoided (Vogel et al., 2010). It is convenient to divide Minkowski functionals by the analysed volume and report them as densities m_{0-3}. Basically, the Minkowski functionals describe the visible porosity, its surface area, convexity and connectivity. Their importance for transport in porous media and mass transfer at interfaces is evident and has been reviewed recently in Armstrong et al. (2019). A very recent development has been to derive wettability (i.e. contact angles, in partially saturated porous media directly from the so-called deficit curvature (Blunt et al., 2020; Sun et al., 2020)), which is related to M_3 of the non-wetting phase, e.g. air-filled pores in wet soil. Minkowski functionals can be calculated with the MorphLibJ plugin for Fiji/ImageJ (Legland et al., 2016).

There are other metrics that can be directly retrieved from binary images of pore space. Fractal properties assess the complexity or irregularity of the pore space and in how far this aspect is similar across spatial scales analysed by changing sample volumes and image resolutions. The fractal dimension can be computed with the Box Counting method that is incorporated in the standard Fiji/ImageJ distribution. In addition, there is a suite of the so-called two-point functions (Jiao et al., 2009; Karsanina et al., 2015; Renard & Allard, 2013; Schlüter & Vogel, 2011) that quantify whether two randomly chosen points at a certain separation distance belong to pores (two-point correlation function, indicator variogram), have a straight connection through the pore space (two-point chord function, lineal path function) or belong to the same pore cluster through any tortuous connection (two-point cluster function). Note that for the latter, a connected components labelling is required as explained below. These two-point functions quantify different aspects of pores like regularity, width and separation distance as well as connectivity. They have been mainly used for stochastic reconstruction of porous media in order to assess what minimum structural information is required to create a stochastic pore structure that reproduces a given soil structure visually or functionally (e.g. solute transport).

Variogram analysis on X-ray CT images are more common on unsegmented, grey scale data to quantify the spatial extent across which features are correlated (Hapca et al., 2015; Quigley et al., 2018; Rawlins et al., 2016). The anisotropy of pore structure is captured, when these two-point relations as a function of distance are analysed separately for different directions and compared with each other. To the best of our knowledge, two-point functions are not yet part of Fiji/ImageJ or associated plugins. Among others, they are implemented in QuantIm (Vogel et al., 2010), which is an alternative, open-source image processing software.

6.2.2 *Metrics Based on Connected Components*

The pore space can be distinguished into individual components such that any connected cluster has a unique label or ID (shown as a colour code in Fig. 6.1). This connected component labelling is implemented in MorphoLibJ (Legland et al., 2016), but can also be retrieved with the 3-D Objects Counter in Fiji/ImageJ. Based on this label image the connection probability of two randomly chosen pore voxels, Γ, can be computed (Renard & Allard, 2013) according to

$$\Gamma = \frac{1}{N_v^2} \sum\nolimits_{i=1}^{N_c} n_i^2$$

where N_v is the number of pore voxels, N_c is the number of individual pore clusters and n_i is the number of voxels in cluster i. If all pores are connected in one cluster, then =1. The value of Γ converges to zero if porosity is fragmented into many clusters of similar size. The two-point cluster function mentioned above determines this connection probability as a function of separation distance.

Percolation is another salient property of pore systems. A pore network percolates when at least one pore cluster extends from one image boundary to the opposite boundary, e.g. top to bottom. In its simplest form this is a Boolean property, but percolation theory comes with a list of related metrics and theoretical predictions of functional behaviour (Hunt et al., 2014). Noteworthy is the critical pore diameter d_c (Koestel et al., 2018), i.e. the narrowest constriction along the percolating pore cluster. The aperture of this bottleneck poses a dominant resistance to flow and has been shown to predict saturated water flow in macropores well (Koestel et al., 2018; Schlüter et al., 2020a). Technically, this critical pore diameter is derived from the Euclidean distance map of the pore space—see next section—(i.e. the shortest distance of each pore voxel to the next matrix voxel). This distance map is segmented at increasing distance values resulting in binary images of receding pore space and each time checked for percolation. The pore surface distance, at which this connectivity is lost, corresponds to the critical pore radius. The critical pore diameter is implemented in the Fiji/ImageJ Plugin SoilJ (Koestel, 2018).

6.2.3 Metrics Based on Distance

There are different ways to express distance in space. The most common one is the Euclidean (or shortest) distance between two points. The Euclidean distance transform is implemented in the standard Fiji/ImageJ distribution. The outcome is displayed in Fig. 6.1 as the Euclidean distance to the closest pore in the soil matrix. Sampling the entire soil matrix results in pore distance histograms. The average pore distance is retrieved as the first central moment of the pore distance histogram. This distance information is helpful to predict the aeration status in soil, in particular when the matric potential in an experiment is adjusted such that all visible pores are air-filled and all unresolved pores are water-filled (Schlüter, Sammartino, & Koestel, 2020b). Then this pore distance essentially reflects the diffusion distance of dissolved oxygen into the wet soil matrix, if the detailed trajectory at the molecular scale is disregarded. A distance threshold can be determined in order to estimate the anaerobic soil volume fraction of soil (Kravchenko et al., 2018; Rohe et al., 2021). Distance maps have also been applied to root distances in soil as a measure for how effectively a root system architecture with a given root length explores the soil (Koebernick et al., 2014; Schlüter et al., 2018a). Euclidean distances are often combined with other structural information in order to quantify property X as a function of distance to material Y, e.g. radial gradients of visible porosity in the rhizosphere as function of root distance (Helliwell et al., 2019; Koebernick et al., 2017; Lucas et al., 2019) or element concentration as a function of pore or root distances (Kölbl et al., 2017; Quigley et al., 2018).

Another common distance metric is geodesic distance, i.e. the distance between two points along a path within the pore space, i.e. around obstacles posed by the soil matrix (Schlüter et al., 2019b). The geodesic distance transform is implemented in the MorhoLibJ plugin (Legland et al., 2016). The discrepancy between geodesic distance and Euclidean distance arises at lengths that are much longer than the typical size of structural features and can be used as a measure for tortuosity along these distances (Schlüter et al., 2019b). An introduction to the use of distance transforms in soil images is given in Holden (2001).

6.2.4 Metrics Based on Pore Size

There are different ways to quantify pore size distribution depending on the definition of what constitutes a pore. The cluster size distribution mentioned above would be one approach, if pore continuity was the defining criterion. However, this measure is more related to pore connectivity than to pore size in terms of pore diameters. Moreover, this measure is highly sensitive to the image resolution since connectivity and cluster size can change dramatically with the smallest pores considered for the analysis. A more robust approach is to employ the maximum inscribed sphere method to compute the local pore diameter or local thickness of the pore space (Vogel et al., 2010). The local thickness method was originally developed in the BoneJ plugin (Doube et al., 2010) and is now also part of the standard Fiji/ ImageJ distribution. Sampling the entire thickness map of the pore space results in the pore size distribution. The average pore diameter of the visible pore space is retrieved as the first central moment of the pore size distribution. The pore size distribution is related to water release curves (Lucas et al., 2020b) since, according to the Young-Laplace law, the local pore radius is directly linked to the entry pressure at which air invades a water-filled pore. Summing over all pore voxels into which a sphere with a given radius does not fit results in the water saturation at a given capillary pressure. It should be noted, however, that this "pore morphology" approach describes the distribution of actual pore diameters while water drainage curves and derived pore size distributions are always biased towards smaller pore diameters because the diameter of narrow bottlenecks are critical for the drainage of larger pores. Moreover, the pore morphology approach assumes spherical fluid interfaces, perfect wettability and thermodynamic equilibrium of phase distribution within the pore space when used to estimate water content at a given water potential which simplifies reality (Pot et al., 2015; Vogel et al., 2005). Nevertheless, this "pore morphology" approach to modelling drainage of porous media is a valuable and well-established method (Hazlett, 1995; Hilpert & Miller, 2001; Vogel et al., 2005) that is much faster than 3-D flow simulations in porous media.

The Minkowski functionals M_{0-3} (volume, surface area, convexity, connectivity, see sect. 6.2) can also be determined as a function of pore size. This is done by repeated analysis of the image after removing pores smaller than an increasing size threshold. As a result, the Minkowski functionals are obtained as a function of the minimum pore diameter (Fig. 6.2) and are then referred to as Minkowski functions $M_{0-3}(d)$ (Vogel et al., 2010). The increasing volume $M_0(d)$ with decreasing d is an alternative approach to quantify the cumulative pore size distribution. Useful applications of $M_{0-3}(d)$ to predict functional behaviour are again reviewed in Armstrong et al. (2019).

6.2.5 Case Study on Tillage-Induced Pore Structure

All the aforementioned metrics were summarized in Table 6.1 for the pore structures obtained with X-ray CT scans of a ploughed and no-till topsoil with a voxel size of 60 μm each (Fig. 6.1). The total volume (6.48 mm^3) is too small to be representative. The comparison is only meant to serve demonstration purposes. For a more comprehensive comparison of the pore structures including implications on saturated and near-saturated hydraulic conductivity, see Schlüter et al. (2020a). Even though the visible porosity of the ploughed soil is only increased by a factor of 1.4, the surface area density is 3.4 times higher. This is because a large fraction of porosity in the no-till pore structure is concentrated in a few, rather smooth biopores, whereas ploughing results in a continuum of homogeneously distributed pores with irregular, rough surfaces. The mean breadth is positive in both cases, implying that both

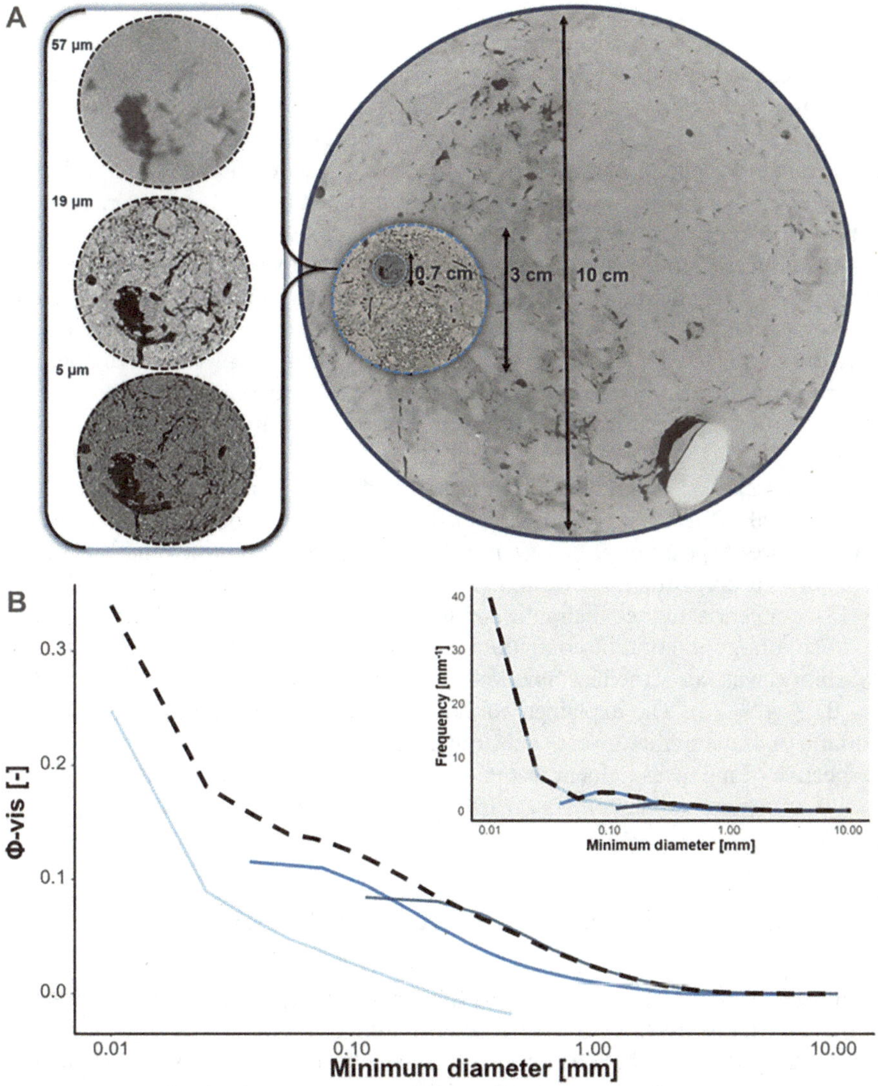

Fig. 6.2 X-ray CT derived pore size distribution within different sample sizes. (**a**) A cross section through one sample at three sample sizes to illustrate the relation in size (right) and the change in resolution (left). (**b**) Cumulative pore size distribution (dashed line) and corresponding distribution of the different sample sizes (decreasing intensity of blue = decreasing sample size; $\phi_{vis} = m_0$). Inset shows the frequency distribution of the corresponding curves. The intersection points were used as transition points between two sample sizes to calculate the joint, cumulative distribution (Lucas et al., 2020b) reprinted with permission

structures are on average convex. A positive (negative) m_2 can be interpreted more intuitively as an increase (decrease) in surface area, when the pore space is slightly dilated. A positive Euler characteristic in the no-till soil points to poor connectivity, with a higher number of isolated pore objects than redundant connections. Note that many of these isolated pores (clearly visible in Fig. 6.1) would be connected if the soil was scanned at a finer resolution. The fragmentation of the intact soil structure by ploughing causes a well-connected pore structure which is reflected in a negative Euler characteristic. The small differences in fractal dimension and mean pore distance between both images are not really indicative of the vastly different pore morphologies. Both pore structures facilitate percolation. However, the connection probability Γ is much higher in ploughed soil due to one dominating pore cluster. Typically, there is a steep increase of Γ in a porosity range of 0.05-0.1 (Jarvis et al., 2017; Schlüter & Vogel, 2016). Note that a limited domain size like in Fig. 6.1 may lead to a bias towards lower Γ, since the biopores in the no-till soil are likely to be connected outside the domain (Lucas et al., 2020b). The mean pore size and critical pore diameter are greater by one order of magnitude because of one approx. 5 mm thick earthworm burrow that stretches from top to bottom of the domain.

The variability in simulated, saturated hydraulic conductivity reported by Schlüter et al. (2020a) was best explained by the critical pore diameter in no-till pore structures (Spearmen $R = 0.89$, $n = 13$) and by visible porosity and connection probability in pore structures resulting from ploughing (Spearmen $R = 0.62\text{-}0.63$, $n = 13$). For near-saturated hydraulic conductivity at a matric potential of $h = -2$ cm, i.e. when a large fraction of continuous biopores is blocked by air, the connection probability was an excellent predictor across both pore structures (Spearmen $R = 0.97$, $n = 26$). The explained variability generally decreased when simulated hydraulic conductivities were substituted by experimentally obtained values irrespective of the measurement technique (hood infiltrometer in the field or tension disc infiltrometers on lab columns) (Schlüter et al., 2020a).

6.3 Scale Issues in Pore Structure Characterization

The image resolution of X-ray CT scans is linked to the diameter of sample through a fixed factor that is given by the detector resolution (i.e. the number of pixels in the horizontal direction of the field of view). Typically for industrial CT scanners this factor is currently in the range of 1000-2000 (Rabot et al., 2018). For instance, the sub-volumes shown in Fig. 6.1 were cut from larger CT scans of cylindrical soil cores with a diameter of 100 mm and a resolution of 0.06 mm, amounting to a factor of 1667. When following the well-founded rule that (a) pore objects need to be at least 2-3 voxels thicker than the voxel resolution to be faithfully detected and (b) that the analysed volume needs to be much larger than the typical size of pore objects, the size range of pores that can be analysed while being sampled representatively is rather below one order of magnitude than above (Vogel et al., 2010). The trade-off between resolved details and representativeness is demonstrated in Fig. 6.2a for intact cores of different diameters taken from the same sample.

To broaden the range of scales detected with X-ray CT it is appealing to combine pore structure information of samples scanned at different resolutions. The methodological approach to generate a joint pore size distribution from the pore size distribution of differently sized samples has been introduced by Vogel et al. (2010) and applied frequently since (Lucas et al., 2020b; Schlüter et al., 2018b; Schlüter et al., 2011). In Fig. 6.2b, the frequency distributions of pore diameters obtained by the local thickness transform are plotted for three different resolutions. The joint pore size distribution is obtained by using the intersection points between the individual pore size distributions as transition point from one resolution to the next (Lucas et al., 2020b). The total porosity detected after this fusion of different scales is always higher than the porosity of each individual scale, while this approach accounts for the fact that there is a considerable overlap in pore sizes detected at various scales. Modifications of this approach are possible. For instance, information on smaller pore diameters does not necessarily have to come from X-ray CT but can be provided through other imaging modes such as electron microscopy (Gerke et al., 2021). The pore size does not have to be determined with the local thickness transform (maximum inscribed sphere method) but could be determined via rescaling towards coarser resolutions.

The change in connectivity metrics (Γ, m_3, d_c) with increasing minimum pore diameter can also be indicative of the underlying pore structure (Koestel et al., 2020; Lucas et al., 2020b) as it reveals the typical size range of salient pore types such as root channels or packing pores that provide long-range pore continuity. Therefore, it would be attractive to also combine connectivity information from different scales in a similar way as described above. This is easily done for the Euler characteristic, whereas the connection probability, Γ, needs to be corrected at the scale transition for the fact that smaller sample sizes always cut off long-distance connections and therefore introduce a bias (Lucas et al., 2020b). Approaches for scale fusion of other metrics listed in Table 6.1 have yet to be developed.

6.4 Correlative Imaging

6.4.1 Practical Issues

Scale issues do not only arise when pore metrics obtained at different voxel sizes are combined in a statistical way. They become even more aggravating when the image data from different sources needs to be merged or aligned with each other. The technical term for this method is image registration, or co-registration. An example of image registration is shown in Fig. 6.2a for the nested subsamples of a larger soil column. This particular example is a rather simple exercise, as the imaging technique is the same, the sample was not deformed internally and the size difference from one scale to another was less than an order of magnitude. Successful image registration can be more challenging when internal deformation of the soil volume occurs during consecutive scans. In some cases, this deformation is the actual research interest and can be quantified by maps of local deformation magnitude or vector fields. Soil science related topics include soil compaction (Peth et al., 2010; Schlüter et al., 2016) or bioturbation (Keyes et al., 2016; Koestel & Schlüter, 2018). In most cases, however, this internal deformation is unintended. It can be induced by subsidence of loosely packed soil, inner erosion through water flow, gas trapping and bubble expansion, desiccation or many other processes that occur during sample preparation, incubation or any other measurements carried out on the sample in between scanning events.

Probably the toughest challenge of image registration arises when 2-D microscopy data on soil sections is registered with 3-D X-ray CT data obtained prior to sectioning. Not only is the imaging modality different (i.e. salient features in both images are not necessarily the same) and the structure perhaps deformed during resin embedding, but also is the field of view of the microscopy image often limited and depth resolution in the third dimension unknown and locally variable. This turns the quest for good registration into a search for the 2-D needle in the 3-D haystack. This challenge is tackled best by splitting it into a series of registration steps. That is, light microscopy (LM) can be applied as a bridging technique on the entire soil section at low cost to increase the chances to capture salient features that are easily found in the X-ray CT image of the 3-D sample. The actual micrograph of interest is usually obtained on a smaller region (e.g. fluorescence microscopy for counting bacteria or electron microscopy for elemental mapping), since imaging the entire soil section would be too costly or time-consuming. The 2-D-2-D registration of this micrograph into the light microscopy image is much easier than the direct 2-D-3-D registration of a small micrograph into X-ray CT image. An example for this sequential procedure is given in Fig. 6.3. In that case study, the combination of physical structure and bacterial ecology revealed that a large majority of bacteria are located in mesopores (<10 μm) (Schlüter et al., 2019a).

6.4.2 Software Implementations

Image registration is required for the joint analysis of consecutive X-ray CT scans, which is sometimes referred to as 4-D or time-lapse X-ray CT. Likewise, it is required for the joint analysis of different tomographic techniques, subsumed under the term correlative imaging. The methodology is by and large the same.

Image registration in Fiji/ImageJ is available in various ways. They all operate with landmarks that have to be set by the user in both the target image and the moving image or by identifying salient features in both images automatically. Deformable registration based on manually defined landmarks is implemented in the BigWarp/BigDataViewer plugin (Bogovic et al., 2016). Descriptor-based registration based on automatic feature detection was implemented in Fiji/ImageJ by Preibisch et al. (2010). Both are included in the standard Fiji/ImageJ distribution. Image registration of 2-D microscopy data of various modalities is, among others,

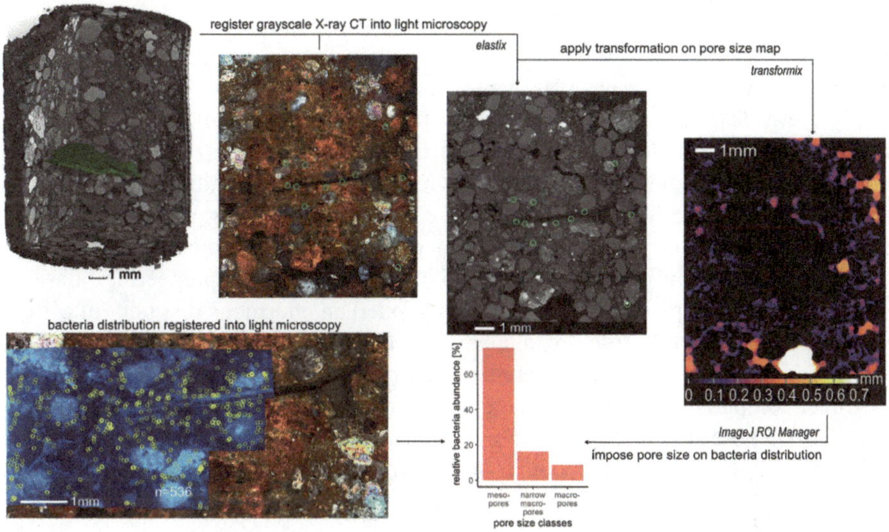

Fig. 6.3 Case study of correlative imaging on a small, repacked, soil core with a leaf fragment depicted in green. First, the X-ray CT data is registered to the light microscopy image with the elastix software by jointly (**a**) minimizing the distance between landmark pairs (green circles) and (**b**) optimizing a mutual information criterion that evaluates the correspondence of intensities everywhere in the overlapping domain. Thus the obtained transformation matrix is employed on the pore size map via a sub-routine called transformix to project 3-D pore diameters into the microscopy plane. The spatial distribution of bacteria (yellow circles, $n = 536$) was manually detected on the fluorescence microscopy image, which was also registered into the light microscopy image. In this way, the relative abundance of bacteria in different pore diameter classes was determined using the ROI Manager of Fiji/ImageJ. Bacteria located in pores below the detection limit of CT (<10 μm) were assigned to mesopores. Visible pores were classified into narrow macropores (<50 μm) and wide macropores (>50 μm). Modified from Schlüter et al. (2019a)

available through the Correlia plugin (Rohde et al., 2020). Elastix is a powerful and very flexible standalone software for image registration (Klein et al., 2010) that is also available as a plugin for Fiji/ImageJ. Many practical implementation tips for image registration with elastix are given in Schlüter et al. (2016).

There is limited functionality in Fiji/ImageJ for 2-D-3-D registration as explained above, due to the different dimensionality of the data. A notable exception is BigWarp, which can handle 2-D microscopy images and 3-D tomography images simultaneously. In addition, with BigWarp it is possible to determine the landmarks on the greyscale CT data, which typically contains the richest details in microstructure, and subsequently employ the registration with these landmarks on any other processed CT data derived from it (segmented data, distance data, etc.). The same two-step strategy has been applied with elastix and a sub-routine called transformix in Fig. 6.3 (Schlüter et al., 2019a).

6.5 Conclusions

We have surveyed frequently used metrics for quantifying pore morphology in binary X-ray CT data of soil and in how far they relate to soil functioning. The trade-off between resolution (detail) and size (representativeness) has been stressed together with strategies to overcome it. In addition to extending the scale, an extension in information content was described by combining different imaging techniques and employing correlative imaging. We are optimistic that the soil science community might be on the verge of leaving behind a pioneering phase in which working groups have developed and applied an enormous diversity of different protocols for pore structure analysis. In the future we may progress into a consolidation phase with more standardized image analysis protocols that allow for direct comparisons of X-ray CT data from different sources or to develop large databases on pore structure information as recently suggested by Rabot et al. (2018). This would open up a novel avenue to identify structural characteristics that are typical for different soil types or different land use and soil management.

References

Andersson, L., Herring, A., Schlueter, S., & Wildenschild, D. (2018). Defining a novel pore-body to pore-throat "morphological aspect ratio" that scales with residual non-wetting phase capillary trapping in porous media. *Advances in Water Resources, 122*, 251–262.

Armstrong, R. T., McClure, J. E., Robins, V., Liu, Z., Arns, C. H., Schlüter, S., & Berg, S. (2019). Porous media characterization using Minkowski Functionals: Theories, applications and future directions. *Transport in Porous Media, 130*(1), 305–335.

Blunt, M. J., Akai, T., & Bijeljic, B. (2020). Evaluation of methods using topology and integral geometry to assess wettability. *Journal of Colloid and Interface Science, 576*, 99–108.

Bogovic, J. A., Hanslovsky, P., Wong, A., & Saalfeld, S. (2016). Robust registration of calcium images by learned contrast synthesis. In *2016 IEEE 13th international symposium on biomedical imaging (ISBI)* (pp. 1123–1126).

Doube, M., Kłosowski, M. M., Arganda-Carreras, I., Cordelières, F. P., Dougherty, R. P., Jackson, J. S., Schmid, B., Hutchinson, J. R., & Shefelbine, S. J. (2010). BoneJ: Free and extensible bone image analysis in ImageJ. *Bone, 47*(6), 1076–1079.

Gerke, K. M., Korostilev, E. V., Romanenko, K. A., & Karsanina, M. V. (2021). Going submicron in the precise analysis of soil structure: A FIB-SEM imaging study at nanoscale. *Geoderma, 383*, 114739.

Hapca, S., Baveye, P. C., Wilson, C., Lark, R. M., & Otten, W. (2015). Three-dimensional mapping of soil chemical characteristics at micrometric scale by combining 2-D SEM-EDX data and 3-D X-ray CT images. *PLoS One, 10*(9), e0137205.

Hazlett, R. D. (1995). Simulation of capillary-dominated displacements in microtomographic images of reservoir rocks. *Transport in Porous Media, 20*(1-2), 21–35.

Helliwell, J. R., Sturrock, C. J., Miller, A. J., Whalley, W. R., & Mooney, S. J. (2019). The role of plant species and soil condition in the structural development of the rhizosphere. *Plant, Cell & Environment, 42*(6), 1974–1986.

Hildebrand, T., & Rüegsegger, P. (1997). Quantification of bone microarchitecture with the structure model index. *Computer Methods in Biomechanics and Biomedical Engineering, 1*(1), 15–23.

Hilpert, M., & Miller, C. T. (2001). Pore-morphology-based simulation of drainage in totally wetting porous media. *Advances in Water Resources, 24*(3-4), 243–255.

Holden, N. M. (2001). Description and classification of soil structure using distance transform data. *European Journal of Soil Science, 52*, 529–545.

Hunt, A., Ewing, R., & Ghanbarian, B. (2014). *Percolation theory for flow in porous media, 880*. Springer.

Jarvis, N., Larsbo, M., & Koestel, J. (2017). Connectivity and percolation of structural pore networks in a cultivated silt loam soil quantified by X-ray tomography. *Geoderma, 287*, 71–79.

Jiao, Y., Stillinger, F. H., & Torquato, S. (2009). A superior descriptor of random textures and its predictive capacity. *Proceedings of the National Academy of Science, 106*(42), 17634–17639.

Karsanina, M. V., Gerke, K. M., Skvortsova, E. B., & Mallants, D. (2015). Universal spatial correlation functions for describing and reconstructing soil microstructure. *PLoS One, 10*(5), e0126515–e0126515.

Keyes, S. D., Gillard, F., Soper, N., Mavrogordato, M. N., Sinclair, I., & Roose, T. (2016). Mapping soil deformation around plant roots using in vivo 4-D X-ray computed tomography and digital volume correlation. *Journal of Biomechanics, 49*(9), 1802–1811.

Klein, S., Staring, M., Murphy, K., Viergever, M. A., & Pluim, J. P. W. (2010). Elastix: A toolbox for intensity-based medical image registration. *Medical Imaging, IEEE Transactions on, 29*(1), 196–205.

Koebernick, N., Daly, K. R., Keyes, S. D., George, T. S., Brown, L. K., Raffan, A., Cooper, L. J., Naveed, M., Bengough, A. G., Sinclair, I., Hallett, P. D., & Roose, T. (2017). High-resolution synchrotron imaging shows that root hairs influence rhizosphere soil structure formation. *The New Phytologist, 216*(1), 124–135.

Koebernick, N., Weller, U., Huber, K., Schlüter, S., Vogel, H.-J., Jahn, R., Vereecken, H., & Vetterlein, D. (2014). In situ visualization and quantification of three-dimensional root system architecture and growth using X-ray computed tomography. *Vadose Zone Journal, 13*(8).

Koestel, J. (2018). SoilJ: An ImageJ plugin for the semiautomatic processing of three-dimensional X-ray images of soils. *Vadose Zone Journal, 17*(1).

Koestel, J., Dathe, A., Skaggs, T. H., Klakegg, O., Ahmad, M. A., Babko, M., Giménez, D., Farkas, C., Nemes, A., & Jarvis, N. (2018). Estimating the permeability of naturally structured soil from percolation theory and pore space characteristics imaged by X-ray. *Water Resources Research, 54*(11), 9255–9263.

Koestel, J., Larsbo, M., & Jarvis, N. (2020). Scale and REV analyses for porosity and pore connectivity measures in undisturbed soil. *Geoderma, 366*, 114206.

Koestel, J., & Schlüter, S. (2018). *Quantification of the structure evolution in a garden soil over the course of two years*. Geoderma.

Kölbl, A., Schweizer, S., Mueller, C., Höschen, C., Said-Pullicino, D., Romani, M., Lugmeier, J., Schlüter, S., & Kögel-Knabner, I. (2017). Legacy of Rice roots as encoded in distinctive microsites of oxides, silicates, and organic matter. *The Soil, 1*(1), 2.

Kravchenko, A. N., Guber, A. K., Quigley, M. Y., Koestel, J., Gandhi, H., & Ostrom, N. E. (2018). X-ray computed tomography to predict soil N2O production via bacterial denitrification and N2O emission in contrasting bioenergy cropping systems. *GCB Bioenergy, 10*(11), 894–909.

Kravchenko, A. N., Guber, A. K., Razavi, B. S., Koestel, J., Blagodatskaya, E. V., & Kuzyakov, Y. (2019). Spatial patterns of extracellular enzymes: Combining X-ray computed microtomography and 2-D zymography. *Soil Biology and Biochemistry, 135*, 411–419.

Legland, D., Arganda-Carreras, I., & Andrey, P. (2016). MorphoLibJ: Integrated library and plugins for mathematical morphology with ImageJ. *Bioinformatics, 32*(22), 3532–3534.

Loo, D. V., Bouckaert, L., Leroux, O., Pauwels, E., Dierick, M., Hoorebeke, L. V., Cnudde, V., Neve, S. D., & Sleutel, S. (2014). Contrast agents for soil investigation with X-ray computed tomography. *Geoderma, 213*(0), 485–491.

Lucas, M., Pihlap, E., Steffens, M., Vetterlein, D., & Kögel-Knabner, I. (2020a). Combination of imaging infrared spectroscopy and X-ray computed microtomography for the investigation of bio- and physicochemical processes in structured soils. *Frontiers in Environmental Science, 8*(42).

Lucas, M., Schlüter, S., Vogel, H.-J., & Vetterlein, D. (2019). Roots compact the surrounding soil depending on the structures they encounter. *Scientific Reports, 9*(1), 16236.

Lucas, M., Vetterlein, D., Vogel, H.-J., & Schlüter, S. (2020b). Revealing pore connectivity across scales and resolutions with X-ray CT. *European Journal of Soil Science, n/a*(n/a).

Ohser, J., & Mücklich, P. (2000). *Statistical Analyisis of microstructures in material science*. Wiley.

Peth, S., Chenu, C., Leblond, N., Mordhorst, A., Garnier, P., Nunan, N., Pot, V., Ogurreck, M., & Beckmann, F. (2014). Localization of soil organic matter in soil aggregates using synchrotron-based X-ray microtomography. *Soil Biology and Biochemistry, 78*(0), 189–194.

Peth, S., Nellesen, J., Fischer, G., & Horn, R. (2010). Non-invasive 3-D analysis of local soil deformation under mechanical and hydraulic stresses by μCT and digital image correlation. *Soil and Tillage Research, 111*(1), 3–18.

Pot, V., Peth, S., Monga, O., Vogel, L. E., Genty, A., Garnier, P., Vieublé-Gonod, L., Ogurreck, M., Beckmann, F., & Baveye, P. C. (2015). Three-dimensional distribution of water and air in soil pores: Comparison of two-phase two-relaxation-times lattice-Boltzmann and morphological model outputs with synchrotron X-ray computed tomography data. *Advances in Water Resources, 84*, 87–102.

Preibisch, S., Saalfeld, S., Schindelin, J., & Tomancak, P. (2010). Software for bead-based registration of selective plane illumination microscopy data. *Nature Methods, 7*(6), 418–419.

Quigley, M. Y., Rivers, M. L., & Kravchenko, A. N. (2018). Patterns and sources of spatial heterogeneity in soil matrix from contrasting long term management practices. *Frontiers in Environmental Science, 6*(28).

Rabot, E., Wiesmeier, M., Schlüter, S., & Vogel, H. J. (2018). Soil structure as an indicator of soil functions: A review. *Geoderma, 314*, 122–137.

Rawlins, B. G., Wragg, J., Reinhard, C., Atwood, R. C., Houston, A., Lark, R. M., & Rudolph, S. (2016). Three-dimensional soil organic matter distribution, accessibility and microbial respiration in macroaggregates using osmium staining and synchrotron X-ray computed tomography. *The Soil, 2*(4), 659–671.

Renard, P., & Allard, D. (2013). Connectivity metrics for subsurface flow and transport. *Advances in Water Resources, 51*(0), 168–196.

Rohde, F., Braumann, U.-D., & Schmidt, M. (2020). Correlia: An ImageJ plug-in to co-register and visualise multimodal correlative micrographs. *Journal of Microscopy, 280*(1), 3–11.

Rohe, L., Apelt, B., Vogel, H. J., Well, R., Wu, G. M., & Schlüter, S. (2021). Denitrification in soil as a function of oxygen availability at the microscale. *Biogeosciences, 18*(3), 1185–1201.

Schindelin, J., Arganda-Carreras, I., Frise, E., Kaynig, V., Longair, M., Pietzsch, T., Preibisch, S., Rueden, C., Saalfeld, S., & Schmid, B. (2012). Fiji: An open-source platform for biological-image analysis. *Nature Methods, 9*(7), 676–682.

Schlüter, S., Albrecht, L., Schwärzel, K., & Kreiselmeier, J. (2020a). Long-term effects of conventional tillage and no-tillage on saturated and near-saturated hydraulic conductivity–can their prediction be improved by pore metrics obtained with X-ray CT? *Geoderma, 361*, 114082.

Schlüter, S., Blaser, S. R. G. A., Weber, M., Schmidt, V., & Vetterlein, D. (2018a). Quantification of root growth patterns from the soil perspective via root distance models. *Frontiers in Plant Science, 9*, 1084.

Schlüter, S., Eickhorst, T., & Mueller, C. W. (2019a). Correlative imaging reveals holistic view of soil microenvironments. *Environmental Science & Technology, 53*(2), 829–837.

Schlüter, S., Großmann, C., Diel, J., Wu, G.-M., Tischer, S., Deubel, A., & Rücknagel, J. (2018b). Long-term effects of conventional and reduced tillage on soil structure, soil ecological and soil hydraulic properties. *Geoderma, 332*, 10–19.

Schlüter, S., Leuther, F., Vogler, S., & Vogel, H.-J. (2016). X-ray microtomography analysis of soil structure deformation caused by centrifugation. *Solid Earth, 7*(1), 129–140.

Schlüter, S., Sammartino, S., & Koestel, J. (2020b). Exploring the relationship between soil structure and soil functions via pore-scale imaging. *Geoderma, 370*, 114370.

Schlüter, S., & Vogel, H.-J. (2011). On the reconstruction of structural and functional properties in random heterogeneous media. *Advances in Water Resources, 34*(2), 314–325.

Schlüter, S., & Vogel, H.-J. (2016). Analysis of soil structure turnover with garnet particles and X-ray microtomography. *PLoS One, 11*(7), e0159948.

Schlüter, S., Weller, U., & Vogel, H.-J. (2011). Soil-structure development including seasonal dynamics in a long-term fertilization experiment. *Journal of Plant Nutrition and Soil Science, 174*(3), 395–403.

Schlüter, S., Zawallich, J., Vogel, H. J., & Dörsch, P. (2019b). Physical constraints for respiration in microbial hotspots in soil and their importance for denitrification. *Biogeosciences, 16*(18), 3665–3678.

Sun, C., McClure, J. E., Mostaghimi, P., Herring, A. L., Berg, S., & Armstrong, R. T. (2020). Probing effective wetting in subsurface systems. *Geophysical Research Letters, 47*(5), e2019GL086151.

Vogel, H.-J., Tölke, J., Schulz, V. P., Krafczyk, M., & Roth, K. (2005). Comparison of a lattice-Boltzmann model, a full-morphology model, and a pore network model for determining capillary pressure-saturation relationships. *Vadose Zone Journal, 4*, 380–388.

Vogel, H.-J., Weller, U., & Schlüter, S. (2010). Quantification of soil structure based on Minkowski functions. *Computers & Geosciences, 36*(10), 1236–1245.

Zheng, H., Kim, K., Kravchenko, A., Rivers, M., & Guber, A. (2020). Testing Os staining approach for visualizing soil organic matter patterns in intact samples via X-ray dual-energy tomography scanning. *Environmental Science & Technology, 54*(14), 8980–8989.

Chapter 7
X-ray Computed Tomography for Studying Solute Transport in Soils

Sandeep Kumar, Poulamee Chakraborty, and Stephen Anderson

7.1 Introduction

The study of pollutant transport through soil is important to protect soil and water resources from contamination (Anderson et al., 2015a, b). Chemicals (including agrochemicals and waste from petroleum, nuclear and several other industries) may appear at elevated concentrations causing major soil and water pollution and have the potential for unintended consequences to human health (Bus & Hammond, 2007). From an agricultural perspective, the usage of mineral fertilizer has risen 10 times since 1960, and pesticide sales have risen from approximately $1 billion to $35 billion per year since 1970 worldwide (FAO, 2018,). In addition, global annual mineral nitrogen fertilizer application to croplands is around 115 million tonnes, and around 4.6 million tonnes of pesticides are sprayed into the environment each year (FAO, 2018). Particularly, soluble nitrate ions (NO_3^-) originating from fertilizers are very mobile and are common contaminants of surface and groundwater causing serious health issues in infants (methemoglobinemia or blue-baby syndrome) as well as eutrophication of water bodies (Hubbard et al., 2004). Most of these pollutants

S. Kumar
Department of Agronomy, Horticulture and Plant Science, South Dakota State University, Brookings, SD, USA
e-mail: sandeep.kumar@sdstate.edu

P. Chakraborty
Department of Agronomy, Horticulture and Plant Science, South Dakota State University, Brookings, SD, USA

School of Natural Resources, University of Missouri, Columbia, MO, USA
e-mail: chakra82@msu.edu

S. Anderson (✉)
Great Lakes Bioenergy Research Center, Michigan State University, East Lansing, MI, USA
e-mail: AndersonS@missouri.edu

S. Jon Mooney et al. (eds.), *X-ray Imaging of the Soil Porous Architecture*,
https://doi.org/10.1007/978-3-031-12176-0_7

may be transported to groundwater via leaching through the vadose zone and can lead to significant ecological implications through soil and water contamination (Dor et al., 2019). Furthermore, these chemicals can persist in the soil for many years and can make once-rich soil unusable for farming, and their bioaccumulation can wipe out living creatures. In addition to these human-induced contaminants and industrial chemicals, several pollutants (e.g., arsenic) are of geological and geohydrological origin (van der Zee & Leijnse, 2013). Significant costs are involved in the remediation of contaminated soil and water sources (Pimentel et al., 1992). Therefore, it is important to understand how these chemicals interact with and are transported through the soil profile (Carvalho, 2006) to reduce their negative impacts on the environment using different management practices (structural and non-structural) leading to increased bio-efficacy and reduced use rates (Grayling et al., 2018).

Traditionally, the concentration of selected chemicals in soil is considered as one of the controlling measures of the degree of soil pollution. However, over the last few decades, the residence time of a contaminant or pollutant in the soil is becoming an important measure, which is largely influenced by a combination of transport, retention, reaction, and transformation processes which a solute undergoes in soil (Govindaraju & Das, 2007). Understanding and parameterizing solute transport through porous media is the first step towards the mitigation of soil and water pollutants. Therefore, this chapter focuses on conventional and recent Computed Tomographic (CT) techniques used in the measurement of solute and water transport through porous media.

7.2 Methods to Study Solute and Water Transport in Soils

In the last two decades, the study of solute transport to groundwater has become of increasing prominence due to the rising awareness about pollution and water quality. Upgrading groundwater preservation techniques are indispensable and are supported by detailed studies on pollutant transportation and hydraulic behaviour of soils (Anderson et al. 2015a). Knowledge of solute and water transport helps to lessen the detrimental impacts of pesticides on the environment and develop various innovative approaches that reduce the application rate and increase the efficiency of agrochemicals (Grayling et al., 2018). There are various methods proposed to measure solute and water transport parameters. Some of these methods are discussed below.

Solute transport can be predicted by performing column transport experiments in the laboratory (outflow experiments) under a variety of water flow conditions (Reddy et al., 1992; Weber et al., 1993) and by lysimeter studies (Byers et al., 1995; Weber et al., 1994). Parameter estimation codes such as CXTFIT (Toride et al., 1995) are used for fitting analytical solutions of the transport equation to experimental breakthrough curves (BTCs). These breakthrough experiments are used to measure the movement and dissipation of pesticides, insecticides, and other harmful chemicals and to determine the potential contamination of groundwater (Redondo et al., 1997). Solute transport parameter estimation for various conditions such as nonlinear adsorption of chemicals or preferential flow, for which closed form analytical solutions are difficult, are obtained through numerical solutions (Šimůnek & Van Genuchten, 1999). The numerical codes enable one to obtain several parameters in the unsaturated soil; these hydraulic functions can be estimated from observed water contents, pressure heads, and/or instantaneous or cumulative boundary conditions using the numerical inversion of the Richards equation. Similarly, the numerical inversion of advection-dispersion equation (ADE) allows the estimation of solute transport and/or reaction parameters using breakthrough curves. However, the heterogeneity of the soil profile often results in anomalous dispersion of solute. Subsequently, the ADE has been continuously modified over the decades; for example, Coats and Smith (1964) proposed a mobile-immobile model by adding a source/sink term to the ADE to take care of the mass exchange between mobile and immobile water. Then after that, various modifications of ADE with the introduction of multiple-rate and multiple-region for solute transport simulation have been proposed (Šimůnek et al., 2003). However, these modifications introduced a large number of parameters that may depend not only on the soil physical properties (Zhang et al., 2007) but also on the water flow (Pang & Hunt, 2001). Moreover, the direct measurements of solute transport using breakthrough curves can be time-consuming (Karup et al., 2016). In addition, traditional methods of bulk ex-situ chemical or eluate analysis can be difficult at estimating the flow of solute in a spatio-temporal manner simultaneously (Grayling et al., 2018). The extrapolation of laboratory-derived data for predicting solute behaviour under field conditions is tenuous considering the many variables that influence solute dynamics. Therefore, a technique providing spatial distributions of contaminant concentrations and groundwater velocities such as the non-destructive method of X-ray CT may be a possible method for studying solute transport through porous media. Some of the key parameters of solute and water transport estimated using the X-ray CT approach are given in Table 7.1, and a comparative analysis between traditional and X-ray CT technique is given in Table 7.2.

Effective evaluation of solute and water transport in soils in the laboratory can be arduous (Anderson et al., 2014). CT imaging techniques, on the contrary, are fast, robust, non-invasive, permit the 3-D visualization of soil structural properties on a micro-metre scale (Carlson et al., 2003; Hapca et al., 2015), provide information on the spatial distribution of soil pores (Rab et al., 2014), and can be used to study transport parameters at a resolution not possible with traditional methods. In addition, CT procedures have been applied to characterize microbial habitat microstructures, root architectures, solute movement, gas exchange, and water flow properties under different soils and crop management systems (Nunan et al., 2006; Tracy et al., 2010; Flavel et al., 2014; Kettridge & Binley, 2008). In numerous examinations, X-ray CT has been utilized in a spatial-temporal manner to explore the hydrophysical attributes of the soil (Taina et al., 2008). Compacted soil layers and their boundaries can also be identified through CT scanning with 3-D images (Garbout et al., 2013; Grevers et al., 1989). Imaging the soil pore space in 3-D and the quantification of pore structure can help in understanding the links between

Table 7.1 Traditional and X-ray Computed Tomography (CT) methods for studying solute and water transport in soils

Studied key parameters	Methods	References
Preferential water flow	Helical CT approach	Sammartino et al. (2012)
Preferential water flow in paddy fields	X-ray CT and break-through curve	Zhang et al. (2015)
Porosity and hydraulic conductivity	X-ray CT-measured solute breakthrough	Anderson et al. (2003)
Soil structural (macroporosity and bulk density) properties	X-ray CT and visible near-infrared spectroscopy	Katuwal et al. (2018a)
Cu and water transport	X-ray CT approach	Paradelo et al. (2013)
Solute transport	Visible–near-infrared spectroscopy	Katuwal et al. (2018b)
Transport of solute and colloid tracers in soils	X-ray CT approach	Soto-Gómez et al. (2019)
Solute transport	Fast X-ray CT	Van Offenwert et al. (2019)
Tracers to visualize solute movement	X-ray CT	Anderson et al. (2003); Clausnitzer and Hopmans (2000); Luo et al. (2008)
Visualization of soil water movement	X-ray CT	Mooney (2002)
Tracing particle movement in soils	X-ray CT	Grayling et al. (2018)
Microstructure due to wetting and drying, and subsequent pesticide mobility	X-ray CT	Dor et al. (2019); Zhang et al. (2015)
Root structure in influencing nitrate leaching	X-ray CT	Galdos et al. (2020)

macropore characteristics and fluid flow and solute transport (Katuwal et al., 2015). In many studies, X-ray CT has been used to investigate the hydro-physical characteristics of the soil in a functional and temporal manner. The following sections describe the procedure for utilizing CT scan techniques for measuring the solute and water transport parameters in soil.

7.3 CT Scanning for Estimating Solute and Water Transport

Solute and water transport in soils are impacted by pore-water velocity, hydraulic conductivity, solute dispersivity, and porosity (Anderson et al., 2014). The pore structure of soils is one of the most important parameters influencing solute and water transport through soil. This structure is closely related to surface runoff and soil permeability and, therefore, is an important determinant of soil water storage for crop production. Pore-structure is a critical parameter that can affect the movement

Table 7.2 The advantages and disadvantages of the traditional and CT approaches for measuring solute and water transport in soils

Methods	Advantages	Disadvantages	References
Traditional (e.g. tracer outflow experiments)	Less expensive	Lack of real-time data especially a 3-dimensional perspective	Grayling et al. (2018)
	Equipment often readily available	Destructive methods	
		Inability to represent spatial distribution of pores and their network	
Computed Tomography	Visualization of the internal fabric of a porous media such as soil.	Expensive	Zhang et al. (2015)
	Non-destructively visualize the internal pore geometry of the structured media with repeating scanning. Characterization of the active processes in situ can be attempted assuming the solute/particulate material can be distinguished from the soil background.	Best suited for the visualization of motionless objects. Resolution-sample size trade off exists. Representative elementary volume may be different for different property.	Grayling et al. (2018) Luo et al. (2008)

and storage of water, nutrients, and microfauna in the soil (Seobi et al., 2005). Traditionally, various methods such as density measurements, gas pycnometry (Danielson & Sutherland, 1986), mercury intrusion (Danielson & Sutherland, 1986), soil bulk density data, soil water characteristic curves, and tension infiltrometer data (Everts & Kanwar, 1993) have been successfully used to characterize soil porosity or pore-size distribution; albeit these methods may not precisely expose the obscure features of soil pores. Within a distance of only a few millimetres, soil porosity can be highly variable, causing water and transport parameters to differ spatially with orders of magnitude. Therefore, the measurement of pore structure at mm to μm scales has been considered to be significant in anticipating micro-scale liquid flow properties (Yang et al., 2018). X-ray CT has been effectively used for its potential to directly study soil-pore networks (Katuwal et al., 2018a, also see Chap. 6).

Anderson et al. (2003) conducted a breakthrough experiment taking intact soil cores and scanning the cores several times before, during, and after solute breakthrough experiments using potassium iodide (KI) (a conservative solute, which does not react with the soil). The soil core was scanned while being saturated with water and while being saturated with dilute KI solution to obtain:

$$\mu_{(p+w)} = \mu_p(1-\phi) + \mu_w\phi \tag{7.1}$$

$$\mu_{(p+w+c)} = \mu_p(1-\phi) + \mu_{(w+c)}\phi \tag{7.2}$$

where μ is the linear attenuation coefficient with the subscript $p + w$ for water-saturated pores, p for dry porous media, $p + w + c$ for saturated KI solution, and ϕ is the porosity of the soil core. Combining the two equations, the following is obtained:

$$\phi = \frac{\mu_{(\mathrm{p+}w\mathrm{+}c)} - \mu_{(\mathrm{p+w})}}{\mu_{(w+c)} - \mu_{(w)}} \tag{7.3}$$

Therefore, porosity, ϕ, for a voxel, centred at x,y,z is given as

$$\phi(x, y, z) = \frac{CTN(x, y, z)_{(\mathrm{p+}w\mathrm{+}c)} - CTN(x, y, z)\mu_{(\mathrm{p+w})}}{\overline{CTN}_{(w+c)} - \overline{CTN}_{(w)}} \tag{7.4}$$

where $CTN(x, y, z)_{(p\,+\,w)}$ is the CT number in the voxel centred at x,y,z containing only water-saturated pores (CT number is the X-ray absorption coefficient in each voxel normalized to the corresponding absorption coefficient for water expressed in Hounsfield units, Lindgren (1991)); $CTN(x, y, z)_{(p\,+\,w\,+\,c)}$ is the CT number in the voxel centred at position x,y,z containing pores saturated with aqueous KI solution; $\overline{CTN}_{(w)}$ is the mean cross-sectional CT number for the cylinder containing only water; and $\overline{CTN}_{(w+c)}$ is the mean cross-sectional CT number for the cylinder containing only aqueous KI solution. The average CT-determined porosity for each soil core was calculated by averaging the porosity values estimated for the entire soil core.

In addition to total porosity, soil macropores can have an important role in transporting solutes, water, and other colloids in the system (Soto-Gómez et al., 2018). These macropores are the pores or structural cracks in the soil with an effective diameter of >1000 μm (Jarvis, 2007; Anderson et al., 2014). As the indirect measure of soil pore size distribution or porosity does not include any detailed information on the spatial distribution of the pores or the nature of the pores (Gantzer & Anderson, 2002), CT scanning can be applied to estimate the numbers, structure (in terms of size and shape), distribution, and 3-D visualization of macropores in soil (Anderson et al., 2003). For example, Fig. 7.1 shows reconstructed 3-D macropore structure for intact soil cores collected from a native grass land, a cropland under long-term no-till, and a conventionally-tilled cropland. Luo et al. (2010) quantified and observed distinct differences in the various characteristics of macropore networks such as continuous macroporosity change along depth, macropore size distribution, and connectivity (path number and node density) for two different soil types (Hagerstown silt loam and Morrison sand) and two land uses (row crop and pasture). Zhang et al. (2019) studied intact soil cores from a plot with long-term fertilizer experiments, analysed the macropores with the help of CT scanning, and characterized the soil cores as bio-pore dominated and matric dominated samples. The pores with pore length to pore radius ratio of more than 20 were considered as bio-pores in their study. They also calculated the air permeability and saturated conductivity for the cores using traditional methods and assessed their relationships with the macropore characteristics. They identified the mean macropore diameter of

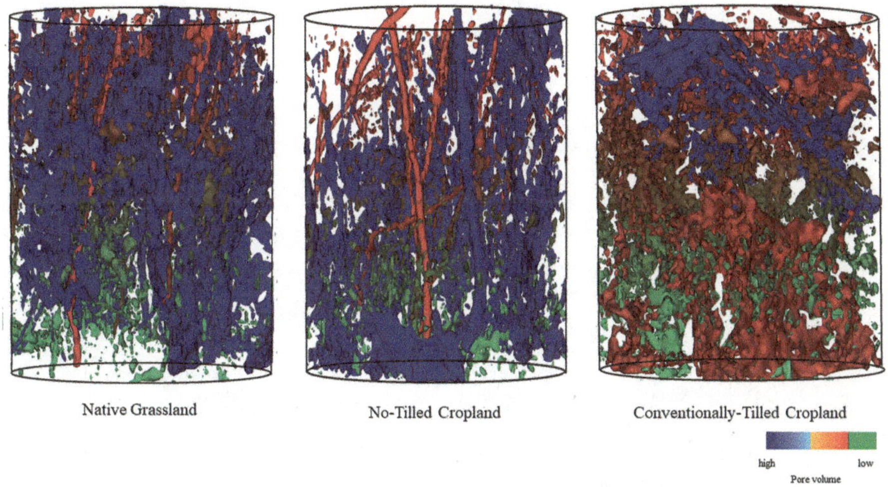

Fig. 7.1 Three-dimensional visualization of soil macropore networks for a native grassland, no-tilled cropland, and conventionally-tilled cropland. Coloured soil macropores are shown in non-porous white background. Source: Created by Dr. Poulamee Chakraborty

the limiting layer (the layer with minimum value of macroporosity) as the best correlated variable.

Hydraulic conductivity is difficult to measure precisely at small scales using conventional laboratory methods as porosity varies over small distances making the conductivity to vary by orders of magnitude. Therefore, averaging is considered, which often fails to yield sufficiently accurate information, particularly for heterogeneous porous media such as soil. The direct chemical transport observation collected using CT scan images, therefore, has emerged as a valuable tool for understanding solute and water transport processes. Anderson et al. (2003) combined X-ray CT imaging with iodide tracer breakthrough experiments and used an inverse approach with CT-determined porosity distributions (Eq. 7.4) to estimate the distribution of soil hydraulic conductivity within undisturbed soil cores. They measured the hydraulic conductivity values on a per voxel basis and observed a positively skewed frequency distribution for the conductivity values for a particular soil core which showed the influence of macropores on the conductivity values. Dal Ferro et al. (2015) used the X-ray CT technique to obtain the saturated hydraulic conductivity of the largest pore, which represents the highly networked large singular pore within soil cores. They utilized a modified Poiseuille equation to estimate water flow through the pores as a function of the CT-measured tortuosity and discrete compactness. Their study combined the 3-D pore-shape aspects (i.e. tortuosity and discrete compactness) with the pore volume to calculate the saturated hydraulic conductivity using Darcy's equation and obtained a good estimate of the conductivity values within the same order of magnitude as that of the experimentally determined ones. The CT technique, therefore, is useful in providing additional information on pore structure to estimate the variations in soil hydraulic parameters.

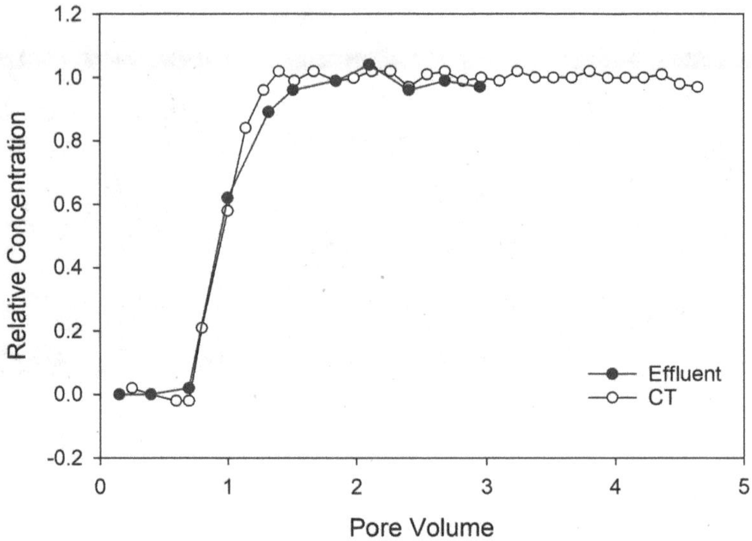

Fig. 7.2 Relative iodide concentration versus pore volume for solute breakthrough using both effluent samples and computed tomography measurements Source: reconstructed from Anderson et al. (2003)

In addition to estimating soil hydraulic properties, CT scanning of a solute breakthrough curve can also be utilized to measure solute transport parameters (Anderson et al., 2014). Transport breakthrough experiments were conducted on intact cores with dilute KI as a solute. The relative concentration of KI solution for each pixel in a core based on attenuation coefficient was, $C(x,y,z)_{rel}$, defined as:

$$C(x, y, z)_{rel} = \frac{CT(x, y, z, t) - CT(x, y, z, t = 0)}{CT(x, y, z, T) - CT(x, y, z, t = 0)} \tag{7.5}$$

where $CT(x,y,z,t)$ is the X-ray attenuation value for a voxel during breakthrough at time t; $CT(x,y,z,t = 0)$ is the X-ray attenuation value when the core is saturated with water; and $CT(x,y,z,T)$ is the X-ray attenuation value when the soil is fully replaced by the dilute KI solution (Anderson et al., 1992). The solute velocity, dispersion coefficient, and dispersivity values were obtained using the CT-measured breakthrough curves. Fig. 7.2 shows the relative iodide concentration versus pore volume for the solute breakthrough curve obtained using effluent outflow measurements and computed tomography measurements (adapted from Anderson et al. (2003)). Anderson et al. (2014) observed lower values for dispersivity using the CT method as compared to the outflow method, which might be attributed to additional dispersion through the column end-plate in the outflow experiments. Their study showed the utility of CT as a useful technique to estimate solute transport parameters.

In addition to using CT-measured breakthrough curves to estimate the water and solute transport parameters, the CT technique has also been used for simulating solute and water transport solving the Navier–Stokes equation directly on the CT derived pore network (lattice Boltzmann method) or using pore network modelling (Bultreys et al., 2015). For example, Köhne et al. (2011) utilized the CT derived pore size distribution, interface area density, and connectivity to set up an equivalent pore network model for predicting the breakthrough curves of Br—and Brilliant Blue FCF at unsaturated and steady-state flux. Similarly, Gharedaghloo et al. (2018) modelled water flow and solute transport in the unstructured pore networks extracted from CT images of peat soil. In addition to these studies, X-ray CT is used to visualize the various structural changes in soil due to anthropogenic or natural causes. For example, soil cracks caused by hysteresis can change the soil structure (Dor et al., 2019). Zhang et al. (2015) used CT to study the impacts of wetting and drying on soil microstructures and the subsequent impact on solute mobility in agricultural fields. Dor et al. (2019) packed sandy clay loam and Geva clay soils, saturated them with distilled water, and obtained CT images. The same soil samples were then subjected to drying followed by rewetting and rescanning to complete the wet-dry cycle. They reported that during the wetting and drying cycle, the structure of clay soil was weakened due to aggregate disintegration as the equivalent diameter of the particle-size distribution modal value was decreased, which enhanced atrazine desorption and leaching. However, the structure of a sandy clay loam soil was strengthened due to $CaCO_3$ cementation and reduced the atrazine desorption and leaching. Similarly, Paradelo et al. (2013) also used CT to study soil pores ($>$180 μm) and their subsequent impact on solute transport. They used CT scanning to measure soil macroporosity and observed it to be the best predictor of air permeability and saturated hydraulic conductivity. They also reported that the relative amount of Cu released was strongly correlated with the CT-measured macroporosity indicating the potential of this visualization technique for predicting contaminant transport through soil. X-ray CT can also be used to investigate the agrochemical movement through various plant systems, for example, Scotson et al. (2021) have recently used CT scan to study the differences in solute movement through ridged and flat plant systems with and without ponding water. They observed lower infiltration of solute in ridged plant system as compared to the flat plant system in absence of ponding water, however, ponding water negated the effect. This indicates that the anticipated rainfall is an important factor to be considered while applying agrochemicals. In addition, CT techniques have been used to study root structure and estimate its influence on solute transport. Galdos et al. (2020) scanned repacked soil cores with maize, ruzigrass, and palisade grass grown in the cores and obtained the CT scans of the cores to visualize the root structure. The images were also used to calculate the total porosity in each image and the relative size, shape, and connection of each pore in 3-D using the BoneJ plugin within *ImageJ*. Nitrate breakthrough curves were obtained for the same cores and the CT scan-derived parameters were combined to observe that a high volume of thin roots was more effective at reducing and delaying leaching in comparison with similar volumes of thicker roots. They reported that plants with fine roots can generate soils with higher microporosity, increased connectivity of the pore network, and increased tortuosity, which can help to retain the nitrate within the soil. Thus, CT imaging techniques are emerging as an important tool to visualize soil structural changes and root networks (see Chap. 9). The visualization of pore networks using CT techniques can further be improved by combining this with other visualization tools, for example, to study the pores in soil that can transport microspheres. Soto-Gómez et al. (2018) utilized fluorescence macrophotography along with CT techniques to trace the colloid transporting pores in intact soil columns. Thus, this technology can be used to assess and refine important management practices to reduce nutrient leaching. The chapter so far has discussed the advantages of CT scan as a straightforward interpreter of transport processes in soil; however, some of the limitations of the technique that can reduce the accuracy of parameter estimation are discussed in the following section.

7.4 Limitations of CT Techniques for Estimating Water and Solute Transport through Soil

Although CT has emerged as an important tool to visualize and estimate water and solute transport through soil, it still has some limitations and requires a continuous improvement in the imaging and image processing techniques to improve the quantification of transport properties. For example, the CT technique generates spatial data of the entire sample over the whole course of the scan, therefore, it requires repeated scanning to track solute transport and is not suitable for tracers moving too rapidly through the soil relative to the time of imaging (Grayling et al., 2018). However, due to minimal disturbance and recent reduction in scanning time, the same soil sample can be made available for repeated scanning in the region of interest for different parameters (Nielsen, 2004). Another limitation of the CT procedure is that a thick object usually absorbs too much energy, resulting in low X-ray flux and poor image quality thereby reducing the sample size that can be scanned. However, with recent advancements in CT technology, larger soil samples are being measured (Mooney et al., 2012). The increase in scanning sample size has also helped in achieving a representative elementary volume (see Chap. 3) for the CT-measured porosity and fluid properties (Gharedaghloo et al., 2018). In addition, as previously stated elsewhere, various CT scanning artefacts, such as beam hardening (the increase in mean X-ray beam energy as it passes through the scanned object as lower-energy X-rays are attenuated more readily by the object than higher-energy X-rays), are reduced by the introduction of filters in modern CT machines (see Chap. 4). Another artefact that is encountered in CT is the partial volume effect that occurs because of the presence of a number of different substances in a voxel resulting in a CT number, which is an average value for all the materials present. This creates difficulty in obtaining a bimodal histogram of the grey level when thresholding (Gackiewicz et al., 2019). However, recent improvements in the resolution of the CT images and image processing techniques have reduced this effect (see Chap. 5).

7.5 Conclusions

Studying the transport of solutes in porous media is important to protect water resources from contamination. The X-ray Computed Tomographic techniques have emerged as a fast, robust, and non-invasive method to provide a unique opportunity to quantify detailed pore morphological parameters and permit 3-D visualization of soil structural properties. Most studies using X-ray CT in soil assessed the 3-D structure of soil macropores because of the strong contrast between the linear attenuation coefficients of the soil matrix and pores. However, this technology is also an important tool for non-destructively imaging and estimating the dynamic water and solute transport properties of porous media, particularly with the recent advancement in the reduction of scanning interval. In addition, CT has made the pore scale modelling of water and solute transport possible by providing the 3-D pore network and allowing the Navier–Stokes equations to be solved directly on a gridded or meshed 3-D image of the pore space geometry. While CT can be expensive, it does provide promising potential for assessing transport processes to help in more accurately predicting contaminant transport through soil which is likely to improve over time.

References

Anderson, S., Heinze, D., & Peyton, R. (2014). Computed tomography-estimated transport velocity and chemical dispersivity in undisturbed geomedia. *Procedia Computer Science, 36*, 643–648.

Anderson, S., Heinze, D., & Peyton, R. (2015a). Assessment of selected methods for estimating chemical transport parameters from computed tomographic imaging. *Procedia Computer Science, 61*, 460–465.

Anderson, S., Holmes, J., & Peyton, R. J. P. C. S. (2015b). Tomography-measured spatial distributions of non-aqueous phase liquids in porous. *Procedia Computer Science, 61*, 466–471.

Anderson, S., Peyton, R., Wigger, J., & Gantzer, C. J. J. G. (1992). Influence of aggregate size on solute transport as measured using computed tomography. *Geoderma, 53*(3-4), 387–398.

Anderson, S., Wang, H., Peyton, R., & Gantzer, C. J. G. S. (2003). Estimation of porosity and hydraulic conductivity from X-ray CT-measured solute breakthrough. *Geological Society, 215*(1), 135–149.

Bultreys, T., Van Hoorebeke, L., & Cnudde, V. (2015). Multi-scale, micro-computed tomography-based pore network models to simulate drainage in heterogeneous rocks. *Advances in Water Resources, 78*, 36–49.

Bus, J. S., & Hammond, L. E. (2007). Regulatory progress, toxicology, and public concerns with 2, 4-D: Where do we stand after two decades? *Crop Protection, 26*(3), 266–269.

Byers, M., Tyess, D., Antonious, G., Hilborn, D., & Jarret, L. (1995). Monitoring herbicide leaching in sustainable vegetable culture using tension lysimeters. *Bulletin of Environmental Contamination and Toxicology, 54*(6), 848–854.

Carlson, W. D., Rowe, T., Ketcham, R. A., & Colbert, M. W. (2003). Applications of high-resolution X-ray computed tomography in petrology, meteoritics and palaeontology. *Geological Society, 215*(1), 7–22.

Carvalho, F. P. (2006). Agriculture, pesticides, food security and food safety. *Environmental Science & Policy, 9*(7-8), 685–692.

Clausnitzer, V., & Hopmans, J. W. (2000). Pore-scale measurements of solute breakthrough using microfocus X-ray computed tomography. *Water Resources Research, 36*(8), 2067–2079.

Coats, K. H., & Smith, B. D. (1964). Dead-end pore volume and dispersion in porous media. *Society of Petroleum Engineers Journal, 4*(01), 73–84.

Dal Ferro, N., et al. (2015). Application of smoothed particle hydrodynamics (SPH) and pore morphologic model to predict saturated water conductivity from X-ray CT imaging in a silty loam Cambisol. *Geoderma, 255*, 27–34.

Danielson, R., & Sutherland, P. (1986). Porosity. Methods of soil analysis: Part 1 physical and mineralogical. *Methods, 5*, 443–461.

Dor, M., Emmanuel, S., Brumfeld, V., Levy, G. J., & Mishael, Y. G. (2019). Microstructural changes in soils induced by wetting and drying: Effects on atrazine mobility. *Land Degradation & Development, 30*(7), 746–755.

Everts, C. J., & Kanwar, R. S. (1993). Interpreting tension-infiltrometer data for quantifying soil macropores: Some practical considerations. *Transactions of the ASAE, 36*(2), 423–428.

FAO. (2018). *Pollutants from agriculture a serious threat to world's water.* http://www.fao.org/news/story/en/item/1141534/icode/

Flavel, R. J., Guppy, C. N., Tighe, M. K., Watt, M., & Young, I. M. (2014). Quantifying the response of wheat (Triticum aestivum L) root system architecture to phosphorus in an Oxisol. *Plant and Soil, 385*(1), 303–310.

Gackiewicz, B., Lamorski, K., & Slawinski, C. (2019). Saturated water conductivity estimation based on X-ray CT images-evaluation of the impact of thresholding errors. *International Agrophysics, 33*(1).

Galdos, M. V., Brown, E., Rosolem, C. A., Pires, L. F., Hallett, P. D., & Mooney, S. J. (2020). Brachiaria species influence nitrate transport in soil by modifying soil structure with their root system. *Scientific Reports, 10*(1), 1–11.

Gantzer, C. J., & Anderson, S. H. (2002). Computed tomographic measurement of macroporosity in chisel-disk and no-tillage seedbeds. *Soil and Tillage Research, 64*(1-2), 101–111.

Garbout, A., Munkholm, L. J., & Hansen, S. B. (2013). Tillage effects on topsoil structural quality assessed using X-ray CT, soil cores and visual soil evaluation. *Soil and Tillage Research, 128*, 104–109.

Gharedaghloo, B., Price, J. S., Rezanezhad, F., & Quinton, W. L. (2018). Evaluating the hydraulic and transport properties of peat soil using pore network modeling and X-ray micro computed tomography. *Journal of Hydrology, 561*, 494–508.

Govindaraju, R. S., & Das, B. S. (2007). *Moment analysis for compounds undergoing sequential decay chain reactions, moment analysis for subsurface hydrologic applications* (pp. 207–222). Springer.

Grayling, K. M., Young, S. D., Roberts, C. J., de Heer, M. I., Shirley, I. M., Sturrock, C. J., & Mooney, S. J. (2018). The application of X-ray micro computed tomography imaging for tracing particle movement in soil. *Geoderma, 321*, 8–14.

Grevers, M., Jong, E. D., & St. Arnaud, R. (1989). The characterization of soil macroporosity with CT scanning. *Canadian Journal of Soil Science, 69*(3), 629–637.

Hapca, S., Baveye, P. C., Wilson, C., Lark, R. M., & Otten, W. (2015). Three-dimensional mapping of soil chemical characteristics at micrometric scale by combining 2-D SEM-EDX data and 3-D X-ray CT images. *PLoS One, 10*(9), e0137205.

Hubbard, R. K., Newton, G. L., & Hill, G. M. (2004). Water quality and the grazing animal. *Journal of Animal Science, 82*(Suppl_13), E255–E263.

Jarvis, N. (2007). A review of non-equilibrium water flow and solute transport in soil macropores: Principles, controlling factors and consequences for water quality. *European Journal of Soil Science, 58*(3), 523–546.

Karup, D., Moldrup, P., Paradelo, M., Katuwal, S., Norgaard, T., Greve, M. H., & de Jonge, L. W. (2016). Water and solute transport in agricultural soils predicted by volumetric clay and silt contents. *Journal of Contaminant Hydrology, 192*, 194–202.

Katuwal, S., Hermansen, C., Knadel, M., Moldrup, P., Greve, M. H., & de Jonge, L. W. (2018a). Combining X-ray computed tomography and visible near-infrared spectroscopy for prediction of soil structural properties. *Vadose Zone Journal, 17*(1), 1–13.

Katuwal, S., Knadel, M., Moldrup, P., Norgaard, T., Greve, M. H., & de Jonge, L. W. (2018b). Visible–near-infrared spectroscopy can predict mass transport of dissolved chemicals through intact soil. *Scientific Reports, 8*(1), 1–9.

Katuwal, S., Norgaard, T., Moldrup, P., Lamandé, M., Wildenschild, D., & de Jonge, L. W. (2015). Linking air and water transport in intact soils to macropore characteristics inferred from X-ray computed tomography. *Geoderma, 237*, 9–20.

Kettridge, N., & Binley, A. (2008). X-ray computed tomography of peat soils: Measuring gas content and peat structure. *Hydrological Processes: An International Journal, 22*(25), 4827–4837.

Köhne, J. M., Schlüter, S., & Vogel, H. J. (2011). Predicting solute transport in structured soil using pore network models. *Vadose Zone Journal, 10*(3), 1082–1096.

Lindgren, L. O. (1991). Medical CAT-scanning: X-ray absorption coefficients, CT-numbers and their relation to wood density. *Wood Science and Technology, 25*(5), 341–349.

Luo, L., Lin, H., & Halleck, P. (2008). Quantifying soil structure and preferential flow in intact soil using X-ray computed tomography. *Soil Science Society of America Journal, 72*(4), 1058–1069.

Luo, L., Lin, H., & Li, S. (2010). Quantification of 3-D soil macropore networks in different soil types and land uses using computed tomography. *Journal of Hydrology, 393*(1-2), 53–64.

Mooney, S. J. (2002). Three-dimensional visualization and quantification of soil macroporosity and water flow patterns using computed tomography. *Soil Use and Management, 18*(2), 142–151.

Mooney, S. J., Pridmore, T. P., Helliwell, J., & Bennett, M. J. (2012). Developing X-ray computed tomography to non-invasively image 3-D root systems architecture in soil. *Plant and Soil, 352*(1), 1–22.

Nielsen, B. D. (2004). *Non-destructive soil testing using X-ray computed tomography*. Montana State University-Bozeman, College of Engineering.

Nunan, N., Ritz, K., Rivers, M., Feeney, D. S., & Young, I. M. (2006). Investigating microbial micro-habitat structure using X-ray computed tomography. *Geoderma, 133*(3-4), 398–407.

Pang, L., & Hunt, B. (2001). Solutions and verification of a scale-dependent dispersion model. *Journal of Contaminant Hydrology, 53*(1-2), 21–39.

Paradelo, M., Moldrup, P., Arthur, E., Naveed, M., Holmstrup, M., López-Periago, J. E., & de Jonge, L. W. (2013). Effects of past copper contamination and soil structure on copper leaching from soil. *Journal of Environmental Quality, 42*(6), 1852–1862.

Pimentel, D., Acquay, H., Biltonen, M., Rice, P., Silva, M., Nelson, J., et al. (1992). Environmental and economic costs of pesticide use. *Bioscience, 42*(10), 750–760.

Rab, M. A., Haling, R. E., Aarons, S. R., Hannah, M., Young, I. M., & Gibson, D. (2014). Evaluation of X-ray computed tomography for quantifying macroporosity of loamy pasture soils. *Geoderma, 213*, 460–470.

Reddy, K. N., Singh, M., & Alva, A. K. (1992). Sorption and leaching of bromacil and simazine in Florida Flatwoods soils. *Bulletin of Environmental Contamination and Toxicology, 48*(5), 662–670.

Redondo, M., Ruiz, M., Font, G., & Boluda, R. (1997). Dissipation and distribution of atrazine, simazine, chlorpyrifos, and tetradifon residues in citrus orchard soil. *Archives of Environmental Contamination and Toxicology, 32*(4), 346–352.

Sammartino, S., Michel, E., & Capowiez, Y. (2012). A novel method to visualize and characterize preferential flow in undisturbed soil cores by using multislice helical CT. *Vadose Zone Journal, 11*(1).

Scotson, C. P., Duncan, S. J., Williams, K. A., Ruiz, S. A., & Roose, T. (2021). X-ray computed tomography imaging of solute movement through ridged and flat plant systems. *European Journal of Soil Science, 72*(1), 198–214.

Seobi, T., Anderson, S., Udawatta, R., & Gantzer, C. (2005). Influence of grass and agroforestry buffer strips on soil hydraulic properties for an Albaqualf. *Soil Science Society of America Journal, 69*(3), 893–901.

Šimůnek, J., Jarvis, N. J., Van Genuchten, M. T., & Gärdenäs, A. (2003). Review and comparison of models for describing non-equilibrium and preferential flow and transport in the vadose zone. *Journal of Hydrology, 272*(1-4), 14–35.

Šimůnek, J., & Van Genuchten, M. T. (1999). Using the HYDRUS-1D and HYDRUS-2-D codes for estimating unsaturated soil hydraulic and solute transport parameters. *Characterization and measurement of the hydraulic properties of unsaturated porous media, 1*, 523–521.

Soto-Gómez, D., Pérez-Rodríguez, P., Vázquez-Juiz, L., López-Periago, J. E., & Paradelo, M. (2018). Linking pore network characteristics extracted from CT images to the transport of solute and colloid tracers in soils under different tillage managements. *Soil and Tillage Research, 177*, 145–154.

Soto-Gómez, D., Pérez-Rodríguez, P., Vázquez Juíz, L., López-Periago, J. E., & Paradelo Pérez, M. (2019). A new method to trace colloid transport pathways in macroporous soils using X-ray computed tomography and fluorescence macrophotography. *European Journal of Soil Science, 70*(3), 431–442.

Taina, I., Heck, R., & Elliot, T. (2008). Application of X-ray computed tomography to soil science: A literature review. *Canadian Journal of Soil Science, 88*(1), 1–19.

Toride, N., Leij, F., & Van Genuchten, M. T. (1995). *The CXTFIT code for estimating transport parameters from laboratory or filed tracer experiments, 2*. US Salinity Laboratory Riverside.

Tracy, S. R., Roberts, J. A., Black, C. R., McNeill, A., Davidson, R., & Mooney, S. J. (2010). The X-factor: Visualizing undisturbed root architecture in soils using X-ray computed tomography. *Journal of Experimental Botany, 61*(2), 311–313.

van der Zee, S., & Leijnse, A. (2013). Solute transport in soil, environmental sciences: Soil processes and current trends in quality assessment. *INTECH*, 33–86.

Van Offenwert, S., Cnudde, V., & Bultreys, T. (2019). Pore-scale visualization and quantification of transient solute transport using fast microcomputed tomography. *Water Resources Research, 55*(11), 9279–9291.

Weber, J., Keller, K., Honeycutt, R., & Schabacker, D. (1994). *Mobility of pesticides in field lysimeters*. Lewis Publishers.

Weber, J. B., Strek, H. J., & Sartori, J. L. (1993). Mobility of fomesafen and atrazine in soil columns under saturated-and unsaturated-flow conditions. *Pesticide Science, 39*(1), 39–46.

Yang, Y., Wu, J., Zhao, S., Han, Q., Pan, X., He, F., & Chen, C. (2018). Assessment of the responses of soil pore properties to combined soil structure amendments using X-ray computed tomography. *Scientific Reports, 8*(1), 1–10.

Zhang, X., Lv, M., Crawford, J. W., & Young, I. M. (2007). The impact of boundary on the fractional advection–dispersion equation for solute transport in soil: Defining the fractional dispersive flux with the Caputo derivatives. *Advances in Water Resources, 30*(5), 1205–1217.

Zhang, Z., Liu, K., Zhou, H., Lin, H., Li, D., & Peng, X. (2019). Linking saturated hydraulic conductivity and air permeability to the characteristics of biopores derived from X-ray computed tomography. *Journal of Hydrology, 571*, 1–10.

Zhang, Z., Peng, X., Zhou, H., Lin, H., & Sun, H. (2015). Characterizing preferential flow in cracked paddy soils using computed tomography and breakthrough curve. *Soil and Tillage Research, 146*, 53–65.

Chapter 8
X-ray Imaging of Mechanical Processes in Soil

Mathieu Lamandé and Lars J. Munkholm

8.1 Introduction

Soils are generally subjected to a range of mechanical stresses. These stresses can be applied by soil biota (e.g. earthworms, roots), during wetting and drying, freezing and thawing, by trampling of animals and humans, or by machinery during agricultural operations (e.g. tillage, traffic). The resulting deformations and their effects on soil functions are complex. They can be quantified through bulk measurements (e.g. water and gas transport parameters), where the interpretation is based on conceptualisation of soil pore space (e.g. Arah & Ball, 1994). An alternative approach, X-ray Computed Tomography (CT) imaging solves the issue of soil opacity and allows three-dimensional observations of the internal changes in soil structure (Taina et al., 2008; Chap. 6). This can be used to evaluate our interpretation of bulk measurements and the concepts for the soil pore system. Our knowledge of the stress-strain relationships of unsaturated structured soils is theoretical, with simplifications of soil intrinsic properties (e.g. Burland, 1990), and empirical, based on bulk observations where the stress-strain relationships are then given by constitutive equations (Koolen & Kuipers, 1983). Developments in X-ray imaging techniques have facilitated our understanding of the mechanical processes in soil. This chapter will present a range of studies reporting these developments and will give perspectives of using X-ray imaging in soil mechanics related research.

M. Lamandé (✉)
Department of Agroecology, Aarhus University, Research Centre Foulum, Tjele, Denmark

Faculty of Environmental Sciences and Natural Resource Management, Norwegian University of Life Sciences, Ås, Norway
e-mail: mathieu.lamande@agro.au.dk

L. J. Munkholm
Department of Agroecology, Aarhus University, Research Centre Foulum, Tjele, Denmark
e-mail: lars.munkholm@agro.au.dk

S. Jon Mooney et al. (eds.), *X-ray Imaging of the Soil Porous Architecture*,
https://doi.org/10.1007/978-3-031-12176-0_8

8.2 X-ray CT in Relation to Tillage

The impact of soil tillage on soil is complex as it directly induces mechanical stresses and indirectly strongly affects the stresses induced by soil biota, wetting/drying and freezing/thawing processes. The mechanical stresses directly induced by tillage may result in soil deformation, ranging from overall loosening (brittle/shear failure) to deformation with, or without, volume change. Soil loosening is normally desired in tillage, but will in many cases also involve compaction or shearing of parts of the disturbed soil volume (e.g. compaction below tines). Further, soil tillage will normally be associated with soil traffic, which may result in traffic-induced soil compaction, as described in the next section of the present chapter. The avoidance of tillage—as in minimal or no-tillage systems—limits the direct mechanical impact from tillage and traffic, meaning that soil structure formation and stabilisation depend strongly on natural physical processes and biotic influences from roots, microorganisms and soil fauna (Bronick & Lal, 2005; Munkholm et al., 2019). X-ray CT has been applied to the soil sciences since the 1980/90s as a powerful tool to visualise and quantify, on the one hand, soil structural properties resulting from tillage (or no-tillage) and associated key soil functions and, on the other hand, the soil structural properties affecting soil behaviour in tillage. In one of the first studies, Werner et al. (1993) used soil from tillage experiments to illustrate the potential of X-ray radiographs to characterise the spatial changes in soil structural properties—both on greyscale images and on binarised and segmented images. In the following years more than 150 peer-reviewed papers were published according to Web of Science (tillage + (X-ray CT or "Computed Tomography" or "CT scanner") + soil; accessed 11 Jan 2021) (Fig. 8.1). It took time before X-ray CT became more broadly used in tillage research: fewer than five papers were published per year until 2012 and fewer than 30 papers were published from 1991 to 2010. The availability of the methodology and the interest in using X-ray CT visualisation for tillage studies have escalated in recent years, and more than half of all the papers (78) have been published in 2017–2020.

8.2.1 Visualisation and Evaluation of Seedling Growing Conditions Produced by Tillage

Most of the early X-ray CT studies focussed on visualisation and evaluation of macro-structure produced by tillage due to the coarse resolution of CT images and simplicity in available image analysis methods. For instance, Munkholm et al. (2003) used a medical CT scanner as a supplementary tool to visualise the short-term impact of two different no-tillage drills on soil physical conditions for germination and seedling growth. For the no-tillage drills, the tine created more disturbance than the disk, though both created less disturbance than the traditionally ploughed system (Fig. 8.2), which was interpreted as beneficial for readily

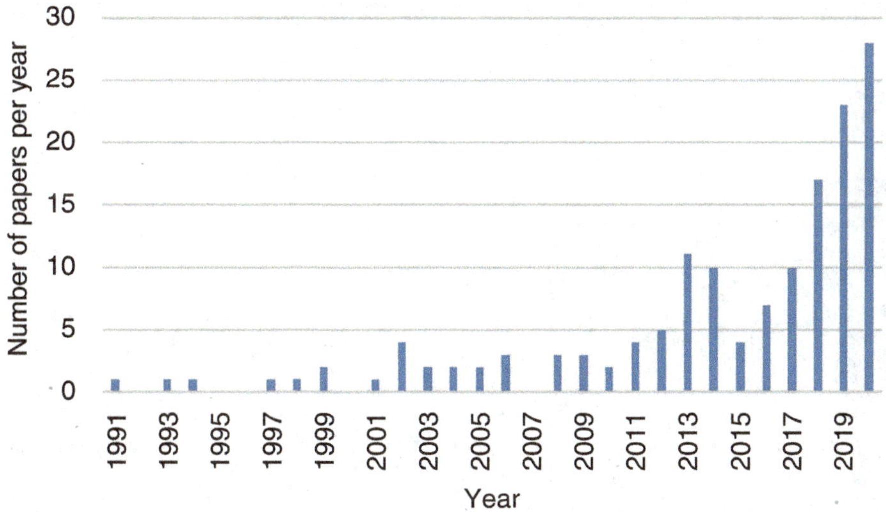

Fig. 8.1 Yearly number of peer-reviewed papers that have been published according to Web of Science using the following topic search (11 January 2021): tillage + (X-ray CT or "computed tomography" or "CT scanner") + soil

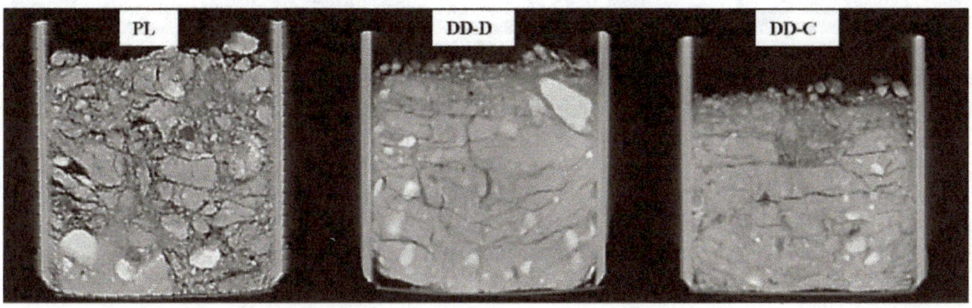

Fig. 8.2 2-D CT-scanning images showing variations in grey value, with the seed furrow at the centre. Bulk density increases with brightness. Each image is from a single undisturbed column (180 mm diameter, 200 mm height) from each treatment taken at T2. PL: conventionally ploughed soil; DD-D: direct-drilled with single-disc coulters; DD-C: direct-drilled with chisel coulters. Reproduced from Munkholm et al. (2003) with permission from Elsevier

compactable sandy loams in a cool and moist climate. More recently, Blunk et al. (2021) significantly advanced the use of X-ray CT for high-resolution analysis of the effects of tillage on soil structural quality of the seedling environment (Fig. 8.3). They showed that X-ray CT was a powerful tool to quantify the variability in soil-seed contact in the field for sugar beet and to evaluate effects of soil type and tillage intensity on seedbed conditions. Additionally, Blunk et al. (2017, 2019) demonstrated the benefits of using X-ray CT for evaluating the impact of seed priming techniques on early seedling growth. Thus, X-ray CT can be a valuable tool to

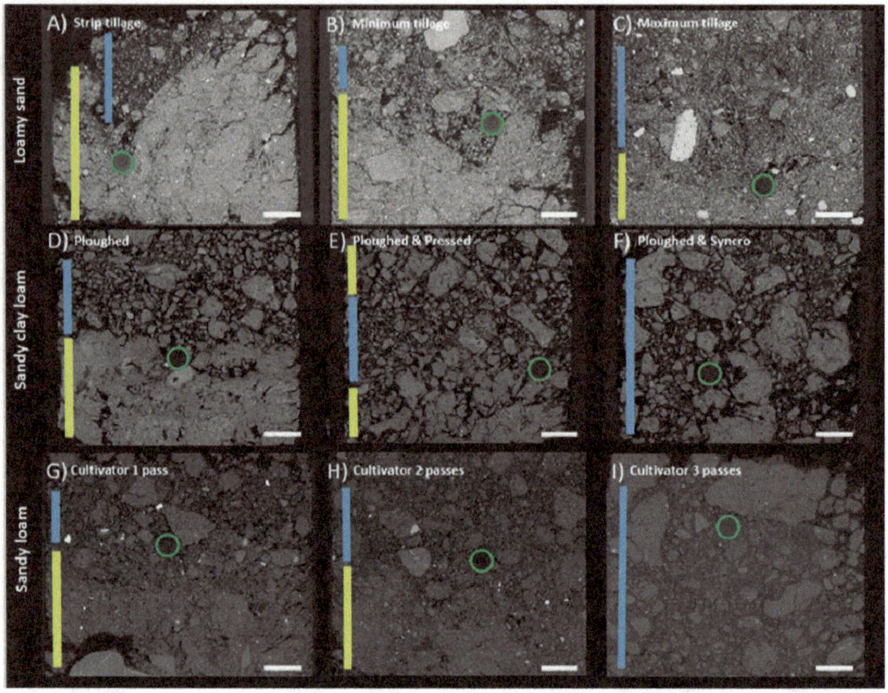

Fig. 8.3 Example 2-D slices of each sampled soil texture and management technique. (**a**) Loamy sand—strip tillage. (**b**) Loamy sand—minimum tillage. (**c**) Loamy sand—maximum tillage. (**d**) Sandy clay loam—ploughed. (**e**) Sandy clay loam—ploughed and pressed. (**f**) Sandy clay loam—ploughed and syncro. (**g**) Sandy loam—cultivator 1 pass. (**h**) Sandy loam—cultivator 2 passes. (**i**) Sandy loam—cultivator 3 passes. Seedbed density characterisation was based on visual assessment: Yellow = compacted layer. Blue = loose layer. Green = seed. Scale bar represents 7 mm. Reproduced from Blunk et al. (2021) with permission from Elsevier

develop management options (precision tillage, seed enhancement technologies) to optimise seeding and seedling growth.

8.2.2 Macropore Structure and Functioning Affected by Tillage/no-Tillage

Tillage strongly influences the volume, size distribution and spatial organisation of soil pores, especially macropores (>80 μm). X-ray CT is a very powerful tool for 3-D soil pore analysis and therefore it has been used to advance knowledge on tillage impact on soil macropores since the method was first applied ca. 30 years ago, e.g., Werner et al. (1993). Early studies focussed on visualisation and analysis of volume, diameter and shape of mainly large macropores (typically >0.3 mm pores) (Werner et al., 1993; Gantzer & Anderson, 2002) due to low-resolution scanning (typically

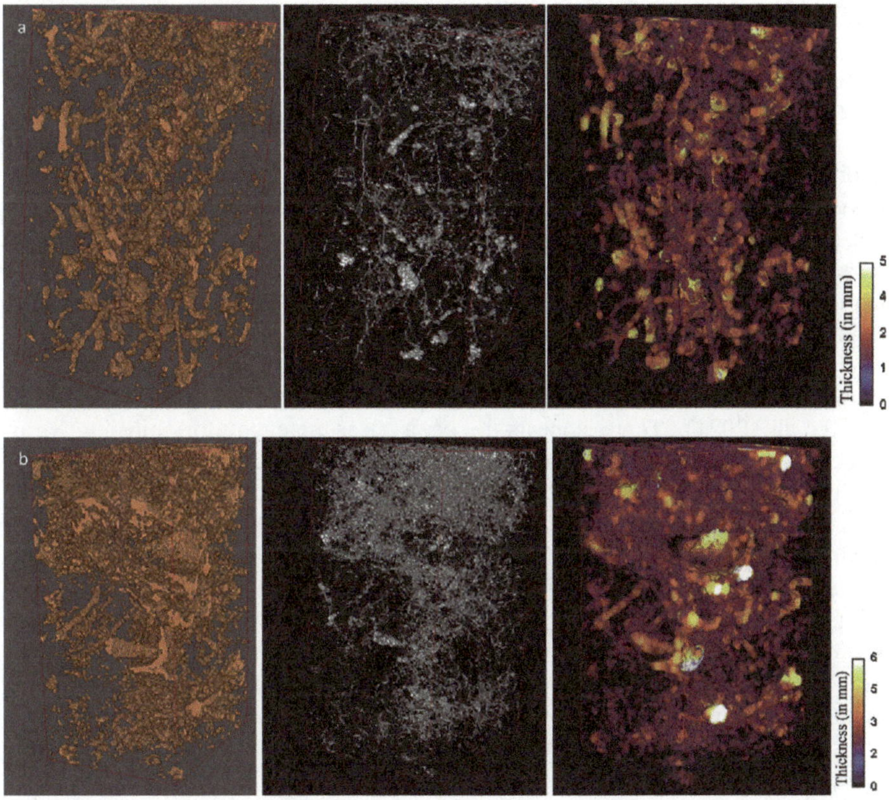

Fig. 8.4 3-D visualisation of the X-ray CT-detected pore system for soil cores (Ø = 20, height = 20 cm) of (**a**) direct-drilled soil, D, and (**b**) ploughed soil, P. The left image is a 3-D view of the pore system. In the middle the soil pore network is obtained after skeletonisation. The right image shows the pore thickness with the calibration thickness in mm. The samples were taken in the same block and scanned using a medical scanner at a resolution of 0.39 mm x 0.39 mm x 0.60 mm. Reproduced from Garbout et al. (2013) with permission from Elsevier

associated with medical X-ray CT scanners). For research applications, most associated CT scanners and image analysis methods have strongly advanced during the last decade. Garbout et al. (2013) visualised and quantified macropore system characteristics from a long-term tillage study (Fig. 8.4) and linked these measurements to visual field evaluations and traditional soil pore results from measurements in the laboratory on soil cores. Their X-ray CT results supported the detailed results from more classical measurements and evaluations in relation to stratification of the topsoil (0–20 cm layer) and showed that the ploughed soil had more macropore networks, branches and junctions than the no-tilled soil. On the other hand, the macropores from the ploughed soil had a lower degree of anisotropy and shorter average branch length than the no-tilled soil. Thus, vertically oriented continuous macropores (biopores), i.e., a more pipe-like system, were more dominant in no-tillage, which is in line with a number of other studies (Munkholm et al., 2013; Schlüter et al., 2020; Pires et al., 2017; Borges et al., 2019). The pore characteristics mentioned all significantly affect the functionality of the pore system for gas and water transport. That is, a vertical oriented pipe-like system indicates good conditions for fast movement of air and water in the vertical direction but limited conditions for gas exchange and availability of nutrients and water in the zones between the major vertical pores. In recent years, high-resolution scanners combined with improved image segmentation and analysis methods have strongly advanced 3 characterisation and quantification of the macropores and the ability to utilise the data for prediction and analysis of soil pore functionality (see Chap. 6). Significant correlations between X-ray CT-derived soil pore system parameters and soil functional properties related to gas and water transport in soil have been shown in a number of recent tillage studies (Muller et al., 2019; Schlüter et al., 2020; Stumpf da Silva et al., 2021). There appears to be a clear potential for utilising X-ray CT-derived parameters (macroporosity, connectivity, critical diameter) to predict and improve current pedotransfer functions used in the prediction of especially convective gas and water transport properties in soil (Schlüter et al., 2020; Stumpf da Silva et al., 2021). X-ray CT also has significant potential for improving our understanding of the impact of tillage on greenhouse gas emissions (Alskaf et al., 2021; Cooper et al., 2021; Mangalassery et al., 2014). These emissions are affected by processes at both micro- and macroscale and therefore call for studies where X-ray CT is used at different scales to study the impact of tillage-induced soil structural changes on emissions of CO_2, N_2O and CH_4 from soil. For instance, soil pore characteristics detected using X-ray CT can be applied to quantify potential of O_2 supply in microsites and the risk of N_2O emissions (Rohe et al., 2021).

8.2.3 Using X-Ray CT to Predict Soil Fragmentation from Tillage

Predicting the impact of tillage on soil structure was stated by Dexter (1979) as a key challenge in soil tillage research and remains a valid and relevant challenge despite 40 years of focussed research. Seedbed preparation and sometimes also seeding operations involve the break-up and fragmentation of soil. This fragmentation results from the opening of cracks and pores and there is therefore a strong link between the 3-D characteristics of air-filled pores and the resulting break-up of soil when exposed to tensile stresses (Munkholm, 2011). Thus, X-ray CT has the potential to significantly improve our understanding of soil fragmentation and may be applied to model the break-up of soil as part of tillage processes. Munkholm et al. (2012) showed a strong positive correlation between CT-derived pore characteristics and the fragmentation of intact soil cores. Soil fragmentation showed strongest correlations with pore volume, surface area and number of junctions per cm^3, indicating the importance of both the abundance and morphological characteristics of the air-filled pore space. Soil aggregate strength plays a key role in the fragmentation of soil in tillage. Munkholm et al. (2016) used X-ray CT to evaluate the role of aggregate pore and shape characteristics in relation to aggregate tensile strength and the fragmentation of intact soil cores. They concluded that bulk soil fragmentation could be predicted by CT-derived aggregate characteristics.

8.3 Soil Compaction Due to Traffic

Two types of soil structure deformation are included in the general definition of soil compaction: the reduction of soil pore space (compression of unsaturated soils, consolidation of saturated soils) and the modification of the soil architecture (deformation, with or without a volume change) (Koolen & Kuipers, 1983). Petrovic et al. (1982) reported an early study showing X-ray imaging techniques were well suited to identifying soil densification. More recent studies, as described below, have used X-ray CT to quantify structure deformation due to traffic in the field and to improve our understanding of the processes leading to soil compaction.

8.3.1 Identification of the Long-term Effects of Soil Compaction

Lamandé et al. (2013) investigated the potential of using medical CT scanning for analysing subsoil structure in a long-term soil compaction experiment in Southwest Sweden. Despite using a coarse resolution (0.6 mm in all three directions for 0.2 m × 0.2 m soil columns), the visualisation techniques could identify persistent effects of subsoil compaction on the CT number of the soil matrix (i.e., the soil matrix density) 14 years after establishment of the experiment (Fig. 8.5). These differences were also identified by the dry bulk density measurements on the same soil cores. One could argue that there is no need to use an advanced technique such as X-ray CT if simple mass-balanced laboratory measurements give the same results. However, one advantage is that the visualisation technique is not destructive, so that soil cores can be used for further analysis of the soil structure (e.g. measurement of hydraulic properties) or for submitting to remediation treatments (wetting/drying or freezing/thawing cycles). In addition, in the same study, visualisation detected core-sampling artefacts (displacement of stones creating cavities) that influenced the evaluation of the effects of soil compaction from mass-balanced total porosity measurements. Most importantly, visualisation of the soil structure in 3D helped interpreting the bulk soil measurements. The comparison of the mass-balanced total porosity and the CT-estimated macroporosity for five slices of the soil columns showed that root exploration was more intense in the compacted plots (Fig. 8.5).

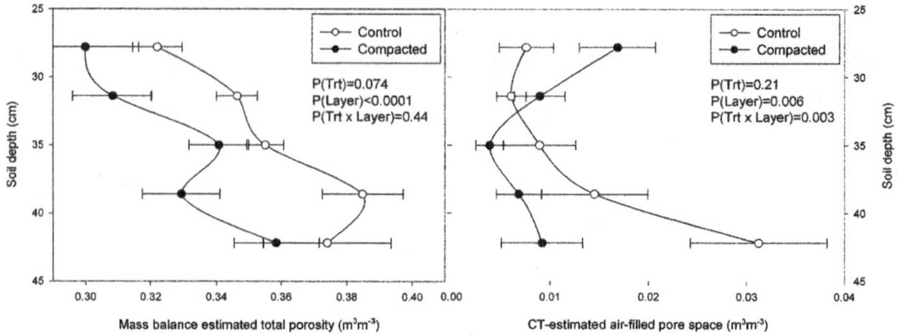

Fig. 8.5 Mass balance-estimated total porosity and X-ray CT-estimated pore fraction (the resolution was approx. 0.6 mm in all three dimensions) as a function of depth for five slices of intact soil columns (0.2 high and 0.2 m in diameter) sampled below the ploughing depth in a long-term soil compaction experiment in Sweden. Reproduced from Lamandé et al. (2013) with permission from Wolters Kluwer

In another long-term experiment, Schjønning et al. (2013) observed lower CT-estimated macroporosity and a smaller number of CT-identified pores for a clayey subsoil in Finland 29 years after compaction. Per definition, soil compaction implies a densification of the soil (Håkansson 1990). The destruction of the largest pores is expected first, but by how much are the "largest" pores affected? This surely depends on the magnitude of the stresses applied to the soil, and on the soil's resistance to compaction, which is itself influenced by intrinsic soil properties and the pore water pressure during deformation. The resolution prevents a complete characterisation of all the soil pore space through visualisation. However, CT data below the threshold of what can be segmented on the images, i.e., the "unseen", can also give interesting information about the densification of soil. For example, Schjønning et al. (2013) reported a surprisingly precise estimation (with a standard deviation of only 0.015 g cm^{-3}) of the soil matrix density measured in the laboratory from the CT number of the soil matrix measured on images with a rather coarse resolution (0.3 mm horizontally and 0.6 mm vertically).

8.3.2 *Segmentation and Compaction*

X-ray imaging allows the detection of materials according to their X-ray attenuation (see Chap. 2). This makes the application of these techniques interesting for the study of soil compaction as a change in soil density (Petrovic et al., 1982). However, due to the influence of soil density in both the detection method (X-ray attenuation is a function of soil bulk density) and the studied object (soil compaction often leads to a densification of the soil structure), detection of the eventual compaction must be done carefully. Lamandé et al. (2013) performed a sensitivity study to evaluate the effect of the chosen segmentation threshold between the pore space and soil matrix on the CT-estimated porosity for compacted and control subsoil cores. They showed

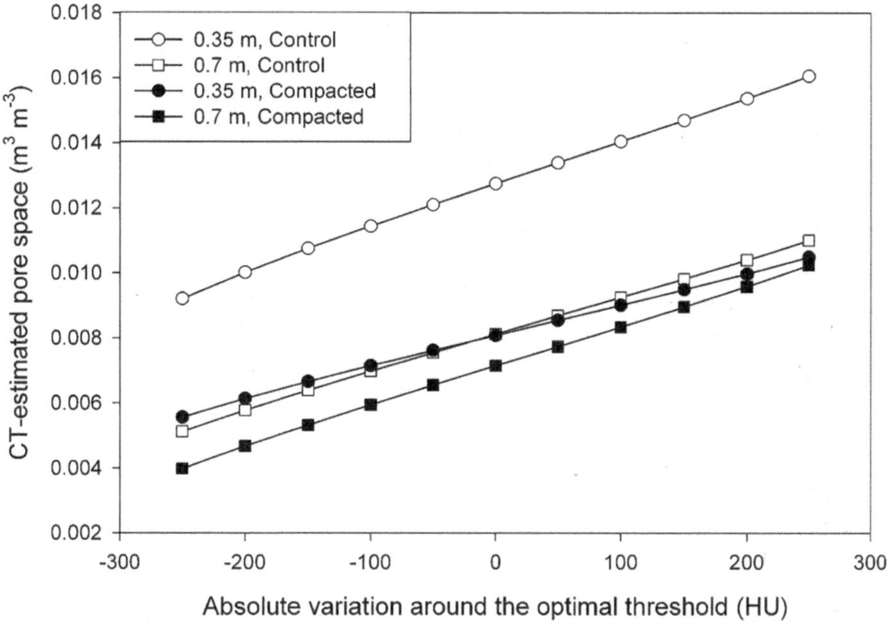

Fig. 8.6 A sensitivity study showed the effect of the chosen segmentation threshold on the differences in air-filled porosity quantified from the CT images between the control and the compacted treatment at two depths in a long-term soil compaction experiment in Sweden. An increase of the segmentation threshold led to a larger residual soil compaction at 0.35 m depth. Reproduced from Lamandé et al. (2013) with permission from Wolters Kluwer

that the choice of threshold influenced the CT-estimated porosity differently for control and compacted subsoil cores. An increase in segmentation threshold (in Hounsfield Units; HU) led to an increase of the difference in CT-estimated porosity between the control and compacted treatments at one of the two depths of the Mollic Endogleyic Luvisol investigated (Fig. 8.6).

8.3.3 From Soil Displacement to Stress Transmission

As X-ray CT imaging is non-destructive and can therefore be applied several times on the same soil core, this presents opportunities in soil mechanics to record observations in time steps during the loading of a soil core. However, this also requires the construction of a loading frame made out of low-density materials to allow the scanning process. Naveed et al. (2016) produced such a frame and were able to observe soil deformation during stepwise loadings of undisturbed topsoil columns by a hydraulic jack in a medical X-ray CT scanner. The deformation of regularly tilled topsoil cores was first quantified by a reduction in the CT-estimated macroporosity and an increase in the CT number of the matrix. At a given loading

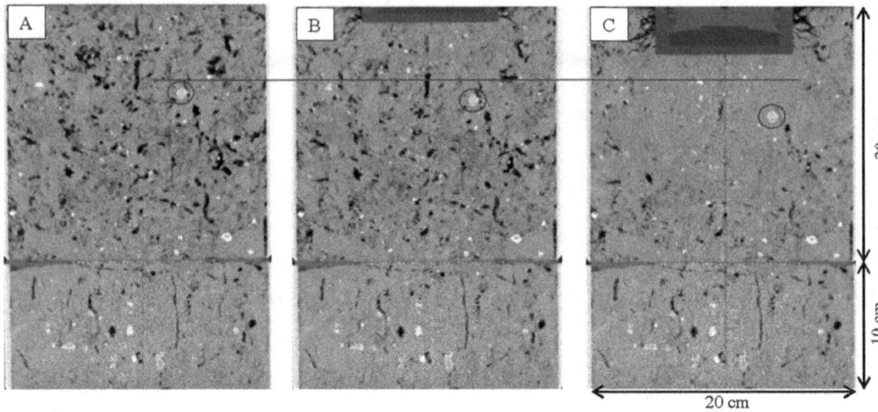

Fig. 8.7 Greyscale images reconstructed from X-ray CT scans of an intact topsoil column (Loam) at a water content corresponding to field capacity (**a**). The same soil column was then subjected to partially confined uniaxial compression at 275 kPa (**b**) and 620 kPa (**c**) applied stress. For this soil column, 506 gravel bits were used to quantify the vertical soil deformation during the compression test. The red circles indicate the vertical displacement of one of these gravels at 0 kPa (**a**), 275 kPa (**b**) and 620 kPa (**c**) applied stress. Reproduced from Naveed et al. (2016) with permission from Elsevier

step, this reduction was largest close to the piston at the top of the soil column and decreased with depth. Aiming at a better understanding of the deformation processes during loadings, Naveed et al. (2016) went further in the analysis of the CT images. They followed the vertical displacement of the hundreds of gravels naturally present in the soil columns, which kept their shape during loading and were therefore easy to recognise on the sequential images (Fig. 8.7). Using these gravels as markers, soil vertical displacement could be quantified in the whole soil column. Displacement could be converted to stress using stress-displacement curves at the contact between the piston and the top of the soil column, and assuming homogeneous mechanical properties throughout the soil column. Results indicated that below an applied stress of 275 kPa, the soil stress propagation pattern corresponds to that of a discrete element (i.e., a collection of resistant aggregates), while at an applied stress of 620 kPa, the stress propagation followed the stress propagation pattern as described by the elasticity theory (Fig. 8.8).

8.4 Visualisation of Soil Deformation

Although X-ray CT has been used in the study of rock mechanics since the late 1980s (e.g., Raynaud et al., 1989), Shi et al. (1999) were probably the first to publish a study on the visualisation of soil-like geomaterials deformation using X-ray CT. They built a test box for loading and shearing in Plexiglas that could be fitted in a medical X-ray CT scanner and could allow acquisition of cross-sectional images

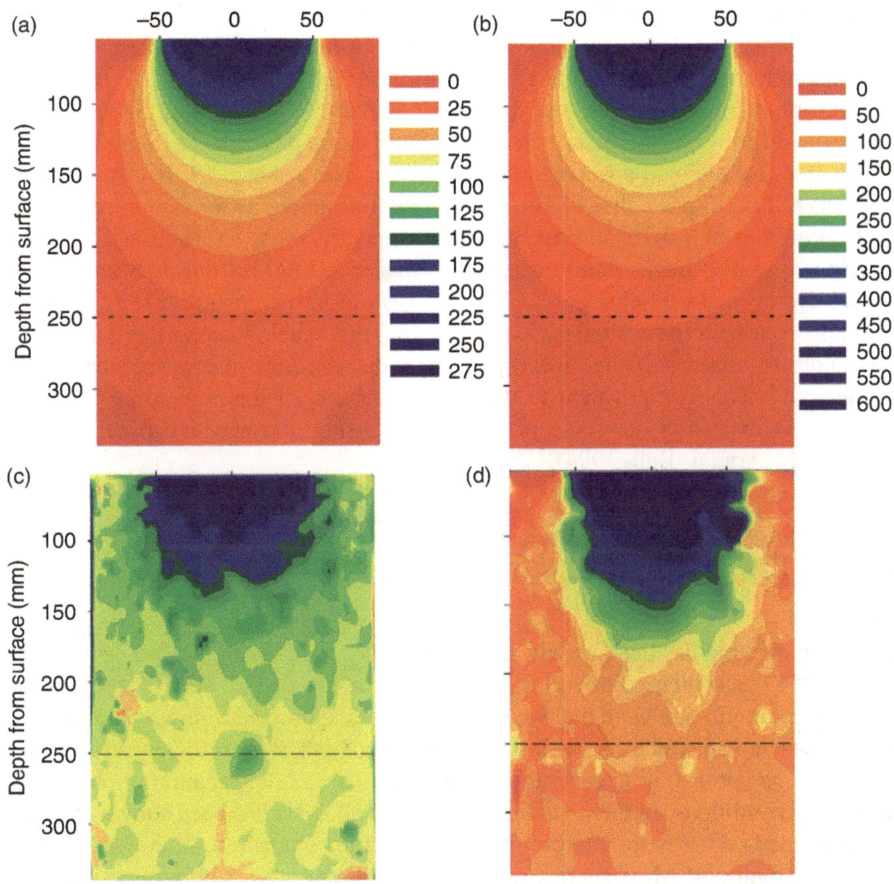

Fig. 8.8 Vertical stress (kPa) in the soil column calculated according to Boussinesq (1885) at 275 kPa (**a**), and at 620 kPa (**b**) applied stress, and measured from the displacement of the gravels during the compression test at 275 kPa (**c**), and at 620 kPa (**d**) applied stress. Reproduced from Naveed et al. (2016) with permission from Elsevier

at several strain levels. About 10 years later, Peth et al. (2010) presented a quantitative study of the local deformation of repacked soil cores (95 cm^2) using X-ray micro-tomography. Soil deformation was due to either hydraulic stresses (swell-shrink tests) or mechanical stresses (confined compression tests on air-dried soil cores). They observed highly heterogeneous deformation during swell-shrink tests in spite of the homogenised structure of the cores prior to the tests. In addition, the cracks formed during shrinkage controlled the deformation pattern of the soil cores during the confined compression tests. Obviously, these kinds of observations are not possible using bulk field or laboratory techniques. The understanding of the mechanisms involved during soil structure formation and deformation needs micro-scale analysis, but the precise description of micro-scale deformation is not straightforward. During deformation, the material moves from voxel to voxel, and advanced image analysis techniques are needed to relate successive images of the soil matrix during deformation (e.g., Crostack et al., 2008). Alternatively, the use of non-deformable markers has been shown to be effective for aligning the volume of interest and following the deformation of some specific macropores (Schäffer et al., 2008a, b) or following the soil matrix deformation (Naveed et al., 2016). Analysis of the bulk soil pore space characteristics during deformation, as for example reported by Bottinelli et al. (2016) during shrinkage of paddy soils, by Zhou et al. (2021) during successive freeze-thaw cycles, by Pires et al. (2020) during wetting and drying cycles, or by Pöhlitz et al. (2018, 2019) during compression tests, is more straightforward. In these studies, X-ray CT imaging enabled successive bulk measurements of macroporosity, macropore size distribution, macropore shape and connectivity. This is equivalent to the characterisation of the pore space by interpretation of measured gas transport parameters using a conceptual pore network model (e.g., Arah & Ball, 1994).

8.5 Conclusions

X-ray CT imaging has been widely and successfully used during the last 40 years to observe the consequences of soil mechanical processes on soil structure. Compaction and fragmentation were the two main processes addressed through comparative studies, where soil structure was investigated before and after traffic events, following tillage operations, or at several time steps during soil deformation in controlled laboratory conditions. In these studies, 3-D image analysis has been shown to be an efficient method to help interpreting soil structure characterisation using other laboratory methods. Most studies used X-ray CT imaging for bulk measurements of the soil structure, in line with other laboratory methods for the determination of capacity and intensity parameters such as bulk density, soil water characteristics, air permeability, gas diffusion, saturated and near-saturated hydraulic conductivity. To date only few promising studies used X-ray CT imaging to observe and quantify the progress of deformation of specific structural features in soil, which seems to be the approach to choose to really understand the mechanics of soil deformation.

References

Alskaf, K., Mooney, S. J., Sparkes, D. L., Wilson, P., & Sjögersten, S. (2021). Short-term impacts of different tillage practices and plant residue retention on soil physical properties and greenhouse gas emissions. *Soil and Tillage Research, 206*, 104803. https://doi.org/10.1016/j.still.2020.104803

Arah, J. R. M., & Ball, B. C. (1994). A functional model of soil porosity used to interpret measurements of gas diffusion. *European Journal of Soil Science, 45*, 135–144.

Blunk, S., Malik, A. H., de Heer, M. I., Ekblad, T., Bussell, J., Sparkes, D., Fredlund, K., Sturrock, C. J., & Mooney, S. J. (2017). Quantification of seed-soil contact of sugar beet (Beta vulgaris) using X-ray computed tomography. *Plant Methods, 13*, 71. https://doi.org/10.1186/s13007-017-0220-4

Blunk, S., de Heer, M. I., Malik, A. H., Fredlund, K., Ekblad, T., Sturrock, C. J., & Mooney, S. J. (2019). Seed priming enhances early growth and improves area of soil exploration by roots. *Environmental and Experimental Botany, 158*, 1–11. https://doi.org/10.1016/j.envexpbot.2018.11.003

Blunk, S., Bussell, J., Sparkes, D., de Heer, M. I., Mooney, S. J., & Sturrock, C. J. (2021). The effects of tillage on seed-soil contact and seedling establishment. *Soil and Tillage Research, 206*, 104757. https://doi.org/10.1016/j.still.2020.104757

Borges, J. A. R., Pires, L. F., Cássaro, F. A. M., Auler, A. C., Rosa, J. A., Heck, R. J., & Roque, W. L. (2019). X-ray computed tomography for assessing the effect of tillage systems on topsoil morphological attributes. *Soil and Tillage Research, 189*, 25–35. https://doi.org/10.1016/j.still.2018.12.019

Bottinelli, N., Zhoua, H., Boivin, P., Zhang, Z. B., Jouquet, P., Hartmann, C., & Peng, X. (2016). Macropores generated during shrinkage in two paddy soils using X-ray micro-computed tomography. *Geoderma, 265*, 78–86.

Boussinesq, J. (1885). *Application des Potentiels à l'Etude de l'Equilibre et des Mouvements des Solides Elastiques* (p. 30). Gauthier-Villars.

Bronick, C. J., & Lal, R. (2005). Soil structure and management: A review. *Geoderma, 124*, 3–22.

Burland, J. B. (1990). Thirtieth Rankine lecture: On the compressibility and shear strength of natural clays. *Geotechnique, 40*, 329–378.

Cooper, H. V., Sjogersten, S., Lark, R. M., & Mooney, S. (2021). To till or not to till in a temperate ecosystem? Implications for climate change mitigation. *Environmental Research Letters, 16*, 054022.

Crostack, H.-A., Nellesen, J., Fischer, G., Weber, U., Schmauder, S., & Beckmann, F. (2008). 3-D analysis of MMC microstructures and deformation by X-ray CT and FE simulations. In S. R. Stock (Ed.), *Developments in X-ray tomography VI* (Vol. 7078, p. 70781I112). SPIE.

Dexter, A. R. (1979). Prediction of soil structures produced by tillage. *Journal of Terramechanics, 16*, 117–127.

Gantzer, C. J., & Anderson, S. H. (2002). Computed tomographic measurement of macroporosity in chisel-disk and no-tillage seedbeds. *Soil and Tillage Research, 64*, 101–111.

Garbout, A., Munkholm, L. J., & Hansen, S. B. (2013). Tillage effects on topsoil structural quality assessed using X-ray CT, soil cores and visual soil evaluation. *Soil and Tillage Research, 128*, 104–109. https://doi.org/10.1016/j.still.2012.11.003

Håkansson, I. (1990). A method for characterizing the state of compactness of the plough layer. *Soil and Tillage Research, 16*, 105–120.

Koolen, A. J., & Kuipers, H. (1983). *Agricultural soil mechanics: Advanced series in agricultural sciences* (Vol. 13, p. 241). Springer.

Lamandé, M., Wildenschild, D., Berisso, F. E., Garbout, A., Marsh, M., Moldrup, P., Keller, T., Hansen, S. B., de Jonge, L. W., & Schjønning, P. (2013). X-ray CT and laboratory measurements on glacial till subsoil cores–assessment of inherent and compaction-affected soil structure characteristics. *Soil Science, 178*, 359–368.

Mangalassery, S., Sjögersten, S., Sparkes, D. L., Sturrock, C. J., Craigon, J., & Mooney, S. J. (2014). To what extent can zero tillage lead to a reduction in greenhouse gas emissions from temperate soils? *Scientific Reports, 4*, 4586.

Muller, K., Dal Ferro, N., Katuwal, S., Tregurtha, C., Zanini, F., Carmignato, S., de Jonge, L. W., Moldrup, P., & Morari, F. (2019). Effect of long-term irrigation and tillage practices on X-ray CT and gas transport derived pore-network characteristics. *Soil Research, 57*, 657–669. https://doi.org/10.1071/sr18210

Munkholm, L. J., Schjønning, P., Rasmussen, K. J., & Tanderup, K. (2003). Spatial and temporal effects of direct drilling on soil structure in the seedling environment. *Soil and Tillage Research, 71*, 163–173.

Munkholm, L. J. (2011). Soil friability: A review of the concept, assessment and effects of soil properties and management. *Geoderma, 167-168*, 236–246. https://doi.org/10.1016/j.geoderma.2011.08.005

Munkholm, L. J., Heck, R. J., & Deen, B. (2012). Soil pore characteristics assessed from X-ray CT derived images and correlations to soil friability. *Geoderma, 181-182*, 22–29. https://doi.org/10.1016/j.geoderma.2012.02.024

Munkholm, L. J., Heck, R. J., & Deen, B. (2013). Long-term rotation and tillage effects on soil structure and crop yield. *Soil and Tillage Research, 127*, 85–91.

Munkholm, L. J., Heck, R. J., Deen, B., & Zidar, T. (2016). Relationship between soil aggregate strength, shape and porosity for soils under different long-term management. *Geoderma, 268*, 52–59. https://doi.org/10.1016/j.geoderma.2016.01.005

Munkholm, L. J., Pulido-Moncada, M., & Obour, P. B. (2019). Soil Tilth and management. In *Oxford research encyclopedia of environmental science*. Oxford University Press. https://doi.org/10.1093/acrefore/9780199389414.013.241

Naveed, M., Schjønning, P., Keller, T., de Jonge, L. W., Moldrup, P., & Lamandé, M. (2016). Quantifying vertical stress transmission and compaction-induced soil structure using sensor mat and X-ray computed tomography. *Soil and Tillage Research, 158*, 110–122.

Peth, S., Nellesen, J., Fischer, G., & Horn, R. (2010). Non-invasive 3-D analysis of local soil deformation under mechanical and hydraulic stresses by mu CT and digital image correlation. *Soil and Tillage Research, 111*, 3–18. https://doi.org/10.1016/j.still.2010.02.007

Petrovic, A. M., Siebert, J. E., & Rieke, P. E. (1982). Soil bulk density analysis in three dimensions by computed tomographic scanning. *Soil Science Society of America Journal, 46*, 445–450.

Pires, L. F., Borges, J. A. R., Rosa, J. A., Cooper, M., Heck, R. J., Passoni, S., & Roque, W. L. (2017). Soil structure changes induced by tillage systems. *Soil and Tillage Research, 165*, 66–79. https://doi.org/10.1016/j.still.2016.07.010

Pires, L. F., Auler, A. C., Roque, W. L., & Mooney, S. J. (2020). X-ray microtomography analysis of soil pore structure dynamics under wetting and drying cycles. *Geoderma, 362*, 114103.

Pöhlitz, J., Ruecknagel, J., Koblenz, B., Schlüter, S., Vogel, H. J., & Christen, O. (2018). Computed tomography and soil physical measurements of compaction behaviour under strip tillage, mulch tillage and no tillage. *Soil and Tillage Research, 175*, 205–216.

Pöhlitz, J., Ruecknagel, J., Schlüter, S., Vogel, H. J., & Christen, O. (2019). Computed tomography as an extension of classical methods in the analysis of soil compaction, exemplified on samples from two tillage treatments and at two moisture tensions. *Geoderma, 346*, 52–62.

Raynaud, S., Fabre, D., Mazerolle, F., Geraud, Y., & Latière, H. (1989). Analysis of the internal structure of rocks and characterization of mechanical deformation by non-destructive method: X-ray tomodensitometry. *Tectonophysics, 159*, 149–159.

Rohe, L., Apelt, B., Vogel, H. J., Well, R., Wu, G. M., & Schluter, S. (2021). Denitrification in soil as a function of oxygen availability at the microscale. *Biogeosciences, 18*, 1185–1201.

Schäffer, B., Stauber, M., Mueller, T. L., Müller, R., & Schulin, R. (2008a). Soil and macro-pores under uniaxial compression. I. Mechanical stability of repacked soil and deformation of different types of macro-pores. *Geoderma, 146*, 183–191.

Schäffer, B., Mueller, T. L., Stauber, M., Müller, R., Keller, M., & Schulin, R. (2008b). Soil and macro-pores under uniaxial compression. II. Morphometric analysis of macro-pore stability in undisturbed and repacked soil. *Geoderma, 146*, 175–182.

Schjønning, P., Lamandé, M., Berisso, F. E., Simojoki, A., Alakukku, L., & Andreasen, R. R. (2013). Gas diffusion, non-Darcy air permeability and CT images of a clay subsoil affected by compaction. *Soil Science Society of America Journal, 77*, 1977–1990.

Schlüter, S., Albrecht, L., Schwärzel, K., & Kreiselmeier, J. (2020). Long-term effects of conventional tillage and no-tillage on saturated and near-saturated hydraulic conductivity–can their prediction be improved by pore metrics obtained with X-ray CT? *Geoderma, 361*, 114082. https://doi.org/10.1016/j.geoderma.2019.114082

Shi, B., Murakami, Y., Wu, Z., Chen, J., & Inyang, H. (1999). Monitoring of internal failure evolution in soils using computerization X-ray tomography. *Engineering Geology, 54*, 321–328. https://doi.org/10.1016/S0013-7952(99)00080-0

Stumpf da Silva, T., Pulido-Moncada, M., Raul Schmidt, M., Katuwal, S., Schlüter, S., Köhne, J. M., Mazurana, M., Munkholm, L. J., & Levien, R. (2021). Soil pore characteristics and gas transport properties of a no-tillage system in a subtropical climate. *Geoderma, 401*, 115222.

Taina, I. A., Heck, R. J., & Elliot, T. R. (2008). Application of X-ray computed tomography to soil science: A literature review. *Canadian Journal of Soil Science, 88*, 1–20.

Werner, D., Winterot, C., Werner, B., & John, K. (1993). Characterization of soil-structure by X-ray-analysis. *Zeitschrift für Pflanzenernährung und Bodenkunde, 156*, 427–434.

Zhou, G., Hu, X., Li, X.-Y., & Li, Z.-C. (2021). Effects of freeze-thaw cycles on soil macropores and its implications on formation of hummocks in alpine meadows in the Qinghai Lake watershed, northeastern Qinghai-Tibet plateau. *Journal of Soils and Sediments, 21*, 245–256. https://doi.org/10.1007/s11368-020-02765-2

Chapter 9
X-ray Imaging of Root–Soil Interactions

Maik Lucas and Doris Vetterlein

9.1 Introduction

Growing roots interact with soil, and its structure, across a range of spatial and temporal scales, and thus adapt to the local environment (Downie et al., 2015). Several factors influence the development of roots, and as such, shape the root system architecture (RSA). Well-known examples of these are water stress, mechanical impedance, pore connectivity and porosity (Bengough et al., 2011; Correa et al., 2019; Lucas et al., 2019a). However, it is not just physical but also biological and chemical factors such as the availability of nitrogen that can change the growth of roots and their RSA (Flavel et al., 2014; Gao et al., 2019b; Nwankwo et al., 2018). Due to all these factors, perhaps not surprisingly root growth in artificial soil, such as gel media and indeed sieved soil, shows significantly different root growth when compared to roots developed in undisturbed soil (Hargreaves et al., 2009).

X-ray Computed Tomography (X-ray CT) and X-ray synchrotron tomography (SR-CT) have the great advantage of non-invasively characterizing the architecture of the root system and the associated plant–soil interactions at biologically relevant (sub-)mm scales in soils. In recent years, X-ray CT offered a variety of new possibilities to characterize RSA and to reveal processes taking place in the vicinity of the root, i.e. the rhizosphere. In this chapter we first provide a brief review on how rhizosphere research with X-ray CT has developed within the last three decades. In sections two and three, root segmentation methods and observable root traits will be

M. Lucas
Helmholtz Centre for Environmental Research–UFZ, Halle, Germany
e-mail: maik.lucas@ufz.de

D. Vetterlein (✉)
Helmholtz Centre for Environmental Research–UFZ, Halle, Germany

Martin-Luther-University Halle-Wittenberg, Halle, Germany
e-mail: doris.vetterlein@ufz.de

S. Jon Mooney et al. (eds.), *X-ray Imaging of the Soil Porous Architecture*,
https://doi.org/10.1007/978-3-031-12176-0_9

evaluated. Finally, this chapter will highlight root research with X-ray CT in different fields of research such as image-based modelling and interaction of roots with soil structure.

9.2 General Overview and Drawbacks

The first serious application of X-ray CT in plant studies emerged during the 1980s. Hainsworth and Aylmore (1983, 1986) examined water extraction by roots by visualizing changes in water content, i.e. attenuation, around radish roots (*Raphanus sativus*). However, these, and also other, imaging approaches in the beginning of the 1990s had quite a coarse resolution with edge length of more than 1 mm (Tollner et al., 1994; Watanabe et al., 1992). Later the development of industrial scanner systems (Jenneson et al., 2003) made it possible to scan and segment fine root systems like wheat at a resolution of 100 μm (Gregory et al., 2003). In recent years, modern industrial X-ray CT scanners have become increasingly affordable and technical developments have resulted in better ratios of sample size vs. resolution (see Chaps. 2 and 4). This is an important development, as small column sizes needed for high-resolution always lead to artificial root growth at column walls and thus to different RSA compared to plants grown under natural field conditions. In addition, the problem of the trade-off between sample size and resolution may partially be evaded by merging multiple scans of cylindrical columns (Flavel et al., 2012; Koebernick et al., 2014).

In recent years, the number of studies using X-ray CT for root research has significantly increased and in addition, image resolution has improved considerably (Fig. 9.1). In the current decade, the average column size for experiments with plants for X-ray CT has fluctuated around 6 cm in diameter, resulting in resolutions of about 30–40 μm. Thus, although the diameter of roots has a physiological lower threshold of approx. 30 μm (Fitter, 1987), fine roots often cannot be segmented properly, as roots smaller than 5 voxels are hard to quantify during image processing and analysis. The reason for this is that objects close to image resolution can be not captured in a representative way due to imaging artefacts (Vogel et al., 2010) (see discussion in Chaps. 4 and 6). However, new image analysis protocols enable classification of roots with diameters near image resolution (Flavel et al., 2017; Gao et al., 2019a). In addition, time-efficient scanning and imaging protocols enable even for rapid phenotyping of undisturbed field samples (Maenhout et al., 2019; Pfeifer et al., 2015) and high-resolution X-ray CT and SR-CT of small samples containing single roots were successfully applied to observe process and image features like root hairs at high-resolution of approx. 1 μm (Daly et al., 2016; Keyes et al., 2017a; Koebernick et al., 2017, 2019). Thus, X-ray CT has developed into a suitable tool to non-invasively visualize the entire root system of different plants as well as rhizosphere processes with μm resolution in 3-D.

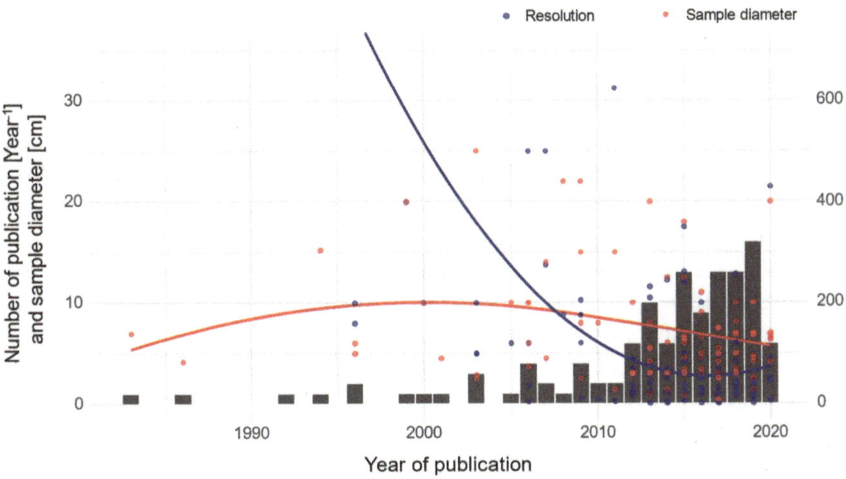

Fig. 9.1 Results of a literature search in web of science, which had 338 hits (4.01.2020, search query: X-ray AND (CT OR tomography) AND (rhizosphere OR plant root) carried out in 2020. From this, 121 publications were about experiments with plants in soil using X-ray CT or SR-CT. Lines show the results of a locally weighted regression computed in R with the plugin ggplot2 (Version 3.30)

9.3 X-ray Dose

During X-ray CT analysis, plants and their surrounding soil are exposed to ionizing radiation. Pioneering experiments on the effects of X-rays on plant cells have shown different effects of the radiation, ranging from none to strong morphologically negative ones associated with the destruction of the growth hormone auxin and the inhibition of the formation of meristematic cells (Evans, 1965; Johnson, 1936; Skoog, 1935). Although plants in these studies were often exposed to much higher doses than required for common experiments, X-ray dose of current X-ray CT studies can influence plant growth as well. This depends on numerous factors such as plant species, growth phase, scan frequency and length, as well as scan settings such as energy and sample distance and the use of filters (Blaser et al., 2018; Zappala et al., 2013a). Blaser et al. (2018) showed in their experiments with faba bean (*Vicia faba)* and barley (*Hordeum vulgare)*, that faba bean reacted to the X-ray dose with shortened root lengths, with lateral root growth being restricted in particular, whereas barley showed no negative effects due to X-ray CT scans. Several factors, such as the soil type and the diameter of the column, determine how much radiation reaches the root. Decomposition of single roots of faba bean incubated in a sandy loam showed higher rates due to X-rays, while there was no effect on decomposition of roots in a light clay soil (Haling et al., 2013). In summary, there is no real threshold value for X-ray CT studies with plants which can yet be assumed and pilot studies on new plants at different energies and scan times might be instructive. In addition, the dose to the plant shoot, and thus the potential impact on root growth, can be significantly reduced by using a lead shield (Lippold et al., 2021). Blaser et al. (2018) concluded from their observations that generally non-scanned controls should be included in X-ray CT studies. These aspects should be especially considered for 4-D imaging of plants, in which a time sequence of the same plant is analysed thus increasing the overall dose (Blaser et al., 2018). So far, such non-scanned controls were used only in very few studies. No effect of X-ray dose of multiple scans was shown for radish (Daly et al., 2018), rice (*Oryza sativa*) (Zappala et al., 2013a) and wheat (*Triticum aestivum*) (Flavel et al., 2012; Colombi & Walter, 2017). In a study by Colombi and Walter (2016), controls showed that despite scanning after 2, 5, 8, 11 and 14 days of sowing, there was no effect of dose on root or shoot biomass of triticale (*x Triticosecale*), soybean (*Glycine max*) and wheat. However, X-rays may not only affect the plant directly but may also impact microorganisms. In a recent work of Ganther et al. (2020) columns with maize (*Zea mays*) were scanned and the corresponding effect on the rhizosphere processes including microbial community composition and root gene expression was analysed 1 h and 24 h after the scanning procedure. It was shown that under the medium dose of 0.81 Gy, which is typically used for scans of plant columns, the composition of the rhizosphere bacterial microbiome was not affected. It was further demonstrated that although a downregulation of cell growth and cell stress related gene expression after 1 h occurred, gene expression recovered 24 h after X-ray CT scanning (Ganther et al., 2020). These partly contradictory findings reveal that more research is needed to evaluate critical thresholds of X-ray dose on various plants and microorganisms in root and rhizosphere.

9.4 Opportunity In Situ over Time

The most unique opportunities arising from the use of X-ray CT in root and rhizosphere research is that it cannot just be used to visualize root structure in 3-D but also in 4-D, i.e. over time. It offers the possibility to investigate specific time-points of root behaviour of the same plant. Thus, it also gives the opportunity to investigate the development of different root types and relate their functions to their age (Vetterlein & Doussan, 2016). One of the drawbacks of repeated measurements with X-ray CT is the increase in X-ray dose per plant as elaborated above. Previous studies used this 4-D analysis for various research questions, e.g. root decay was observed for individual roots (Haling et al., 2013) (Fig. 9.2a), the interaction of roots with granular fertilizer application was shown (Ahmed et al., 2016; Gao et al., 2019b) (Fig. 9.2b) or the shrinkage or roots during drought and associated loss of contact (Carminati et al., 2009, 2013; Koebernick et al., 2018).

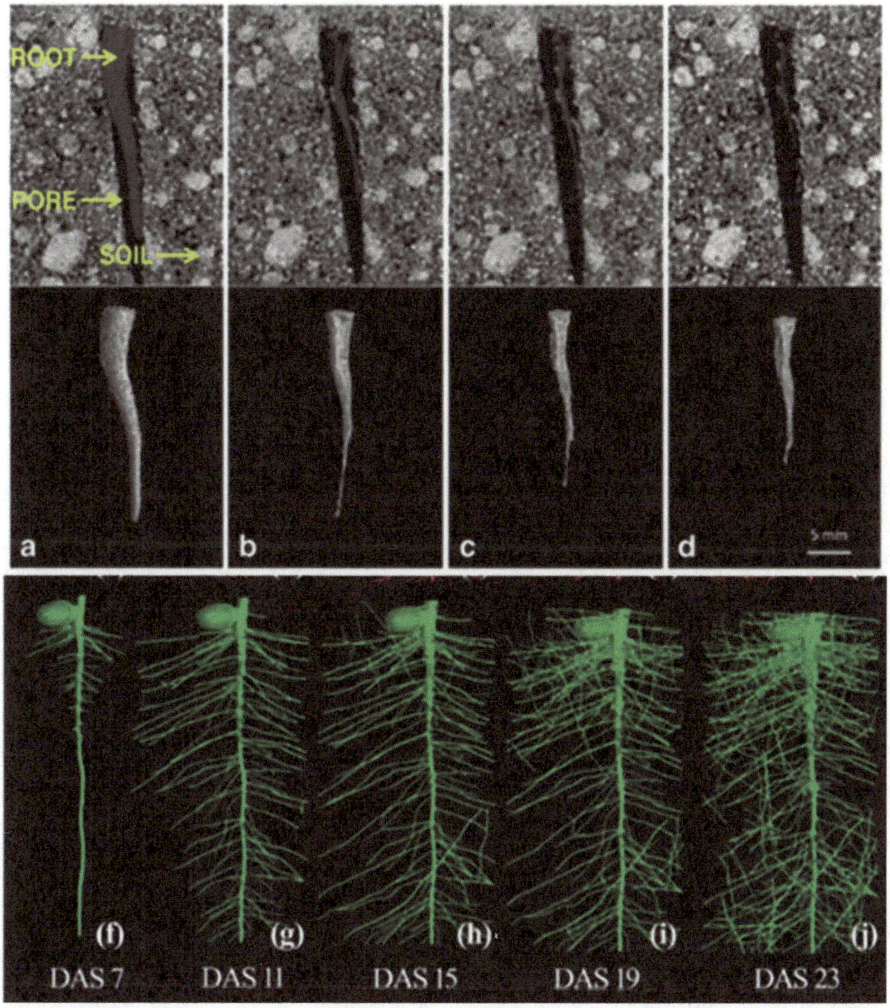

Fig. 9.2 Above: Degradation of a faba bean root after incubation in a sandy loam for (**a**) 13 (**b**) 22 (**c**) 33 and (**d**) 46 days from (Haling et al., 2013). Below: Growth of faba bean with localized phosphor application (**f**) 7, (**g**) 11 (**h**) 15 (**i**) 19 and (**j**) 23 days after seeding (DAS) from (Gao et al., 2019b)

9.5 Root Segmentation

A crucial step during X-ray CT analysis, which describes the root system, but also rhizosphere processes, is the segmentation of the root system. This implies the separation of all image voxels into classes, in this case root voxels and soil matrix (including pores filled with air or water). One of the easiest segmentation methods is based on the attenuation density (i.e. by applying simple thresholds for grey values, see Chap. 5). This fast segmentation based on the histogram was often applied in early X-ray CT studies (Heeraman et al., 1996; Lontoc-Roy et al., 2005, 2006; Pierret et al., 1999). However, this simple thresholding causes major obstacles, as roots, consisting mainly of water, have a similar attenuation as water or other organic material. Zappala et al. (2013b) showed that saturated conditions do not allow a sufficient segmentation of rice roots whereas at field capacity roots can be segmented most accurately (Zappala et al., 2013b). This is reasonable as scans of whole columns lead to resolutions of >30 μm, i.e. only air filled pores are visualized and water is only visible as mixed pixels with soil matrix. Also, partial-volume effects, which are created by mixed voxel of different image features below image resolution (see Chap. 6) lead to many misclassifications.

Thus, many different approaches have been developed to handle these issues of accurate root segmentation. Many of them involve pre- and post-processing of the image. These are used to enhance the contrast between roots and soil matrix and to reduce noise by applying different filters like Gaussian and Non-local means filters (Flavel et al., 2012; Kaestner et al., 2006; Koebernick et al., 2014; Lucas et al., 2019b). In addition, post-processing filters can be applied to remove misclassified features, e.g. by applying size or median filters (Gao et al., 2019a; Lucas et al., 2019b). More details about image processing to enhance the results of segmentation methods can be found in Chap. 5 and in Schlüter et al. (2014).

To overcome the drawbacks of thresholding based on histograms, more complex imaging approaches were developed and used, which make use of features inherent in root systems such as connectivity or their cylindrical shape. Region Growing approaches, i.e. adaptive local thresholding methods, are based on the connectivity of the root system. Region growing classifies voxels based on a seed point or region, usually set manually. Adjacent voxels are added to these regions until they do not match a pre-determined similarity criterion (e.g. grey value intensity, greyscale texture). One of the first who successfully applied such an approach was Gregory et al. (2003), who segmented the roots of wheat and rapeseed (*Brassica napus)* with an underestimation of around 10%. Similarly, the root tracking software RootViz3-D® (http://www.rootviz3-D.org/) determines a probability function on the grey values of roots and tracks the root based on a starting point. It was shown to be successfully in segmenting wheat roots (Tracy et al., 2012b). Flavel et al. (2017) provide an overview of the use of X-ray CT for root segmentation.

Various studies have used the 'region growing' method implemented in the widely used software VGStudio MAX. It was successfully used to segment tap root systems like tomato (S*olanum lycopersicum*) *(*Helliwell et al., 2017, 2019; Tracy et al., 2013*),* pea (*P. sativum*) (Keyes et al., 2017b) and faba bean (Blaser et al., 2018; Koebernick et al., 2014, 2015) and for fibrous root systems like wheat (Atkinson et al., 2020; Flavel et al., 2012; Helliwell et al., 2019), maize (Galdos et al., 2020), barley (Pfeifer et al., 2014), rice (Fang et al., 2018; Kirk et al., 2019) as well for ruzi grass (*Brachiaria ruziziensis*) and palisade grass (*Brachiaria brizantha*) (Galdos et al., 2020). Hudek et al. (2017) successfully applied region growing to segment the root system of 10 pioneer plants in the fore-field of a glacier. This broad field of application for region growing methods shows that region growing indeed is suitable to be used for most root systems. However, it is only semi-automated and therefore needs a lot of user interaction and thus it is time-consuming (Flavel et al., 2012).

Hysteresis thresholding is another local adaptive thresholding method, which can be applied to segment roots (Lontoc-Roy et al., 2005, 2006; Nwankwo et al., 2018). It uses two thresholds, a global stringent threshold to segment the roots and a second threshold which is in an iterative approach only considered if the voxels are connected to segmented root volumes.

Mairhofer et al. (2012) developed the free software RooTrak for the segmentation of roots based on level-sets. It starts with an initial appearance model based on grey values which are manually defined as seed points of the root system. The software tracks the roots through 2-D image slices based on connectivity requirements and updates the model. Since the original version extensions have been made by the developers and it now allows to extract plagiotropic roots, i.e. it also tracks roots growing upwards (Mairhofer et al., 2013) and can be used to segment and separate different root systems in one sample (Mairhofer et al., 2015a, 2016). This allows examining the spatial interaction of different plants. RooTrak was successfully applied to segment wheat, maize, tomato, tillage radish, vetch and black oat (Burr-Hersey et al., 2017; Daly et al., 2018; Mairhofer et al., 2012, 2017).

Root1 is a Fiji/ImageJ (Schindelin et al., 2012) macro developed by Flavel et al. (2017). It first increases the scale of the images to avoid root fragmentation by following imaging steps and then uses a combination of different filters (sobel, median, erosion) for pre- and post-processing of a manually set bi-level threshold. After this a connected region extraction by a user-defined seed point is applied. For large image/file sizes this step is best achieved in the bespoke X-ray CT software (Flavel et al., 2017) where Root 1 successfully extracted the roots of chickpea (*Cicer arietinum*, 10% overestimation, wheat (1% underestimation) and barley (8% underestimation—Flavel et al., 2017; Rabbi et al., 2018a, 2018b).

A recently developed root segmentation tool by Gao et al. (2019a) called Rootine is like Root1 as it is a macro for the software Fiji/ImageJ. However, different to other root segmentation methods, it does not directly rely on grey values and is shape based. The main component of the macro is the use of the Tubeness plugin (https://imagej.net/Tubeness). In short, it finds roots by their salient feature of tubular structure. This is implemented by computing the eigenvalues of the Hessian matrix derived from the second derivative of grey values (Frangi et al., 1998). It is combined with a scale space approach using Gaussian smoothing of different standard deviations σ to segment roots of different diameters. The Tubeness-results are then segmented using hysteresis thresholding, which preserves the connectivity of the roots. This allows filtering of the biggest connected component, i.e. the root system, to remove the noise in the end. Rootine was successfully applied to segment the fibrous root system of maize in different studies (Gao et al., 2019a, 2019b; Lucas et al., 2019a). However, an important advantage of Rootine is that it does not necessarily rely on connectivity of the root system. Thus also roots in subsamples can be segmented, which are not fully connected any more (Lucas et al., 2019a). In addition, Tubeness can also be used to segment biopores, i.e. cylindrical shaped pores, which have the same shape, but not the attenuation value of roots (Lucas et al., 2019b). Due to the high parallelisation of the Fiji/ImageJ plugins used, it was shown to outperform region growing in VGStudio, Root1 and RooTrak in terms of computational time and user interactions (Gao et al., 2019a). Most recently Phalempin et al. (2021) developed an improved version of Rootine, which increased the computational demand, however, the number of user-defined parameters could be further reduced. In a benchmark test, a "worst-case" scenario (i.e. images with poor quality and rather low image resolution and of soils with high heterogeneity) was compared with images of a "best-case" scenario, i.e. with respective image qualities. With 114% in the "best-case" scenario and 73% for the "worst-case" scenario, root recovery has been drastically improved compared to the first version of Rootine, especially for the "worst-case" scenario.

Future root segmentation will include machine learning algorithms (e.g. like the deep learning approach recently developed (Douarre et al., 2016; Soltaninejad et al., 2019)). These give the opportunity to train models based on more than one feature image (e.g. together on grey values, Hessian matrix (structure) and gradient masks (structure)). Thus, more than one salient feature of roots can be combined with minimal user interaction. These deep learning models need to be trained on ground truth images of various root types in different soil types (Xu et al., 2018). A collaborative database of different laboratories working on root imaging with X-ray CT would be a powerful tool to achieve this.

However, no matter which segmentation method is applied, the result needs to be evaluated. This is often done by comparing them with traditional root washing and analysing methods (Flavel et al., 2012, 2017; Gao et al., 2019a; Gregory et al., 2003). In addition, artificial images with similar characteristics as a real X-ray CT images or real X-ray CT images from roots segments can be used for validation (Mairhofer et al., 2015b). These systems are well defined in terms of their length, volumes and areas and in addition offer the possibility to validate a segmentation method under changing signal-to-noise ratios (Mairhofer et al., 2015b).

9.6 Root System Architecture and Root Traits

Root system architecture (RSA) plays an important role in the plants ability to acquire water and nutrients and hence for stabilizing yields under adverse environmental conditions. This has raised interest in characterizing RSA traits in order to use them for breeding strategies related to crop improvement (Tracy et al., 2020). X-ray CT is able to provide all RSA traits in great detail for young plants, however, currently, due to the low throughput it cannot compete with simple methods such as shovelomics (Trachsel et al., 2011). Yet, it can be used to investigate specific changes in RSA of different plants.

The possibility to observe the root system in situ and in 3-D offers new valuable parameters to be characterized. This is important as the heterogeneous nature of soils in time and space leads to highly plastic responses of plant roots. These are governed by changes in elongation rate, branching, surface area and angle, which shape the final RSA (Morris et al., 2017). In the following typical root traits, which were analysed using X-ray CT imaging will be discussed and are summarized in Table 9.1.

Numerous studies have described the root structure by means of classical root traits like total root length (Ahmed et al., 2016; Blaser et al., 2018; Gao et al., 2019b; Jenneson et al., 2003) and length densities (Flavel et al., 2014; Lucas et al., 2019a), root volume (Fang et al., 2019; Tracy et al., 2012b), surface area (Daly et al., 2018) and by root diameter classes (Bauerle & Centinari, 2014; Colombi & Walter, 2016). As elaborated above, these traits can easily be evaluated by comparing them to results obtained with root washing and hence they are also accessible with simple, however, destructive methods. Most of these traits are easily derived during image analysis. Estimating the root volume is simply done by counting the number of voxels classified as root. Root diameters are often described by the using the maximum inscribed ball method (more details can be found in Chap. 6), for instance by using the local thickness tool in Fiji/ImageJ as described in Lucas et al. (2019b). Most research analysing the length of the root system uses the skeletonize approach (Flavel et al., 2017; Blaser et al., 2018; Gao et al., 2019b). Skeletonize reduces the diameter of the roots by removing the border pixel without breaking the connectivity of a root system. The result is a voxel thick root system from which the total amount of voxel represents the total length of the root system. However, as skeletonization can be slow (especially for thick roots which need a high number of iterations) other methods can also be used to measure the length of a root system. Koebernick et al. (2014) used one of the Minkowski functionals (see Chap. 6) to describe the length of the total root system. Lucas et al. (2019a, b) used the local thickness approach for describing the diameter of the roots and in addition for measuring the length of different diameter classes by assuming a perfect cylindrical form of the roots.

Comparable to classical approaches, root length or root length densities can be described for different depths. However, during destructive sampling and root washing, careful decisions need to be made about the depth at which samples are taken and the number of segments into which a soil profile or plant pot will be divided. In contrast, X-ray CT offers the possibility to (1) increase the vertical resolution, which can be estimated in a slice-by-slice manner (i.e. it is only limited by the spatial image resolution) and (2) decide to summarize and compare different depth based on first results. In addition, X-ray CT allows the description of (3) root length density in 3-D (i.e. difference in density within a depth – horizontal resolution) which can also be visualized (Flavel et al., 2017).

In addition, more complex traits like branching angle (Burr-Hersey et al., 2017; Flavel et al., 2014) or number and length of different root orders (Flavel et al., 2012; Blaser et al., 2018; Carminati et al., 2009; Colombi & Walter, 2016) can be estimated. However, these traits are often measured manually within software like Fiji/ImageJ or VGStudio and thus are time-consuming to be analysed. Root branching length can be estimated (e.g. with the Analyse Skeleton function in Fiji/ImageJ), however, this function estimates the length for all branches (Flavel et al., 2017) but cannot distinguish between root branches of different orders.

Table 9.1 X-ray CT derived root traits and the relevant literature examples

	Root trait		Example literature
Classical 2-D traits	Root length, volume, surface and diameter		Blaser et al. (2018); Daly et al. (2018); Koebernick et al. (2014); Lazarević et al. (2016)
	Root length (density) profile and maximum depth / width	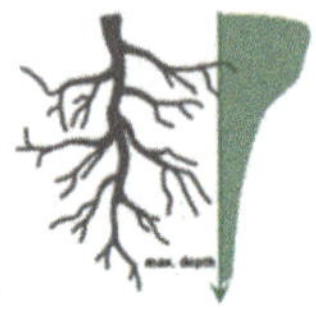	Flavel et al. (2014); Gao et al. (2019b); Tracy et al. (2012b)
	Tuber volume, number of rhizomes		Davey et al. (2011); Perez-Torres et al. (2015)
	Branch angle	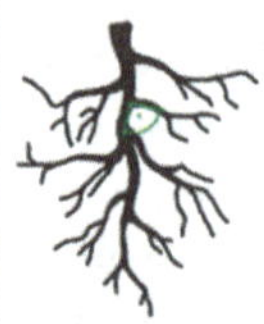	Burr-Hersey et al. (2017); Flavel et al. (2014); Hargreaves et al. (2009); Pierret et al. (1999); Tracy et al. (2012b)
	Number and length of branches	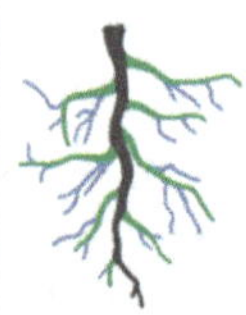	Burr-Hersey et al. (2017); Carminati et al. (2009); Colombi and Walter (2016); Flavel et al. (2012)
Only in 3-D	Euclidean distance	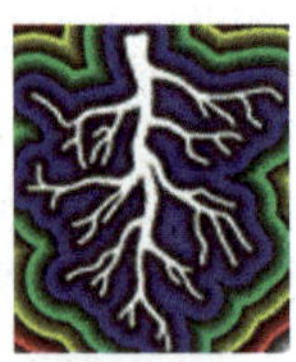	Koebernick et al. (2014); Lucas et al. (2019b); Schlüter et al. (2018)
	Convex hull	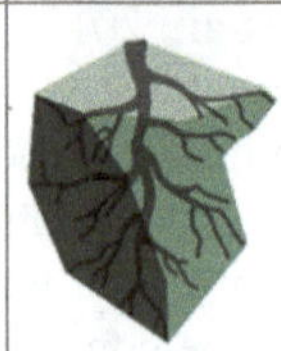	Burr-Hersey et al. (2017); Fry et al. (2018); Helliwell et al. (2019); Hudek et al. (2017); Zappala et al. (2013b)

(continued)

Table 9.1 (continued)

	Root trait		Example literature
	Minimum distances between root systems	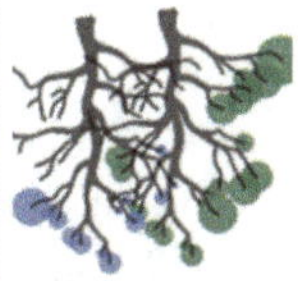	Mairhofer et al. (2017)
	Fractal dimension and tortuosity of root system		Fang et al. (2019); Subramanian et al. (2015); Tracy et al. (2013)
	Center of mass, point of substantial morphological changes	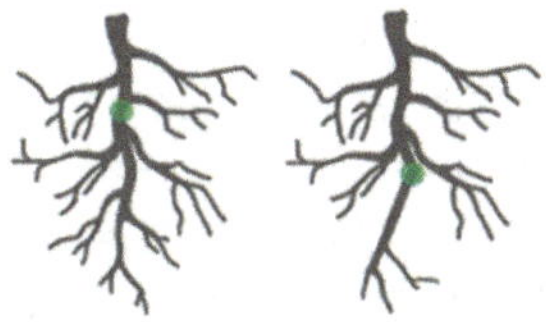	Flavel et al. (2014); Mairhofer et al. (2012, 2017); Tracy et al. (2012b)

Over the past two decades, major advances in new measurements were made, which describe specific root traits in 3-D. One of these is the convex hull, describing the root volume, which is occupied by the root system (Burr-Hersey et al., 2017; Helliwell et al., 2019; Mairhofer et al., 2012; Zappala et al., 2013b). The centre of mass/centroid (Mairhofer et al., 2012; Tracy et al., 2012b) and the point of substantial morphological changes (Flavel et al., 2014) are two measures, which help to reduce the information of the complex 3-D root system and describe changes under changing environmental factors with a single number.

The Euclidean distance to roots is another measure, which describes the root system from the soil perspective (also see Chap. 6). The Euclidean distance image includes the minimum length of a straight line from each soil voxel of a 3-D image to the next root. Comparable to root length densities in 3-D, as shown in Flavel et al. (2017) this allows the description of how a certain root system occupies the soil volume. Schlüter et al. (2018) showed how to fit a model to the frequency distribution of Euclidean distances, which can be used to describe the rhizosphere volume and which can be adapted to specific interaction distances for rhizodeposition and soil water, etc. (Schlüter et al., 2018).

All these measures still cannot reduce the major drawback of X-ray CT, i.e. sample size: resolution trade-off. This means results of these measures obtained from images of growing plants in small columns need to be interpreted with caution, as the root system under natural field conditions profit of much larger available soil volumes.

9.7 Mutual Interaction of Roots and Soil Structure

When roots grow into soil, they reorganize the spatial arrangements of soil particles and after they die and decay, they leave behind cylindrical shaped pores, which are known as biopores (Lucas et al., 2019b). However, these biopores and other macropores can be also used by roots to elongate into soil and bypass zones of high resistance (Jin et al., 2013; Lucas et al., 2019a; White & Kirkegaard, 2010). X-ray CT and SR-CT offer the possibility to reveal these mutual interactions of roots and soil structure. Different to other research fields, elaborated in this chapter, changes in soil structure induced by plants are often observed under natural field conditions in undisturbed field samples (Jarvis et al., 2017; Koestel & Schlueter, 2019; Kuka et al., 2013; Lucas et al., 2019b). Hu and co-workers showed in several publications the influence of roots on macropore structure of different land types, revealing the positive correlation of root density and macroporosity (Hu et al., 2016, 2019a, 2019b, 2020). In an extensive field study of a chronosequence, Lucas et al. (2019b) observed soil structure formation on a reclamation site by describing the porosity with a scale approach. By scanning three different sample sizes extracted from fields of different ages they examined the changes in macroporosity down to a resolution of 5 μm. Unlike Hu et al., Lucas et al. (2019b) argued that roots do not necessarily change the macroporosity over time, but they can rearrange the exiting macropores and thus create a dense system of biopores. For the specific agricultural chronosequence, maximum biopore density close to 20 cm cm^{-3} was already achieved after 6 years in the topsoil and comparable high values after 12 years in 40–60 cm depth (Fig. 9.3a) (Lucas et al., 2019b). In a follow-up study they showed that by creating this highly connected biopore system, roots are able to increase the connectivity of the pore system down to pore diameters of 0.1 mm, which reduces the percolation threshold and thus has a positive effect on air permeability and water conductivity near saturation (Lucas et al., 2020b).

The effect of root growth on pore structure is highly depending on the plant species, its RSA and the soil texture. Galdos et al. (2020) analysed breakthrough curves in addition to X-ray CT images to measure the effect of three different plants on leaching of nitrate. They showed that palisade grass having a fine root system increased porosity most but decreased leaching as it did not increase the widest macropores, but only finer and well-connected ones, compared to maize and ruzi grass (Galdos et al., 2020). Bacq-Labreuil et al. (2019) showed in pot experiments that different cover crops affect macroporosity and pore connectivity differently in two different soil textures (clay and sandy soil). In the same vein, field sample analysis showed that plants increased connectivity and porosity in clay soil but decreased these parameters in sandy soil and perennial plants manifest a wider range of pore sizes (Bacq-Labreuil et al., 2018).

Other research about the changes in soil deformation by roots focused on visualizing root growth with high temporal resolution. Keyes et al. (2016) scanned growing roots over 20 h with 1 scan h^{-1} and visualized soil deformation of a growing root tip over time to quantify soil deformation by applying a digital volume

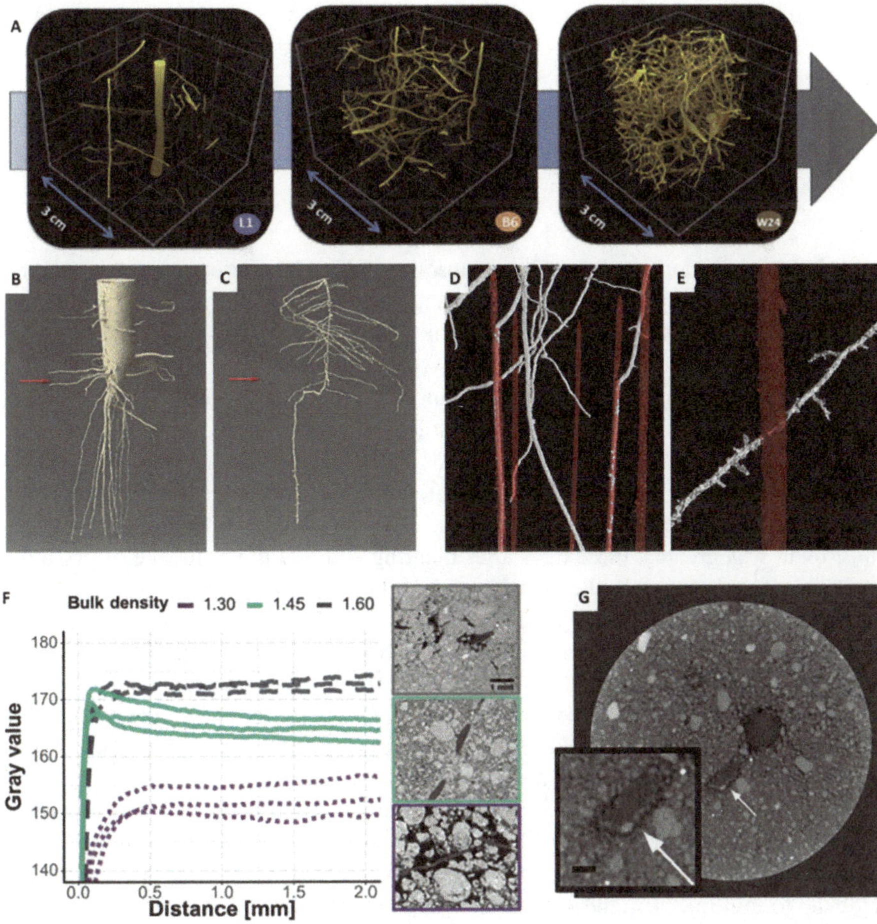

Fig. 9.3 Examples from various for the mutual interaction of soil structure and plant growth. (**a**) Development of biopores segmented from X-ray CT images of 3 cm samples of 50 cm depth from a chronosequence 1, 3 and 24 years after segmentation (Lucas et al., 2019b). Root system of radish (**b**) and vetch (**c**) after 58 days of growth. Red arrow indicates location of bulk density changes with a low bulk density above (1.2 g cm^{-3}) and 1.4 g cm^{-3} below (Burr-Hersey et al., 2017). (**d**) Roots (white) colonising artificial macropores in soil with high bulk densities (1.6 g cm^{-3}) and (**e**) root crossing a pore in low bulk density soil (1.2 g cm^{-3}) (Atkinson et al., 2020). (**f**) Density gradients around roots growing in soil of three different bulk densities (left) and corresponding examples of CT image slices (Lucas et al., 2019a). (**g**) Horizontal slice of faba bean after 17 days of drying. Arrows points to gap around particles adhering to the root surface after root shrinkage (Koebernick et al., 2018)

correlation approach. In a latter study, they increased temporal and spatial resolution using SR-CT and observed growth and displacement by roots over 42 min (8 time steps, 6 min each). This allowed them to visualize root reconfiguration of primary mineral grains with a resolution <2 μm and thus to parameterize kinematic analyses.

Their results revealed that roots without an intact root cap induced a significantly greater degree of soil deformation than intact roots in a dry and compacted soil (Keyes et al., 2017a).

9.8 Compacted Soils

As agricultural management becomes more intensified, including heavier machinery, an increasing impact of soil compaction on plant growth has been witnessed (Tracy et al., 2011). X-ray CT has proven a valuable tool to estimate the impact of soil compaction on root development. A number of studies have observed that under high compaction, plants root diameters are increased and length densities and surface area tend to be reduced and thus also the soil volume explored by roots is reduced (Colombi & Walter, 2016; Fang et al., 2018, 2019; Lucas et al., 2019a; Tracy et al., 2012a). In 4-D X-ray CT analyses of tomato plants in bulk densities varying from 1.2–1.6 Mg m^{-3} Tracy et al. (2013) showed that the impact of dry bulk density on root growth was greatest three days after planting and that the influence of dry bulk densities differs between soil types. In a later study, they revealed that endogenous abscisic acid concentration plays an important role in mediating the root response to soil compaction and has a positive influence on RSA in compacted soil (Tracy et al., 2015). A phenotyping study with 14 different genotypes of wheat, subjected to three different bulk densities, revealed that under moderate soil compaction the capacity to maintain the number of axial and lateral roots plays an important role in maintaining shoot biomass production (Colombi & Walter, 2016).

To determine the effects of compacted layers in columns imitating compacted subsoil, Burr-Hersey et al. (2017) compared three cover crops, which showed different reactions to subsoil compaction. It was shown that radish switched from single tap to multiple perpendicular root system at the compacted layer, while vetch primary root showed less lateral root growth (Fig. 9.3b, c) and black oats exhibited no deviation at compacted layers. Recently, the importance of ethylene signalling was demonstrated to be crucial in triggering root growth responses to soil compaction (Pandey et al., 2021). Among other analysis of these authors, X-ray CT analyses of wild-type rice and ethylene-insensitive mutants growing in uncompacted soil (1.1 g cm^{-3}) or highly compacted soil (1.6 g cm^{-3}) revealed that reduced root length and increased root diameters are rather a response to changed ethylene diffusion than a direct response of increased penetration resistance (Pandey et al., 2021). This mechanism was presumably evolved to avoid growth in compacted soils (Pandey et al., 2021), which, on the other hand, would also further explain how roots preferentially grow in soil regions with low density as, e.g., observed in Lucas et al. (2019a).

At the same time, connected biopores have an important role helping roots overcome limitations of plant growth in compacted soil, as roots can grow into them and thus explore larger soil depths. These interactions of biopores and roots in compacted soils can be visualized with X-ray CT too. It was shown that artificial macropores facilitate root growth (Fig. 9.3d, e) especially under high bulk densities (Atkinson et al., 2020; Colombi et al., 2017; Pfeifer et al., 2014). As soil penetration resistance increases with soil depth, roots are especially in subsoils restricted to growth within macropores. Thus, to investigate small scale root macropore interactions in the field with X-ray CT, many small samples at different depths would be needed to provide the possibility for high-resolution for root segmentation and to cover the whole soil profile, respectively. Another possibility was provided by Zhou et al. (2021), who used X-ray CT images from large soil cores of 100 cm height to segment macropores and combined them with results of root counting by the traditional core break method. They showed soil macroporosity explained 59% of total root depth variation of six different wheat lines, while there were no differences between wheat varieties. Especially at depths greater than 35 cm, most roots were associated with macropores (Zhou et al., 2021).

9.8.1 *Compaction of the Rhizosphere*

While roots grow into the soil as described above, the physical, chemical and biological properties in their surroundings (i.e. in the rhizosphere) are also being modified (Hinsinger et al., 2009; Young, 1998). Root-induced structural changes of the rhizosphere do not only affect microbial habitats, but likewise the transport of water and nutrients. In model calculations supported by X-ray CT imaging, Aravena et al. (2011, 2014) showed that increased rhizosphere compression (reduced porosity) has a positive effect on root water uptake. In an early stage, it was assumed that growing roots, mostly compress the rhizosphere, as their root diameters are often larger than the existing pores (Bruand et al., 1996; Dexter, 1987). In a model by Dexter, compaction is defined as an exponential decrease in porosity towards the roots, which is determined by a constant multiple of the root diameter (Dexter, 1987). The aforementioned development of X-ray CT enables the in-situ observation of the root–soil interface and hence the soil porosity gradient could be visualized and analysed directly in recent years (Feeney et al., 2006; Helliwell et al., 2017; Vollsnes et al., 2010). Confirmations of Dexter's model have been made, as for young maize it was shown that the compaction around roots in sand is exponential and depends on the root radius (Lucas et al., 2019a; Vollsnes et al., 2010). Other researchers, however, found contradictory results (i.e. an increase in porosity in the rhizosphere) (Helliwell et al., 2017, 2019; Rabbi et al., 2018a, 2018b; Zhang et al., 2020). X-ray CT analysis of aggregates from the rhizosphere of ryegrass (*Lolium perenne*) with a resolution of 4.4 μm showed a significant increase in porosity compared to bulk soil samples (Feeney et al., 2006). This was also observed by Helliwell et al. (2017) in the rhizosphere of tomato plants, which the authors attributed to processes such as shrinkage and swelling and thigmotropism that occur more intensively in the rhizosphere. In subsequent work, these authors showed that the porosity (density) gradients depend on the plant species and texture (Helliwell et al., 2019). Furthermore, Rabbi et al. (2018a, 2018b) showed that drought tolerant chickpea genotypes form roots with a larger diameter and a larger, more porous rhizosheath, resulting in increased water absorption capacity. In a similar way, Zhang et al. (2020) showed that the porosity of the rhizosheath near the root surface of highland rice is higher than that of lowland rice. Koebernick et al. (2019) described an increased porosity in the immediate vicinity of the epidermis of barley roots followed by a zone of compaction. They assigned this to a "surface/wall effect" and were able to describe the pattern of porosity around the root using Dexter's model with an extension that takes into account the packing geometry of spherical soil particles (Koebernick et al., 2019). A broader perspective has been adopted by Lucas et al. (2019a); they demonstrated in a study using undisturbed field samples as well as in an experiment with maize grown at different bulk densities under controlled conditions that the initial soil structure has a decisive influence on the porosity of the rhizosphere (Fig. 9.3f). The mechanistic physical effect of the roots growing into the soil structure leads to the compression of the rhizosphere soil as described in Dexter (1987) in combination with a low porosity at the surface of the epidermis as described in Koebernick et al. (2019). This is especially true for soils containing few and unconnected macropores in the size of the root segment diameters. However, besides the formation of new root channels, roots can also penetrate into existing macropores. This leads to low rhizosphere porosity in soils with high macroporosity, due to the interaction of the plant with the existing pore structure (Lucas et al., 2019a). An important aspect arising from this study for future research in the field of physical processes in the rhizosphere is that loosely filled cylinders (1.3 g cm^{-3}), as typically used for laboratory experiments, lead to very porous rhizospheres and not to homogenous physical properties as often assumed.

The differences in rhizosphere densities may also change the relative distribution of organic matter in biopore/rhizosphere walls, as physical protection of organic matter is influenced (Lucas et al., 2020a). In the same vein, Van Veelen et al. (2020) were able to reveal in a correlative analysis of synchrotron X-ray fluorescence microscopy and X-ray absorption near-edge structure with X-ray CT that iron, sulphur and phosphate chemistry are strongly influenced by the spatial variation in porosity. In the direct porous vicinity of the root, dissolution processes are increasingly occurring due to optimal conditions for microbial activity, while the subsequent densification zone is associated with reduced permeability and increased binding and reactions on particle surfaces, e.g. resulting in the formation of stable Fe oxides (van Veelen et al., 2020).

9.8.2 Root-Soil Contact

As discussed above, the space directly at the epidermis is often more porous than the rhizosphere, which tremendously affects the ability of roots to take up water and nutrients. X-ray CT enables not only the determination of gradients in porosity with distance to the root, but also the quantification of the direct contact area of root epidermis with soil particles. In addition, also the seed-soil contact can be visualized and segmented (Blunk et al., 2017, 2019). In experiments with maize and lupin, it was shown that root-soil-contact varies greatly with small changes in macroporosity and increased with decreasing aggregate size (Schmidt et al., 2012).

In pot experiments with lupine and faba bean, it was observed that root-soil contact changes dynamically (Carminati et al., 2013; Koebernick et al., 2018). If a soil dries out, gaps occur around roots due to root shrinkage (Fig. 9.3g) but are closed again after roots swelled due to irrigation. By observing gap formation and transpiration under wetting and drying cycles it was shown that first transpiration decreases then gaps are formed, thus Carminati et al. (2013) concluded that gaps are a result and not a cause of reduced water availability. However, it is important to note that the wall effect described by Koebernick et al. (2019) and shrinkage dynamic of roots can be only differentiated with 4-D X-ray CT image analysis at high image resolution.

9.9 Interaction with (Micro)Organisms

X-ray CT analysis may be combined with analysis of soil (micro-)biota to analyse the soil–plant–microbe complex. Grose et al. (1996) showed that X-ray CT can also indirectly be used to classify soil habitats. Their measurements implied that spatial heterogeneity of soil water in the rhizosphere is biologically significant, that is favourable and less favourable sites can be found in the rhizosphere for the growth of two pathogenic fungi (Grose et al., 1996). However, a drawback of this study is that the authors assumed a constant bulk density for calculating water contents, and thus did not allow for local changes in soil density effecting the X-ray attenuation. However, X-ray CT may be used to describe soil habitats, which can be combined with soil ecological analyses. For instance, Feeney et al. (2006) revealed a complex interplay of structural changes in the rhizosphere of plants with fungal communities. There are a number of other X-ray CT studies focusing on soil pathogens. The influence of a scab-inducing pathogen could be successfully measured with X-ray CT by analysing density gradients of the below ground organs of growing potato plants (Han et al., 2009). Sturrock et al. (2015) coupled real-time PCR analysis with X-ray CT to study the interaction of the plant pathogenic fungus *Rhizoctonia solani* with wheat and rapeseed. They found a strong correlation of pathogen DNA and changes measured in root traits over time and higher tolerance of wheat plants compared to rapeseed (Sturrock et al., 2015). By using split-columns for the growth of a mycorrhiza-defective tomato mutant and a wild-type tomato the effect of arbuscular mycorrhizal fungi (AMF) could be disentangled and revealed much higher impact of roots on porosity and soil stability compared to fungal hyphae (Hallett et al., 2009). Feng et al. (2013) revealed that metal nanoparticles such as AgNPs can have negative effects on AMF and thus change their positive effect on plant growth. Wireworms are known for their damage they cause on crops. 4-D analysis of the interaction of barley and maize with them revealed that the volume of wireworm borrows was different between the two plants but was not correlated with root volume (Booth et al., 2020). Similarly, Johnson et al. (2004) visualized the movement of S. lepidus larvae and their interactions with white clover. Further details on the use of X-ray CT for the analysis of soil biology can be found in Chap. 10.

9.10 The Impact of Nutrition on Root Growth

To efficiently assimilate and utilize nutrients, plants have developed a range of adaptive responses (Hodge, 2004). X-ray CT has proven to be a useful tool to analyse the morphological and physiological responses to the amount and distribution of nutrients. It was shown that phosphorus bands induced high local root length densities and an increase in branch density within the band compared to homogenized phosphorous application for wheat roots (Flavel et al., 2014; Fig. 9.4a, b). Similarly, comparing the influence of localized phosphorus (P) application to maize and faba bean roots, Gao et al. (2019b) showed maize reacts already 11 days after sowing with higher root length density compared to uniform P application, while faba bean showed no response (Fig. 9.4c-f). Nwankwo et al. (2018) showed high root length densities and a higher number of fine roots around seedballs, which can be easily made by hand and include also fertilizer. These changes in root behaviour are important to consider for fertilizer application or more specifically, its location. Concentrated root growth towards nutrients, for example, may not be beneficial under conditions where roots need to explore large or deep soil volumes as in the event of drought.

However, an important aspect to cover the response of RSA to nutrients is the use of additional techniques to cover and analyse nutrient distribution. Brackin et al. (2017) used X-ray CT analysis to position microdialysis probes at known distances from roots. Their results revealed nitrate accumulated close to roots whereas ammonium was depleted as expected, which may be explained by nitrate fluxes arriving at

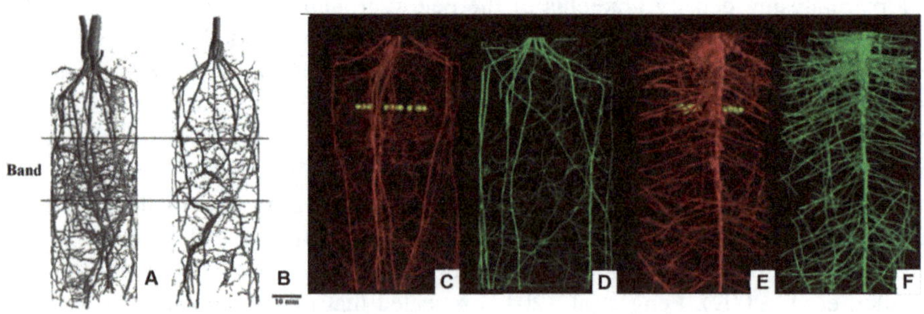

Fig. 9.4 The effect of localized phosphorus application on wheat (Flavel et al., 2014, (**a**, **b**)), maize (**c**, **d**) and faba bean (Gao et al. 2019a, b, (**e**, **f**)). A, C, E show plants growing into soil with localized phosphorus application, while images B, D and F show images of plant roots growing into soil mixed with phosphorus fertilizer

the root surface exceeding the rate of uptake. This combination of techniques allows to reveal new insights of nutrient physiology in the soil–plant continuum and thus may provide new possibilities for nutrient efficient fertilizer application. Micro suction cups at different depth in a soil column can be used to analyse various nutrient concentration and pH-values and link them to root length distribution over time (Gao et al., 2019b). In addition, the correlative imaging of X-ray CT and elemental mapping (SEM-EDS) can be used for numerical diffusion models and calculate root-uptake of nutrients like phosphorus (McKay Fletcher et al., 2019). However, again limiting pot size is restricting the options of investigations.

9.11 Image-Based Modelling

9.11.1 Explicit Models

X-ray CT imaging offers the possibility to process data for image-based models as extensively elaborated in Chap. 11. However, especially SR-CT can be also used to model root related processes on high-resolution images (Daly et al., 2017; Keyes et al., 2013; Koebernick et al., 2017). At present, these can have a resolution down to approx. 1 μm and thus offer the possibility to segment root hairs and be used to model their contribution to P uptake (Fig. 9.5b) (Daly et al., 2016; Keyes et al., 2013) or their role in soil structure formation (Fig. 9.5a) (Koebernick et al., 2017). However, it is important to consider that for these explicit image-based models until now only very small soil volumes have been used (< 1 cm^3), due to the high-resolution required and due to the limitation of computational time. However, although image volumes affect modelling results (Fig. 9.5a), often volumes, which cannot be considered as representative elementary volumes, are analysed. This is in particular the case for macropores, and hence detrimental for soil hydrological processes linked to macropore—connectivity like (near) saturated hydraulic conductivity, which cannot be covered within small samples (Lucas et al., 2020b).

9.11.2 Root System Architecture (RSA) as Input for Models

Models using the RSA as input can cover larger number of scenarios compared to experiments. In addition, RSA with similar traits but different realizations can be modelled and used for contrasting scenarios. Most of these studies rely on the RSA created by models, however, this can be replaced by using 3-D X-ray CT images as an input. In such a way, X-ray CT can be used to calculate water distribution and water uptake by plants (Daly et al., 2018; Koebernick et al., 2015). Koebernick et al. (2015) used a split root experiment with split layers of paraffin to show that despite the hydraulic barrier water redistribution happens.

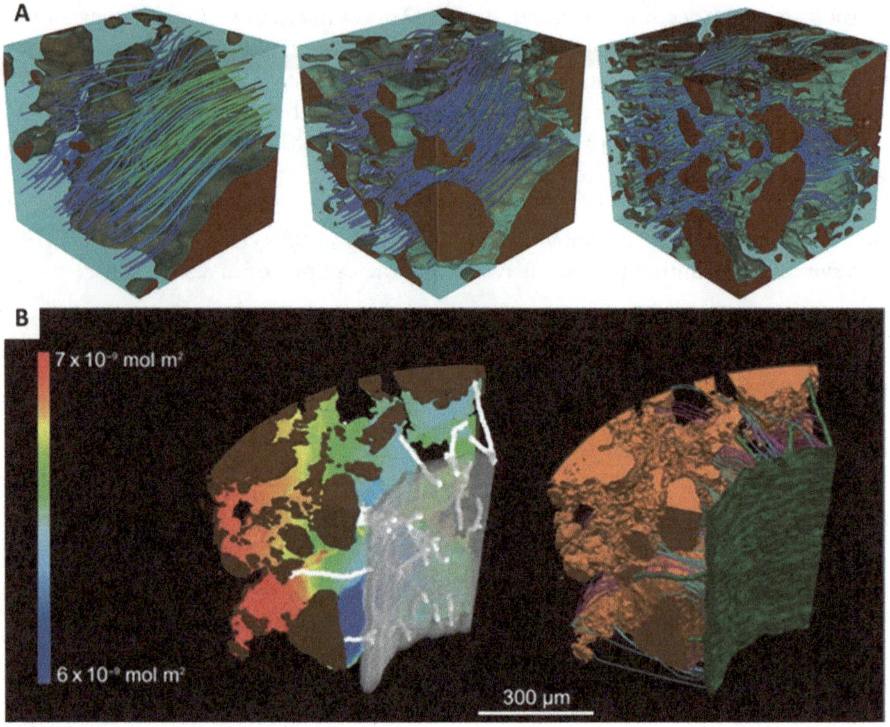

Fig. 9.5 (**a**) Flow streamlines of local Darcy velocities simulated by image-based modelling in the barley rhizosphere with increasing site length (left: 0.16 mm; middle: 0.25 mm; right: 0.4 mm) by Koebernick et al. (2017). Warmer colours indicate greater relative velocity (**b**) Modelled surface concentrations of phosphate on the soil particle surface after 10 h of uptake by a root and its hairs of wheat (left) and streamlines showing phosphate transport paths from soil surface to root hairs ((magenta) and root (blue). (Keyes et al., 2013)

9.12 Conclusions

This chapter has focused on recent advances in X-ray imaging of plant root interaction, particularly in the field of root segmentation and analysis of RSA. In addition, a broad overview of specific research applications for root scientists was given. It was shown that in recent years several segmentation methods were published, which all have their strength and drawbacks. However, these methods all focus on specific features of roots, like attenuation values, connectivity, or their shape. We foresee the next challenges will take place using machine learning algorithms, which can be trained by a joined database of various laboratory data along with root imaging via X-ray CT. In addition to the advances in image processing and image analysis protocols, this will strengthen the possibility of X-ray CT to be used for process understanding in rhizosphere research, as well it will offer the possibility for phenotyping studies with X-ray CT. As extensively reviewed, X-ray CT allows the examination of various processes involved in soil–root interactions as it offers the major advantage of undisturbed visualisation of soil and root structure in 3-D and even 4-D. The next challenge will be to minimize the impact of the trade-off between sample size and resolution and thus not only focus on pot experiments with limiting pot sizes but also on samples taken from field experiments.

References

Ahmed, S., Klassen, T. N., Keyes, S., Daly, M., Jones, D. L., Mavrogordato, M., Sinclair, I., & Roose, T. (2016). Imaging the interaction of roots and phosphate fertiliser granules using 4-D X-ray tomography. *Plant and Soil, 401*, 125–134. https://doi.org/10.1007/s11104-015-2425-5

Aravena, J. E., Berli, M., Ghezzehei, T. A., & Tyler, S. W. (2011). Effects of root-induced compaction on rhizosphere hydraulic properties–X-ray microtomography imaging and numerical simulations. *Environmental Science & Technology, 45*, 425–431. https://doi.org/10.1021/es102566j

Aravena, J. E., Berli, M., Ruiz, S., Suárez, F., Ghezzehei, T. A., & Tyler, S. W. (2014). Quantifying coupled deformation and water flow in the rhizosphere using X-ray microtomography and numerical simulations. *Plant and Soil, 376*, 95–110. https://doi.org/10.1007/s11104-013-1946-z

Atkinson, J. A., Hawkesford, M. J., Whalley, W. R., Zhou, H., & Mooney, S. J. (2020). Soil strength influences wheat root interactions with soil macropores. *Plant, Cell & Environment, 43*, 235–245. https://doi.org/10.1111/pce.13659

Bacq-Labreuil, A., Crawford, J., Mooney, S. J., Neal, A. L., Akkari, E., McAuliffe, C., Zhang, X., Redmile-Gordon, M., & Ritz, K. (2018). Effects of cropping systems upon the three-dimensional architecture of soil systems are modulated by texture. *Geoderma, 332*, 73–83. https://doi.org/10.1016/j.geoderma.2018.07.002

Bacq-Labreuil, A., Crawford, J., Mooney, S. J., Neal, A. L., & Ritz, K. (2019). Cover crop species have contrasting influence upon soil structural genesis and microbial community phenotype. *Scientific Reports, 9*, 7473. https://doi.org/10.1038/s41598-019-43937-6

Bauerle, T. L., & Centinari, M. (2014). Assessment of root system development among four ornamental tree species through time via X-ray computed tomography. *horts, 49*, 44–50. https://doi.org/10.21273/HORTSCI.49.1.44

Bengough, A. G., McKenzie, B. M., Hallett, P. D., & Valentine, T. A. (2011). Root elongation, water stress, and mechanical impedance: A review of limiting stresses and beneficial root tip traits. *Journal of Experimental Botany, 62*, 59–68. https://doi.org/10.1093/jxb/erq350

Blaser, S. R. G. A., Schlüter, S., & Vetterlein, D. (2018). How much is too much?-influence of X-ray dose on root growth of faba bean (Vicia faba) and barley (Hordeum vulgare). *PLoS One, 13*, e0193669. https://doi.org/10.1371/journal.pone.0193669

Blunk, S., Malik, A. H., de Heer, M. I., Ekblad, T., Bussell, J., Sparkes, D., Fredlund, K., Sturrock, C. J., & Mooney, S. J. (2017). Quantification of seed-soil contact of sugar beet (Beta vulgaris) using X-ray computed tomography. *Plant Methods, 13*, 71. https://doi.org/10.1186/s13007-017-0220-4

Blunk, S., de Heer, M. I., Malik, A. H., Fredlund, K., Ekblad, T., Sturrock, C. J., & Mooney, S. J. (2019). Seed priming enhances early growth and improves area of soil exploration by roots. *Environmental and Experimental Botany, 158*, 1–11. https://doi.org/10.1016/j.envexpbot.2018.11.003

Booth, S., Kurtz, B., de Heer, M. I., Mooney, S. J., & Sturrock, C. J. (2020). Tracking wireworm burrowing behaviour in soil over time using 3-D X-ray computed tomography. *Pest Management Science, 76*(8), 2653–2662. https://doi.org/10.1002/ps.5808

Brackin, R., Atkinson, B. S., Sturrock, C. J., & Rasmussen, A. (2017). Roots-eye view: Using microdialysis and microCT to non-destructively map root nutrient depletion and accumulation zones. *Plant, Cell & Environment, 40*, 3135–3142. https://doi.org/10.1111/pce.13072

Bruand, A., Cousin, I., Niccoullaud, B., Duval, O., & Begon, J. C. (1996). Backscattered electron scanning images of soil porosity for analyzing soil compaction around roots. *Soil Science Society of America Journal, 60*, 895–901. https://doi.org/10.2136/sssaj1996.03615995006000030031x

Burr-Hersey, J. E., Mooney, S. J., Bengough, A. G., Mairhofer, S., & Ritz, K. (2017). Developmental morphology of cover crop species exhibit contrasting behaviour to changes in soil bulk density, revealed by X-ray computed tomography. *PLoS One, 12*, e0181872. https://doi.org/10.1371/journal.pone.0181872

Carminati, A., Vetterlein, D., Weller, U., Vogel, H.-J., & Oswald, S. E. (2009). When roots lose contact. *Vadose Zone Journal, 8*, 805–809. https://doi.org/10.2136/vzj2008.0147

Carminati, A., Vetterlein, D., Koebernick, N., Blaser, S., Weller, U., & Vogel, H.-J. (2013). Do roots mind the gap? *Plant and Soil, 367*, 651–661. https://doi.org/10.1007/s11104-012-1496-9

Colombi, T., & Walter, A. (2016). Root responses of triticale and soybean to soil compaction in the field are reproducible under controlled conditions. *Functional Plant Biol., 43*, 114. https://doi.org/10.1071/FP15194

Colombi, T., & Walter, A. (2017). Genetic diversity under soil compaction in wheat: Root number as a promising trait for early plant vigor. *Frontiers in Plant Science, 8*. https://doi.org/10.3389/fpls.2017.00420

Colombi, T., Braun, S., Keller, T., & Walter, A. (2017). Artificial macropores attract crop roots and enhance plant productivity on compacted soils. *Sci Total Environ, 574*, 1283–1293. https://doi.org/10.1016/j.scitotenv.2016.07.194

Correa, J., Postma, J. A., Watt, M., & Wojciechowski, T. (2019). Soil compaction and the architectural plasticity of root systems. *Journal of Experimental Botany, 70*, 6019–6034. https://doi.org/10.1093/jxb/erz383

Daly, K. R., Keyes, S. D., Masum, S., & Roose, T. (2016). Image-based modelling of nutrient movement in and around the rhizosphere. *Journal of Experimental Botany, 67*, 1059–1070. https://doi.org/10.1093/jxb/erv544

Daly, K. R., Cooper, L. J., Koebernick, N., Evaristo, J., Keyes, S. D., van Veelen, A., & Roose, T. (2017). Modelling water dynamics in the rhizosphere. *Rhizosphere, 4*, 139–151. https://doi.org/10.1016/j.rhisph.2017.10.004

Daly, K. R., Tracy, S. R., Crout, N. M. J., Mairhofer, S., Pridmore, T. P., Mooney, S. J., & Roose, T. (2018). Quantification of root water uptake in soil using X-ray computed tomography and image-based modelling. *Plant, Cell & Environment, 41*, 121–133. https://doi.org/10.1111/pce.12983

Davey, E., Wigand, C., Johnson, R., Sundberg, K., Morris, J., & Roman, C. T. (2011). Use of computed tomography imaging for quantifying coarse roots, rhizomes, peat, and particle densities in marsh soils. *Ecological Applications, 21*, 2156–2171. https://doi.org/10.1890/10-2037.1

Dexter, A. R. (1987). Compression of soil around roots. *Plant and Soil, 97*, 401–406.

Douarre Clement, Schielein, R., Frindel, C., Gerth, S., Rousseau, D. (2016). *Deep learning based root-soil segmentation from X-ray tomography.*

Downie, H. F., Adu, M. O., Schmidt, S., Otten, W., Dupuy, L. X., White, P. J., & Valentine, T. A. (2015). Challenges and opportunities for quantifying roots and rhizosphere interactions through imaging and image analysis. *Plant, Cell & Environment, 38*, 1213–1232. https://doi.org/10.1111/pce.12448

Evans, H. J. (1965). Effects of radiations on meristematig cells. *Radiation Botany, 5*, 171–182. https://doi.org/10.1016/s0033-7560(65)80036-9

Fang, H., Zhou, H., Norton, G. J., Price, A. H., Raffan, A. C., Mooney, S. J., Peng, X., & Hallett, P. D. (2018). Interaction between contrasting rice genotypes and soil physical conditions induced by hydraulic stresses typical of alternate wetting and drying irrigation of soil. *Plant and Soil, 430*, 233–243. https://doi.org/10.1007/s11104-018-3715-5

Fang, H., Rong, H., Hallett, P. D., Mooney, S. J., Zhang, W., Zhou, H., & Peng, X. (2019). Impact of soil puddling intensity on the root system architecture of rice (Oryza sativa L.) seedlings. *Soil and Tillage Research, 193*, 1–7. https://doi.org/10.1016/j.still.2019.05.022

Feeney, D. S., Crawford, J. W., Daniell, T., Hallett, P. D., Nunan, N., Ritz, K., Rivers, M., & Young, I. M. (2006). Three-dimensional microorganization of the soil-root-microbe system. *Microbial Ecology, 52*, 151–158. https://doi.org/10.1007/s00248-006-9062-8

Feng, Y., Cui, X., He, S., Dong, G., Chen, M., Wang, J., & Lin, X. (2013). The role of metal nanoparticles in influencing arbuscular mycorrhizal fungi effects on plant growth. *Environmental Science & Technology, 47*, 9496–9504. https://doi.org/10.1021/es402109n

Fitter, A. (1987). An architectural approach to the comparative ecology of plant root systems. *New Phytologist, 106*, 61–77. https://doi.org/10.1111/j.1469-8137.1987.tb04683.x

Flavel, R. J., Guppy, C. N., Tighe, M., Watt, M., McNeill, A., & Young, I. M. (2012). Non-destructive quantification of cereal roots in soil using high-resolution X-ray tomography. *Journal of Experimental Botany, 63*, 2503–2511. https://doi.org/10.1093/jxb/err421

Flavel, R. J., Guppy, C. N., Tighe, M. K., Watt, M., & Young, I. M. (2014). Quantifying the response of wheat (Triticum aestivum L) root system architecture to phosphorus in an Oxisol. *Plant and Soil, 385*, 303–310. https://doi.org/10.1007/s11104-014-2191-9

Flavel, R. J., Guppy, C. N., Rabbi, S. M. R., & Young, I. M. (2017). An image processing and analysis tool for identifying and analysing complex plant root systems in 3-D soil using non-destructive analysis: Root1. *PLoS One, 12*, e0176433. https://doi.org/10.1371/journal.pone.0176433

Frangi, A. F., Niessen, W. J., Vincken, K. L., & Viergever, M. A. (1998). Multiscale vessel enhancement filtering. In *International conference on medical image computing and computer-assisted intervention* (pp. 130–137).

Fry, E. L., Evans, A. L., Sturrock, C. J., Bullock, J. M., & Bardgett, R. D. (2018). Root architecture governs plasticity in response to drought. *Plant and Soil, 433*, 189–200. https://doi.org/10.1007/s11104-018-3824-1

Galdos, M. V., Brown, E., Rosolem, C. A., Pires, L. F., Hallett, P. D., & Mooney, S. J. (2020). Brachiaria species influence nitrate transport in soil by modifying soil structure with their root system. *Scientific Reports, 10*, 5072. https://doi.org/10.1038/s41598-020-61986-0

Ganther, M., Yim, B., Ibrahim, Z., Bienert, M. D., Lippold, E., Maccario, L., Sørensen, S. J., Bienert, G. P., Vetterlein, D., Heintz-Buschart, A., Blagodatskaya, E., Smalla, K., & Tarkka, M. T. (2020). Compatibility of X-ray computed tomography with plant gene expression, rhizosphere bacterial communities and enzyme activities. *Journal of Experimental Botany, 71*, 5603–5614. https://doi.org/10.1093/jxb/eraa262

Gao, W., Schlüter, S., Blaser, S. R. G. A., Shen, J., & Vetterlein, D. (2019a). A shape-based method for automatic and rapid segmentation of roots in soil from X-ray computed tomography images: Rootine. *Plant and Soil, 441*, 643–655. https://doi.org/10.1007/s11104-019-04053-6

Gao, W., Blaser, S. R. G. A., Schlüter, S., Shen, J., & Vetterlein, D. (2019b). Effect of localised phosphorus application on root growth and soil nutrient dynamics in situ–comparison of maize (Zea mays) and faba bean (Vicia faba) at the seedling stage. *Plant and Soil, 441*, 469–483. https://doi.org/10.1007/s11104-019-04138-2

Gregory, P. J., Hutchison, D. J., Read, D. B., Jenneson, P. M., Gilboy, W. B., & Morton, E. J. (2003). Non-invasive imaging of roots with high-resolution X-ray micro-tomography. *Plant and Soil, 255*, 351–359. https://doi.org/10.1023/A:1026179919689

Grose, M. J., Gilligan, C. A., Spencer, D., & Goddard, B. V. D. (1996). Spatial heterogeneity of soil water around single roots: Use of CT-scanning to predict fungal growth in the rhizosphere. *The New Phytologist, 133*, 261–272. https://doi.org/10.1111/j.1469-8137.1996.tb01893.x

Hainsworth, J. M., & Aylmore, L. A. (1983). The use of computer assisted tomography to determine spatial distribution of soil water content. *Soil Res., 21*, 435. https://doi.org/10.1071/SR9830435

Hainsworth, J. M., & Aylmore, L. A. (1986). WATER EXTRACTION BY SINGLE PLANT-ROOTS. *Soil Science Society of America Journal, 50*, 841–848.

Haling, R. E., Tighe, M. K., Flavel, R. J., & Young, I. M. (2013). Application of X-ray computed tomography to quantify fresh root decomposition in situ. *Plant and Soil, 372*, 619–627. https://doi.org/10.1007/s11104-013-1777-y

Hallett, P. D., Feeney, D. S., Bengough, A. G., Rillig, M. C., Scrimgeour, C. M., & Young, I. M. (2009). Disentangling the impact of AM fungi versus roots on soil structure and water transport. *Plant and Soil, 314*, 183–196. https://doi.org/10.1007/s11104-008-9717-y

Han, L., Dutilleul, P., Prasher, S. O., Beaulieu, C., & Smith, D. L. (2009). Assessment of density effects of the common scab-inducing pathogen on the seed and peripheral organs of potato during growth using computed tomography scanning data. *Transactions of the ASABE, 52*, 305–311. https://doi.org/10.13031/2013.25924

Hargreaves, C. E., Gregory, P. J., & Bengough, A. G. (2009). Measuring root traits in barley (Hordeum vulgare ssp. vulgare and ssp. spontaneum) seedlings using gel chambers, soil sacs and X-ray microtomography. *Plant and Soil, 316*, 285–297. https://doi.org/10.1007/s11104-008-9780-4

Heeraman, D. A., Hopmans, J. W., & Clausnitzer, V. (1996). Three dimensional imaging of plant roots in situ with X-ray computed tomography. *Plant and Soil, 189*, 167–179.

Helliwell, J. R., Sturrock, C. J., Mairhofer, S., Craigon, J., Ashton, R. W., Miller, A. J., Whalley, W. R., & Mooney, S. J. (2017). The emergent rhizosphere: Imaging the development of the porous architecture at the root-soil interface. *Scientific Reports, 7*, 14875. https://doi.org/10.1038/s41598-017-14904-w

Helliwell, J. R., Sturrock, C. J., Miller, A. J., Whalley, W. R., & Mooney, S. J. (2019). The role of plant species and soil condition in the structural development of the rhizosphere. *Plant, Cell & Environment, 42*, 1974–1986. https://doi.org/10.1111/pce.13529

Hinsinger, P., Bengough, A. G., Vetterlein, D., & Young, I. M. (2009). Rhizosphere: Biophysics, biogeochemistry and ecological relevance. *Plant and Soil, 321*, 117–152. https://doi.org/10.1007/s11104-008-9885-9

Hodge, A. (2004). The plastic plant: Root responses to heterogeneous supplies of nutrients. *The New Phytologist, 162*, 9–24. https://doi.org/10.1111/j.1469-8137.2004.01015.x

Hu, X., Li, Z., Li, X., & Liu, L.-Y. (2016). Quantification of soil macropores under alpine vegetation using computed tomography in the Qinghai Lake watershed, NE Qinghai-Tibet plateau. *Geoderma, 264*, 244–251. https://doi.org/10.1016/j.geoderma.2015.11.001

Hu, X., Li, X., Wang, P., Liu, Y., Wu, X.-C., Li, Z., Zhao, Y.-D., Cheng, Y.-Q., Guo, L.-L., Lyu, Y.-L., & Liu, L.-Y. (2019a). Influence of exclosure on CT-measured soil macropores and root architecture in a shrub-encroached grassland in northern China. *Soil and Tillage Research, 187*, 21–30. https://doi.org/10.1016/j.still.2018.10.020

Hu, X., Li, X., Guo, L.-L., Liu, Y., Wang, P., Zhao, Y.-D., Cheng, Y.-Q., Lyu, Y.-L., & Liu, L.-Y. (2019b). Influence of shrub roots on soil macropores using X-ray computed tomography in a shrub-encroached grassland in northern China. *Journal of Soils and Sediments, 19*, 1970–1980. https://doi.org/10.1007/s11368-018-2218-6

Hu, X., Li, X., Li, Z., Gao, Z., Wu, X.-C., Wang, P., Lyu, Y.-L., & Liu, L.-Y. (2020). Linking 3-D soil macropores and root architecture to near saturated hydraulic conductivity of typical meadow soil types in the Qinghai Lake watershed, northeastern Qinghai-Tibet plateau. *Catena, 185*. https://doi.org/10.1016/j.catena.2019.104287

Hudek, C., Sturrock, C. J., Atkinson, B. S., Stanchi, S., & Freppaz, M. (2017). Root morphology and biomechanical characteristics of high altitude alpine plant species and their potential application in soil stabilization. *Ecological Engineering, 109*, 228–239. https://doi.org/10.1016/j.ecoleng.2017.05.048

Jarvis, N., Forkman, J., Koestel, J., Katterer, T., Larsbo, M., & Taylor, A. (2017). Long-term effects of grass-clover leys on the structure of a silt loam soil in a cold climate. *Agriculture Ecosystems & Environment, 247*, 319–328. https://doi.org/10.1016/j.agee.2017.06.042

Jenneson, P. M., Gilboy, W. B., Morton, E. J., & Gregory, P. J. (2003). An X-ray micro-tomography system optimised for the low-dose study of living organisms. *Applied Radiation and Isotopes, 58*, 177–181. https://doi.org/10.1016/S0969-8043(02)00310-X

Jin, K., Shen, J., Ashton, R. W., Dodd, I. C., Parry, M. A. J., & Whalley, W. R. (2013). How do roots elongate in a structured soil? *Journal of Experimental Botany, 64*, 4761–4777. https://doi.org/10.1093/jxb/ert286

Johnson, E. L. (1936). Susceptibility of seventy species of flowering plants to X-radiation. *Plant Physiology, 11*, 319–342.

Johnson, S. N., Read, D. B., & Gregory, P. J. (2004). Tracking larval insect movement within soil using high-resolution X-ray microtomography. *Ecological Entomology, 29*, 117–122. https://doi.org/10.1111/j.0307-6946.2004.00567.x

Kaestner, A., Schneebeli, M., & Graf, F. (2006). Visualizing three-dimensional root networks using computed tomography. *Geoderma, 136*, 459–469. https://doi.org/10.1016/j.geoderma.2006.04.009

Keyes, S. D., Daly, K. R., Gostling, N. J., Jones, D. L., Talboys, P., Pinzer, B. R., Boardman, R., Sinclair, I., Marchant, A., & Roose, T. (2013). High-resolution synchrotron imaging of wheat root hairs growing in soil and image based modelling of phosphate uptake. *The New Phytologist, 198*, 1023–1029. https://doi.org/10.1111/nph.12294

Keyes, S. D., Gillard, F., Soper, N., Mavrogordato, M. N., Sinclair, I., & Roose, T. (2016). Mapping soil deformation around plant roots using in vivo 4-D X-ray computed tomography and digital volume correlation. *Journal of Biomechanics, 49*, 1802–1811. https://doi.org/10.1016/j.jbiomech.2016.04.023

Keyes, S. D., Cooper, L., Duncan, S., Koebernick, N., McKay Fletcher, D. M., Scotson, C. P., van Veelen, A., Sinclair, I., & Roose, T. (2017a). Measurement of micro-scale soil deformation around roots using four-dimensional synchrotron tomography and image correlation. *J R Soc Interface, 14*. https://doi.org/10.1098/rsif.2017.0560

Keyes, S. D., Zygalakis, K. C., & Roose, T. (2017b). An explicit structural model of root hair and soil interactions parameterised by synchrotron X-ray computed tomography. *Bulletin of Mathematical Biology, 79*, 2785–2813. https://doi.org/10.1007/s11538-017-0350-x

Kirk, G. J. D., Boghi, A., Affholder, M.-C., Keyes, S. D., Heppell, J., & Roose, T. (2019). Soil carbon dioxide venting through rice roots. *PLANT CELL AND ENVIRONMENT., 42*(12), 3197–3207. https://doi.org/10.1111/pce.13638

Koebernick, N., Weller, U., Huber, K., Schlüter, S., Vogel, H.-J., Jahn, R., Vereecken, H., & Vetterlein, D. (2014). In situ visualization and quantification of three-dimensional root system architecture and growth using X-Ray computed tomography. *Vadose Zone Journal, 13*, vzj2014.03.0024. https://doi.org/10.2136/vzj2014.03.0024

Koebernick, N., Huber, K., Kerkhofs, E., Vanderborght, J., Javaux, M., Vereecken, H., & Vetterlein, D. (2015). Unraveling the hydrodynamics of split root water uptake experiments using CT scanned root architectures and three dimensional flow simulations. *Frontiers in Plant Science, 6*, 370. https://doi.org/10.3389/fpls.2015.00370

Koebernick, N., Daly, K. R., Keyes, S. D., George, T. S., Brown, L. K., Raffan, A., Cooper, L. J., Naveed, M., Bengough, A. G., Sinclair, I., Hallett, P. D., & Roose, T. (2017). High-resolution synchrotron imaging shows that root hairs influence rhizosphere soil structure formation. *The New Phytologist, 216*, 124–135. https://doi.org/10.1111/nph.14705

Koebernick, N., Schlüter, S., Blaser, S. R. G. A., & Vetterlein, D. (2018). Root-soil contact dynamics of Vicia faba in sand. *Plant and Soil, 431*, 417–431. https://doi.org/10.1007/s11104-018-3769-4

Koebernick, N., Daly, K. R., Keyes, S. D., Bengough, A. G., Brown, L. K., Cooper, L. J., George, T. S., Hallett, P. D., Naveed, M., Raffan, A., & Roose, T. (2019). Imaging microstructure of the barley rhizosphere: Particle packing and root hair influences. *The New Phytologist, 221*, 1878–1889. https://doi.org/10.1111/nph.15516

Koestel, J., & Schlueter, S. (2019). Quantification of the structure evolution in a garden soil over the course of two years. *Geoderma, 338*, 597–609. https://doi.org/10.1016/j.geoderma.2018.12.030

Kuka, K., Illerhaus, B., Fox, C. A., & Joschko, M. (2013). X-ray computed microtomography for the study of the soil-root relationship in grassland soils. *Vadose Zone Journal, 12*. https://doi.org/10.2136/vzj2013.01.0014

Lazarević, B., Sturrock, C. J., Poljak, M., & Mooney, S. J. (2016). Quantification of aluminum-induced changes in wheat root architecture by X-ray microcomputed tomography. *Communications in Soil Science and Plant Analysis, 47*, 263–274. https://doi.org/10.1080/00103624.2015.1122800

Lippold, E., Kleinau, P., Blaser, S. R. G. A., Schlüter, S., Phalempin, M., & Vetterlein, D. (2021). In soil measurement of radiation dose caused by X-ray computed tomography. *Journal of Plant Nutrition and Soil Science, 184*, 343–345. https://doi.org/10.1002/jpln.202000276

Lontoc-Roy, M., Dutilleul, P., Prasher, S. O., Han, L., & Smith, D. L. (2005). Computed tomography scanning for three-dimensional imaging and complexity analysis of developing root systems. *Canadian Journal of Botany, 83*, 1434–1442. https://doi.org/10.1139/b05-118

Lontoc-Roy, M., Dutilleul, P., Prasher, S. O., Han, L., Brouillet, T., & Smith, D. L. (2006). Advances in the acquisition and analysis of CT scan data to isolate a crop root system from the soil medium and quantify root system complexity in 3-D space. *Geoderma, 137*, 231–241. https://doi.org/10.1016/j.geoderma.2006.08.025

Lucas, M., Schlüter, S., Vogel, H.-J., & Vetterlein, D. (2019a). Roots compact the surrounding soil depending on the structures they encounter. *Scientific Reports, 9*, 16236. https://doi.org/10.1038/s41598-019-52665-w

Lucas, M., Schlüter, S., Vogel, H.-J., & Vetterlein, D. (2019b). Soil structure formation along an agricultural chronosequence. *Geoderma, 350*, 61–72. https://doi.org/10.1016/j.geoderma.2019.04.041

Lucas, M., Pihlap, E., Steffens, M., Vetterlein, D., & Kögel-Knabner, I. (2020a). Combination of imaging infrared spectroscopy and X-ray computed microtomography for the investigation of bio- and physicochemical processes in structured soils. *Frontiers in Environmental Science, 8*, 573. https://doi.org/10.3389/fenvs.2020.00042

Lucas, M., Vetterlein, D., Vogel, H.-J., & Schlüter, S. (2020b). Revealing pore connectivity across scales and resolutions with X-ray CT. *European Journal of Soil Science, 72*, 546–560. https://doi.org/10.1111/ejss.12961

Maenhout, P., Sleutel, S., Xu, H., van Hoorebeke, L., Cnudde, V., & de Neve, S. (2019). Semi-automated segmentation and visualization of complex undisturbed root systems with X-ray μCT. *Soil and Tillage Research, 192*, 59–65. https://doi.org/10.1016/j.still.2019.04.025

Mairhofer, S., Zappala, S., Tracy, S. R., Sturrock, C., Bennett, M., Mooney, S. J., & Pridmore, T. (2012). RooTrak: Automated recovery of three-dimensional plant root architecture in soil from x-ray microcomputed tomography images using visual tracking. *Plant Physiology, 158*, 561–569. https://doi.org/10.1104/pp.111.186221

Mairhofer, S., Zappala, S., Tracy, S., Sturrock, C., Bennett, M. J., Mooney, S. J., & Pridmore, T. P. (2013). Recovering complete plant root system architectures from soil via X-ray μ-computed tomography. *Plant Methods, 9*, 8. https://doi.org/10.1186/1746-4811-9-8

Mairhofer, S., Sturrock, C. J., Bennett, M. J., Mooney, S. J., & Pridmore, T. P. (2015a). Extracting multiple interacting root systems using X-ray microcomputed tomography. *Plant Journal, 84*, 1034–1043. https://doi.org/10.1111/tpj.13047

Mairhofer, S., Sturrock, C., Wells, D. M., Bennett, M. J., Mooney, S. J., & Pridmore, T. P. (2015b). On the evaluation of methods for the recovery of plant root systems from X-ray computed tomography images. *Functional Plant Biol., 42*, 460–470. https://doi.org/10.1071/FP14071

Mairhofer, S., Johnson, J., Sturrock, C. J., Bennett, M. J., Mooney, S. J., & Pridmore, T. P. (2016). Visual tracking for the recovery of multiple interacting plant root systems from X-ray $\upmu $ μ CT images. *Machine Vision and Applications, 27*, 721–734. https://doi.org/10.1007/s00138-015-0733-7

Mairhofer, S., Pridmore, T., Johnson, J., Wells, D. M., Bennett, M. J., Mooney, S. J., & Sturrock, C. J. (2017). X-Ray computed tomography of crop plant root systems grown in soil. *Current Protocols in Plant Biology, 2*, 270–286. https://doi.org/10.1002/cppb.20049

McKay Fletcher, D. M., Keyes, S. D., Daly, K. R., van Veelen, A., & Roose, T. (2019). A multi image-based approach for modelling plant-fertiliser interaction. *Rhizosphere, 10*, 100152. https://doi.org/10.1016/j.rhisph.2019.100152

Morris, E. C., Griffiths, M., Golebiowska, A., Mairhofer, S., Burr-Hersey, J., Goh, T., von Wangenheim, D., Atkinson, B., Sturrock, C. J., Lynch, J. P., Vissenberg, K., Ritz, K., Wells, D. M., Mooney, S. J., & Bennett, M. J. (2017). Shaping 3-D root system architecture. *Current Biology, 27*, R919–R930. https://doi.org/10.1016/j.cub.2017.06.043

Nwankwo, C. I., Blaser, S. R. G. A., Vetterlein, D., Neumann, G., & Herrmann, L. (2018). Seedball-induced changes of root growth and physico-chemical properties-a case study with pearl millet. *Journal of Plant Nutrition and Soil Science, 181*, 768–776. https://doi.org/10.1002/jpln.201800059

Pandey, B. K., Huang, G., Bhosale, R., Hartman, S., Sturrock, C. J., Jose, L., Martin, O. C., Karady, M., Voesenek, L. A. C. J., Ljung, K., Lynch, J. P., Brown, K. M., Whalley, W. R., Mooney, S. J., Zhang, D., & Bennett, M. J. (2021). Plant roots sense soil compaction through restricted ethylene diffusion. *Science, 371*, 276–280. https://doi.org/10.1126/science.abf3013

Perez-Torres, E., Kirchgessner, N., Pfeifer, J., & Walter, A. (2015). Assessing potato tuber diel growth by means of X-ray computed tomography. *Plant Cell and Environment, 38*, 2318–2326. https://doi.org/10.1111/pce.12548

Pfeifer, J., Kirchgessner, N., & Walter, A. (2014). Artificial pores attract barley roots and can reduce artifacts of pot experiments. *Journal of Plant Nutrition and Soil Science, 177*, 903–913. https://doi.org/10.1002/jpln.201400142

Pfeifer, J., Kirchgessner, N., Colombi, T., & Walter, A. (2015). Rapid phenotyping of crop root systems in undisturbed field soils using X-ray computed tomography. *Plant Methods, 11*, 41. https://doi.org/10.1186/s13007-015-0084-4

Phalempin, M., Lippold, E., Vetterlein, D., & Schlüter, S. (2021). An improved method for the segmentation of roots from X-ray computed tomography 3-D images: Rootine v.2. *In Plant methods, 17*(1), 39. https://doi.org/10.1186/s13007-021-00735-4

Pierret, A., Capowiez, Y., Moran, C. J., & Kretzschmar, A. (1999). X-ray computed tomography to quantify tree rooting spatial distributions. *Geoderma, 90*, 307–326.

Rabbi, S. M. F., Tighe, M. K., Flavel, R. J., Kaiser, B. N., Guppy, C. N., Zhang, X., & Young, I. M. (2018a). Plant roots redesign the rhizosphere to alter the three-dimensional physical architecture and water dynamics. *The New Phytologist, 219*, 542–550. https://doi.org/10.1111/nph.15213

Rabbi, S. M. F., Tighe, M. K., Knox, O., & Young, I. M. (2018b). The impact of carbon addition on the organisation of rhizosheath of chickpea. *Scientific Reports, 8*. https://doi.org/10.1038/s41598-018-36958-0

Schindelin, J., Arganda-Carreras, I., Frise, E., Kaynig, V., Longair, M., Pietzsch, T., Preibisch, S., Rueden, C., Saalfeld, S., Schmid, B., Tinevez, J.-Y., White, D. J., Hartenstein, V., Eliceiri, K., Tomancak, P., & Cardona, A. (2012). Fiji: An open-source platform for biological-image analysis. *Nature Methods, 9*, 676–682. https://doi.org/10.1038/nmeth.2019

Schlüter, S., Sheppard, A., Brown, K., & Wildenschild, D. (2014). Image processing of multiphase images obtained via X-ray microtomography: A review. *Water Resources Research, 50*, 3615–3639. https://doi.org/10.1002/2014WR015256

Schlüter, S., Blaser, S. R. G. A., Weber, M., Schmidt, V., & Vetterlein, D. (2018). Quantification of root growth patterns from the soil perspective via root distance models. *Frontiers in Plant Science, 9*, 1084. https://doi.org/10.3389/fpls.2018.01084

Schmidt, S., Bengough, A. G., Gregory, P. J., Grinev, D. V., & Otten, W. (2012). Estimating root-soil contact from 3-D X-ray microtomographs. *European Journal of Soil Science, 63*, 776–786. https://doi.org/10.1111/j.1365-2389.2012.01487.x

Skoog, F. (1935). The effect of x-irradiation on auxin and plant growth. *Journal of Cellular and Comparative Physiology, 7*, 227–270. https://doi.org/10.1002/jcp.1030070206

Soltaninejad, M., Sturrock, C. J., Griffiths, M., Pridmore, T. P., & Pound, M. P. (2019). Three dimensional root CT segmentation using multi-resolution encoder-decoder networks. *IEEE Transactions on Image Processing, 55*, 1.

Sturrock, C. J., Woodhall, J., Brown, M., Walker, C., Mooney, S. J., & Ray, R. V. (2015). Effects of damping-off caused by Rhizoctonia solani anastomosis group 2-1 on roots of wheat and oil seed rape quantified using X-ray computed tomography and real-time PCR. *Frontiers in Plant Science, 6*, 461. https://doi.org/10.3389/fpls.2015.00461

Subramanian, S., Han, L., Dutilleul, P., & Smith, D. L. (2015). Computed tomography scanning can monitor the effects of soil medium on root system development: An example of salt stress in corn. *Frontiers in Plant Science, 6*, 256. https://doi.org/10.3389/fpls.2015.00256

Tollner, E. W., Ramseur, E. L., & Murphy, C. (1994). *Techniques and approaches for documenting plant root development with X-ray computed tomography*. SSSA special publication (USA).

Trachsel, S., Kaeppler, S. M., Brown, K. M., & Lynch, J. (2011). Shovelomics: High throughput phenotyping of maize (Zea mays L.) root architecture in the field. *Plant and Soil, 341*, 75–87. https://doi.org/10.1007/s11104-010-0623-8

Tracy, S. R., Black, C. R., Roberts, J. A., & Mooney, S. J. (2011). Soil compaction: A review of past and present techniques for investigating effects on root growth. *Journal of the Science of Food and Agriculture, 91*, 1528–1537. https://doi.org/10.1002/jsfa.4424

Tracy, S. R., Black, C. R., Roberts, J. A., McNeill, A., Davidson, R., Tester, M., Samec, M., Korošak, D., Sturrock, C., & Mooney, S. J. (2012a). Quantifying the effect of soil compaction on three varieties of wheat (Triticum aestivum L.) using X-ray micro computed tomography (CT). *Plant and Soil, 353*, 195–208. https://doi.org/10.1007/s11104-011-1022-5

Tracy, S. R., Black, C. R., Roberts, J. A., Sturrock, C., Mairhofer, S., Craigon, J., & Mooney, S. J. (2012b). Quantifying the impact of soil compaction on root system architecture in tomato (Solanum lycopersicum) by X-ray micro-computed tomography. *Annals of Botany, 110*, 511–519. https://doi.org/10.1093/aob/mcs031

Tracy, S. R., Black, C. R., Roberts, J. A., & Mooney, S. J. (2013). Exploring the interacting effect of soil texture and bulk density on root system development in tomato (Solanum lycopersicum L.). *Environmental and Experimental Botany, 91*, 38–47. https://doi.org/10.1016/j.envexpbot.2013.03.003

Tracy, S. R., Black, C. R., Roberts, J. A., Dodd, I. C., & Mooney, S. J. (2015). Using X-ray computed tomography to explore the role of abscisic acid in moderating the impact of soil compaction on root system architecture. *Environmental and Experimental Botany, 110*, 11–18. https://doi.org/10.1016/j.envexpbot.2014.09.003

Tracy, S. R., Nagel, K. A., Postma, J. A., Fassbender, H., Wasson, A., & Watt, M. (2020). Crop improvement from phenotyping roots: Highlights reveal expanding opportunities. *Trends in Plant Science, 25*, 105–118. https://doi.org/10.1016/j.tplants.2019.10.015

van Veelen, A., Koebernick, N., Scotson, C. S., McKay-Fletcher, D., Huthwelker, T., Borca, C. N., Mosselmans, J. F. W., & Roose, T. (2020). Root-induced soil deformation influences Fe, S and P: Rhizosphere chemistry investigated using synchrotron XRF and XANES. *The New Phytologist, 225*, 1476–1490. https://doi.org/10.1111/nph.16242

Vetterlein, D., & Doussan, C. (2016). Root age distribution: How does it matter in plant processes? A focus on water uptake. *Plant and Soil, 407*, 145–160. https://doi.org/10.1007/s11104-016-2849-6

Vogel, H.-J., Weller, U., & Schlüter, S. (2010). Quantification of soil structure based on Minkowski functions. *Computers & Geosciences, 36*, 1236–1245. https://doi.org/10.1016/j.cageo.2010.03.007

Vollsnes, A. V., Futsaether, C. M., & Bengough, A. G. (2010). Quantifying rhizosphere particle movement around mutant maize roots using time-lapse imaging and particle image velocimetry. *European Journal of Soil Science, 61*, 926–939. https://doi.org/10.1111/j.1365-2389.2010.01297.x

Watanabe, K., Mandang, T., Tojo, S., Ai, F., & Huang, B. K. (1992). *Nondestructive root-zone analysis with X-ray CT scanner*. Paper-American Society of Agricultural Engineers (USA). no. 923018.

White, R. G., & Kirkegaard, J. A. (2010). The distribution and abundance of wheat roots in a dense, structured subsoil–implications for water uptake. *Plant, Cell & Environment, 33*, 133–148. https://doi.org/10.1111/j.1365-3040.2009.02059.x

Xu, Z., Valdes, C., & Clarke, J. (2018). Existing and potential statistical and computational approaches for the analysis of 3-D CT images of plant roots. *AGRONOMY-BASEL, 8*. https://doi.org/10.3390/agronomy8050071

Young, I. M. (1998). Biophysical interactions at the root–soil interface: A review. *The Journal of Agricultural Science, 130*, 1–7. https://doi.org/10.1017/S002185969700498X

Zappala, S., Helliwell, J. R., Tracy, S. R., Mairhofer, S., Sturrock, C. J., Pridmore, T., Bennett, M., & Mooney, S. J. (2013a). Effects of X-Ray dose on rhizosphere studies using X-Ray computed tomography. *PLoS One, 8*, e67250. https://doi.org/10.1371/journal.pone.0067250

Zappala, S., Mairhofer, S., Tracy, S., Sturrock, C. J., Bennett, M., Pridmore, T., & Mooney, S. J. (2013b). Quantifying the effect of soil moisture content on segmenting root system architecture in X-ray computed tomography images. *Plant and Soil, 370*, 35–45. https://doi.org/10.1007/s11104-013-1596-1

Zhang, Y., Du, H., Gui, Y., Xu, F., Liu, J., Zhang, J., & Xu, W. (2020). Moderate water stress in rice induces rhizosheath formation associated with abscisic acid and auxin responses. *Journal of Experimental Botany, 71*(9), 2740–2751. https://doi.org/10.1093/jxb/eraa021

Zhou, H., Whalley, W. R., Hawkesford, M. J., Ashton, R. W., Atkinson, B., Atkinson, J. A., Sturrock, C. J., Bennett, M. J., & Mooney, S. J. (2021). The interaction between wheat roots and soil pores in structured field soil. *Journal of Experimental Botany, 72*, 747–756. https://doi.org/10.1093/jxb/eraa475

Chapter 10
X-ray Computed Tomography Imaging & Soil Biology

Sasha Kravchenko

10.1 Introduction

Since C. Darwin (1809–1882), V. Dokuchaev (1846–1903) and H. Jenny (1899–1992), soil organisms have been well-known as the key element of soil formation (Schaetzl & Anderson, 2005), particularly important for development and properties of soil structure (Meurer et al., 2020). Yet, the opaque nature of the soil and near impossibility of observing soil micro- and macro-fauna activities in their natural habitat, without destroying the habitat itself, have made studies of the soil organisms' behaviours and their subsequent influence on the soil extremely challenging. The development of X-ray Computed Tomography (CT) provided a long-needed capability to look inside soil samples, even though of relatively small (<10–20 cm) sizes, while keeping them intact. X-ray CT enables visualization of intact soil structure with organic objects, such as plant roots or animals, located within it. Thus, it is particularly useful in assessing the role of soil biota, including microorganisms, on soil structure formation and properties and in informing studies of biology and behaviour of soil meso- and macro-fauna, especially earthworms and larger insects. X-ray CT has been readily employed by soil scientists, generating a large body of knowledge concerning interactions between soil physical (e.g. structural) components and soil organisms. Improvements in CT image quality and resolution over the past couple of decades have expanded its implementation in soil studies from viewing relatively large organisms, such as insects, to exploring micro-habitats relevant for functioning of soil microorganisms, to visualizing soil organic matter.

This chapter begins with an overview of the CT use in the studies of soil meso-fauna, primarily earthworms, followed by a review of the current works on soil

S. Kravchenko (✉)
Michigan State University, East Lansing, MI, USA
e-mail: kravche1@msu.edu

S. Jon Mooney et al. (eds.), *X-ray Imaging of the Soil Porous Architecture*,
https://doi.org/10.1007/978-3-031-12176-0_10

microorganisms and their activities that use CT directly as well as indirectly. It then considers the potential effects of scanning on activities and functioning of soil organisms that might influence the interpretation of the experimental results and concludes with a review of current efforts in using CT for visualizing soil particulate and non-particulate organic matter.

10.2 Soil Meso-Fauna and X-ray CT

It is now 140 years since Charles Darwin (Darwin, 1881) highlighted the significance of earthworms as agents of soil development. Yet, despite wide recognition of its importance, the progress in understanding the belowground activities and engineering effects of soil meso-fauna has remained justifiably slow. It is needless to say that soil is an opaque media. But unlike other types of opaque media where burrowing invertebrates live, e.g. wood, soil is also very fragile. Moreover, burrowing of soil meso-fauna lacks consistent structure and is opportunistic in nature, exploiting inherent variations in soil characteristics, presence of existing pores, and low-density spots. These features make it next to impossible to effectively explore the burrowing structures of soil mesa-fauna. Cutting through a soil burrowing system inevitably destroys it, while pre-installed study devices or glass walls alter the burrowing activity and structure construction. Thus, studies of belowground faunal architectures have been few and limited (Oades, 1993) and had to rely on very creative but extremely time-consuming approaches, including casting that involves pouring fast setting liquids/resins into earthworm channels (Garner, 1953), microscopy combined with X-ray stereo-radiography (Rogaar & Boswinkel, 1978), or a complex process of soil thin sectioning (Wendt and Larink, 1990; Ligthart et al., 1993). Utilization of CT offered unprecedented opportunities for progress in understanding and quantifying the invertebrate burrowing effects on soil functioning. The pioneering work of Warner et al. (1989) and Joschko et al. (1991) demonstrated CT capabilities in visualizing macropores of biological origin and earthworm burrows.

Observing temporal dynamics, spatial patterns, and burrowing outcomes from earthworm and insect activities in environments close to their natural habitats and with minimal disturbance became a reality, creating an abundant and actively growing research field. The studies related to activities and biology of soil meso-fauna include works on burrowing behaviours (Joschko et al., 1993; Booth et al., 2020), characteristics of burrow systems of individual earthworm species (Jegou et al., 1999; Bastardie et al., 2003; Capowiez et al., 2003, 2011), species interactions (Capowiez & Belzunces, 2001; Capowiez et al., 2001), effects of different land uses, management practices, and soil pollution on earthworm burrows and activity (Jegou et al., 2002; Auclerc et al., 2013; Pelosi et al., 2017; Mombo et al., 2018). A detailed comprehensive review of the earlier literature on using CT in earthworm studies can be found in Taina et al. (2008).

For soil meso- and macro-organisms, CT can be used to image not just changes in the soil structure stemming from their activities, but also movements of the organisms themselves. One of the latest inspiring examples is the work of Booth et al. (2020) who visualized locations of the freshly made burrows and positions of introduced insects within containers with different plant species. The study enabled analysis of plant–insect interactions and explorations of how root architecture affects insects' burrowing patterns and activities.

X-ray CT has also become instrumental for quantitative characterization of earthworms' effects on formation of soil macropore structure. Earthworms are classified into three broad ecological groups, which differ in their burrowing activities (Lee & Foster, 1991), namely, epigeic (those that do not build permanent burrows and live close to soil surface), endogeic (those that built semi-permanent burrowing systems in the topsoil with a complex network of filled and refilled macropores, in both horizontal and vertical directions), and anecic (those that build permanent vertical burrowing systems up to 1 m deep). The use of CT enabled quantitative descriptions of the burrowing patterns, including such characteristics as diameter, branching, connectivity, verticality, bioturbation, etc. (Capowiez et al., 2015). But quantification also revealed that the variability in burrowing characteristics of individual species even within the same ecotype can be very substantial, leaving behind soils with drastically different macropore systems (Fig. 10.1 from (Capowiez et al., 2015)). However, even though observations from single species studies indicate that different earthworm species make markedly distinct burrow systems, it is very difficult to identify to which species a particular burrow belongs in multi-species studies or in intact soil columns (Jegou et al., 1999; Bastardie et al., 2003; Capowiez et al., 2003, 2015). Smaller worms such as enchytraeids also modify soil pore characteristics, especially increasing presence of biopores with diameters comparable to their sizes (500–700 μm). Yet still smaller organisms, e.g., mites, have not been observed to influence pore volumes and size distributions for pores visible in CT images at 50 μm resolution (Porre et al., 2016).

Being a non-invasive technique CT offers unique opportunities for observing temporal trends in development of pore characteristics as an outcome of soil faunal activities. It is particularly useful for exploring the fate of the macropores, as many of them change their characteristics, e.g., when they are refilled by earthworm casts (Capowiez et al., 2014a). Among the macropore characteristics that can change are open diameter (Whalen et al., 2015), hydrological properties, and ability to transport water and gases. The earthworms not only generate new macropores but also modify their characteristics by making wider pores narrower due to deposition of casts and application of slime on the pore boundaries (Pagenkemper et al., 2015). The macropores created by earthworms or the root-created macropores subsequently populated by earthworms have their walls affected by earthworm presence, including the blockage of pores connected with the macropore. Combining CT with endoscopy allowed observation of decreases in diameters of macropores due to earthworm generated lining as well as closings of laterally interconnected pores which open into vertical macropores by clogging them with coating lining (Kautz et al., 2013; Pagenkemper et al., 2015). The time sufficient for different species to

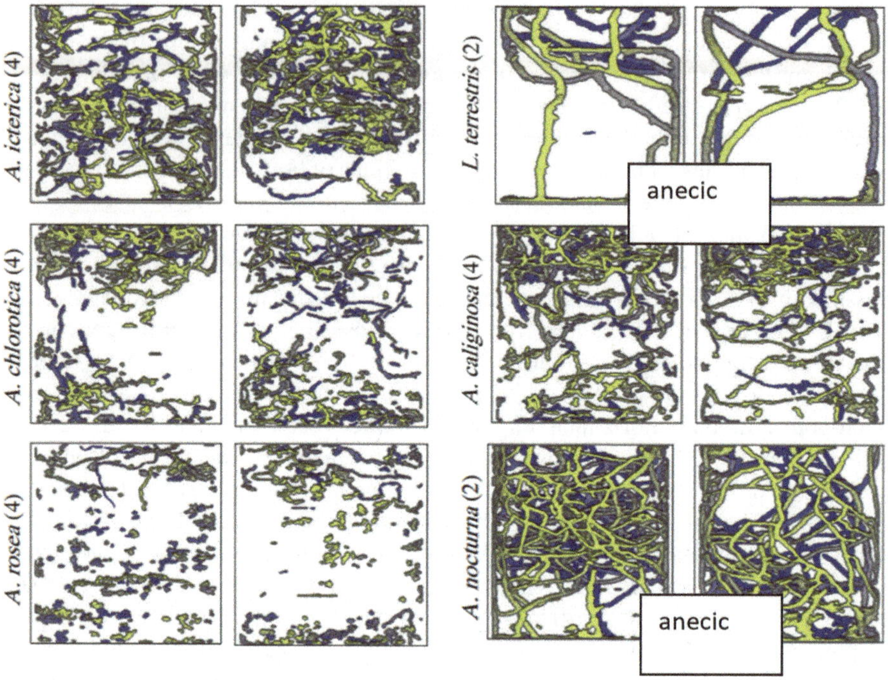

Fig. 10.1 Examples of macropore structures generated by earthworms of different species from different ecotypes (from Capowiez et al., 2015). The systems were built by the earthworms in 6 weeks after being introduced into repacked soil cores

build contrasting macropore networks in the originally sieved and homogenized soil is just 4–6 weeks (Capowiez et al., 2015; Bottinelli et al., 2017). Other recent examples of monitoring of temporal dynamic in pore characteristics in relation to root and insect activities are the work of Koestel and Schluter (2019), who monitored pore architecture evolution in an intact soil core buried into topsoil in the field by taking it out for occasional CT scanning for 2 years, as it was naturally populated by sequences of plants and insects; and Le Mer et al. (2021), who explored how aging of the earthworm burrows affects their connectivity to the surface.

The macropore structure can have a very strong effect on soil hydraulic properties (Capowiez et al., 2014b); for example, the infiltration rate can be 2–4 times higher in soil columns populated by anecic species than by endogeic species (Capowiez et al., 2015), while saturated hydraulic conductivity can be >3 times different among soils where the macropore systems were built by different species (Bottinelli et al., 2017). A notable benefit from using CT in such studies is that it allows identification of the particular features of earthworm burrows and species characteristics that are the most likely drivers of changes in soil hydrology. For example, burrow length and a total volume of percolating burrows appeared to be the characteristics best at describing

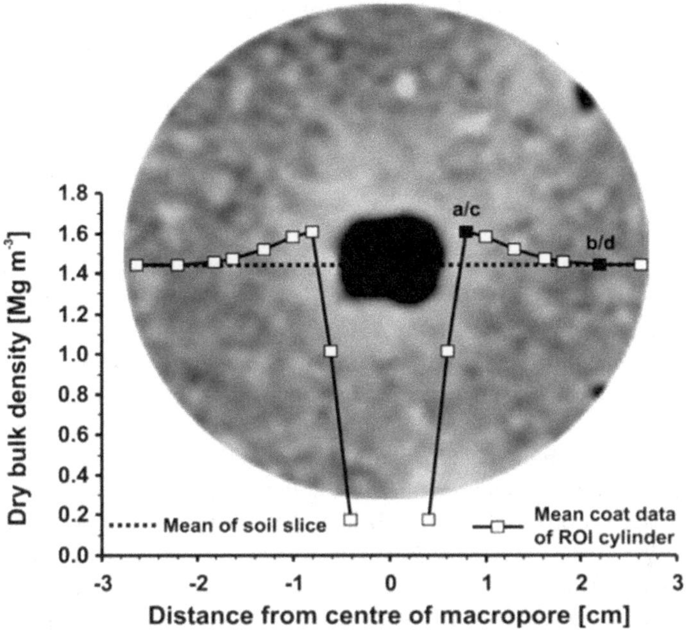

Fig. 10.2 Spatial patterns in soil density in vicinity of a macropore generated by earthworms (from Schrader et al., 2007)

variability in soil hydrological properties (Capowiez et al., 2015; Bottinelli et al., 2017).

A distinct benefit of CT is the capabilities it provides for in-detail exploration of how fine-scale soil structure characteristics, e.g., pore size distributions, spatial patterns in pore locations, and in the soil density, etc., are affected by earthworm burrows. For example, Schrader et al. (2007) used CT to identify spatial patterns in soil density in vicinity of the earthworm burrow, i.e. the drilosphere. They demonstrated that the compacted area surrounding a ~ 5 mm Ø burrow measured ~5–6 mm in width and quantified density distribution in vicinity of the burrow (Fig. 10.2).

Another unique aspect in estimating the contribution of earthworm activities to soil functioning that can be obtained by using CT is the assessment of the volumes of the earthworm created macropores that are connected to the soil surface and thus can serve as preferential flow paths (Bastardie et al., 2005). Such macropores can comprise a surprisingly small portion of overall burrows present within the soil matrix (Bastardie et al., 2005), yet they can play a major role in soil hydrology and in distribution of dissolved organics and nutrients to deep soil layers.

Earthworm casts are an important component of soil structure (Meurer et al., 2020). Endogenic earthworm species ingest soil and then egest it as casts. Under favourable conditions a single earthworm can ingest in a day a volume of soil that can be 4–6 times greater than its own weight (Curry & Schmidt, 2007). The casts thus can constitute a major element of soil structure, an element that has physical and chemical characteristics substantially different from those of the bulk soil or from soil aggregates originated due to physical and biological processes unrelated to earthworms. X-ray CT has been shown to be a valuable tool to study physical characteristics of soil aggregates of earthworm origin in detail at just few micron resolutions. Le Bayon et al. (2020) characterized physical properties of soil aggregates originating from casts of three earthworm species. The use of CT enabled the authors not only to distinguish cast-based aggregates from the aggregates in a control soil, but also to detect differences in cast-aggregate characteristics among the earthworm species.

While significant progress has been made in conducting the tasks of segmenting, identifying, and characterizing soil pores using CT images (see Chap. 6), soil pore analyses relevant to studies of soil fauna remain a challenge. Even obtaining a clear separation of pores of biological origin from the non-biological pores is not a trivial task (Leue et al., 2019), while distinguishing between biological pores originating from root vs. earthworm origin is a substantial challenge (Pelosi et al., 2017). This separation is performed based on size (Capowiez et al., 2011; Pagenkemper et al., 2015) and shape (Zhang et al., 2015, 2018; Leue et al., 2019) characteristics.

It also should be noted that while significant knowledge gains have been achieved in a course of many laboratory studies, extrapolation of the findings to the field must be undertaken with caution. A number of issues limit the universality of laboratory experiment results. For example, the size of containers that can be CT scanned with sufficiently fine resolution is usually limited to diameters <10–50 cm. This limits the size of the colonies that could be studied, altering the fauna's behavioural patterns and generating boundary effects (Booth et al., 2020). Potential negative effects of radiation on the animals also need to be considered.

10.3 Soil Microorganisms and X-ray CT

One of the key advantages of using CT in the studies focusing on soil microbiology is an opportunity to obtain information on physical structure of the micro-habitats where microbes reside (Nunan et al., 2006). Availability of oxygen, water, and nutrients, accessibility to predators are just a few of the environmental parameters that can vary within the soil matrix at spatial scales of a few microns to a few millimetres creating a highly variable mosaic of suitability for microbial functioning (Keiluweit et al., 2017). The flow and transport processes driving the micro-environmental conditions and their variations take place within soil pores. Adhered thin biofilms on pore surfaces are where the active microorganisms live (bar those inside other biological materials) or travel through (Or et al., 2007). The great benefit of CT is that visualization of soil pores is what it does best, with scanning resolutions and methods of pore identification and pore analyses continuously and rapidly improving (see Chap. 6).

However, while there is clear evidence that spatial patterns of microorganisms are related to pore properties and locations (Nunan et al., 2003), the experimental works that employ CT in the studies of soil microbial activities and microbially driven processes are far from being abundant. Overall, the microscale patterns in microbial processes within intact soils are much less explored as compared to the patterns in physical and even biochemical processes (Baveye et al., 2018).

That is not because of the lack of appreciation of the role of physical microhabitats for microbial functioning. In fact, there are a great number of studies that attempt modelling microbial spread and activity as a function of micro-environmental conditions (see a comprehensive list of recent works compiled by Baveye et al. (2018)). Such modelling efforts often utilize the X-ray CT information from scanned intact soils. For example, Kravchenko et al. (2011) modelled fungal growth taking place within soils under long-term contrasting management by visualizing pores from CT images of intact cores. There have been innovative attempts of recreating artificial soil matrixes, via, for example, 3-D printing of CT soil images, with realistic pore architectures reflecting soils of different types and under different land use and management practices (Otten et al., 2012; Lamande et al., 2020). This approach can potentially enable a clear separation of the effects of physical characteristics from those of chemical and biological soil properties. In future, it will offer exciting opportunities for experimental assessments of the roles of pore characteristics on microbial functioning and related soil processes.

Yet, at the moment, the experimental work aimed at demonstrating associations between physical micro-environmental conditions, as visualized via CT, and microbial functioning remains extremely challenging. The greatest issue for CT-based studies is the mismatch between the achievable scanning resolution and the size of microorganisms: bacteria and archaea are too small to be visible on most X-ray CT systems. The CT scanning resolution in most soil studies ranges from 5 to 50 μm. Only recently 1–2 μm resolution-based studies became possible. However, they are plagued by the CTs inverse relationship between the resolution and the sample size, where fine resolution scanning is possible only in relatively small soil samples, limiting representativeness and spatial scale of the findings. The second issue currently limiting the effectiveness of CT use is that most microorganisms consist of at least 80% water, which makes their attenuation coefficient insufficiently distinct from that of water, making them hard to distinguish from the water within the soil matrix.

Biofilms, microbial colonies embedded in self-protecting matrix of extracellular polymeric substances, are a special case. Biofilms, even though thin, have sufficiently large areal extent to be detected on CT images. However, with water being the main component of the biofilms there remains a strong limitation, making it very difficult to distinguish a biofilm from surrounding soil pore water in CT images. The solution implemented to solve this problem usually consists of applying contrast or staining agents, e.g., $BaSO_4$ (Davit et al., 2011; Iltis et al., 2011; Peszynska et al., 2016; Carrel et al., 2017). Yet, even with that enhancement, most of the successful

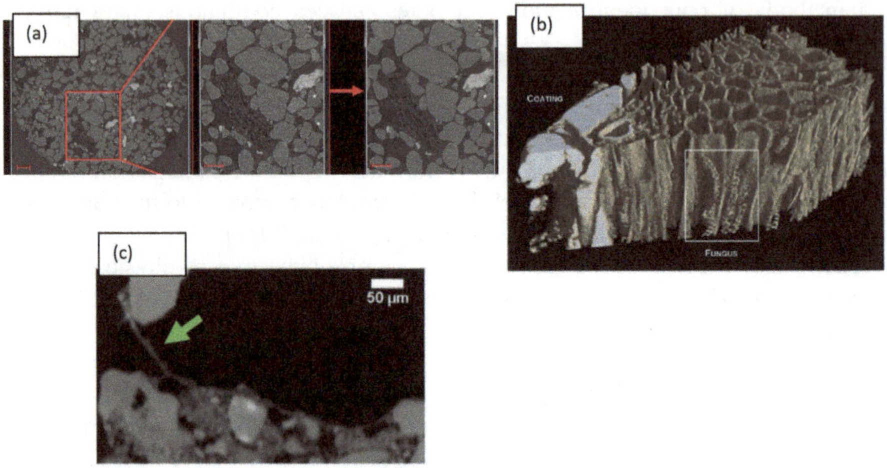

Fig. 10.3 Examples of microorganism visualizations using CT: cyanobacteria bundle in a soil crust (from Couradeau et al., 2018) with red scale bars representing 300 μm on the whole sample image and 150 μm on the zoomed-in cut-outs (**a**), hyphae of *Aureobasidium pullulans* within pine sapwood (from Van den Bulcke et al., 2008) (**b**), a likely fungal hyphae observed in a soil sample previously stained with $AgNO_3$ (from Lammel et al., 2019) (**c**)

biofilm studies up to date were conducted in artificially created porous media, e.g., glass or plastic beads, or washed sand (Davit et al., 2011; Iltis et al., 2011; Peszynska et al., 2016; Carrel et al., 2017), with no work in intact soil reported so far. A recent success story is a study of cyanobacteria in biological soil crusts (Couradeau et al., 2018) (Figure 10.3a), yet it is the cyanobacteria assembled into >100 μm bundles that was possible to visualize and characterize with CT.

Detecting fungi on CT scans has proved to be even more challenging than studying biofilms and at present there are hardly any reports of fungi reliably visualized using CT in the intact undisturbed soil. The main issue is that the scanning resolution to visualize fungal hyphae must be quite small: from 1–3 μm to, preferably, sub-micron range. As has been demonstrated before in studies of CT-based identification and characterization of soil pores, a reliable visualization of an object, say a pore, requires that the scanning resolution be at least 2–5 times smaller than the diameter of the object (Vogel et al., 2010; Koestel et al., 2018). While such assessments have not been yet conducted for fungi visualization tasks, one can expect that similar image analyses and segmentation procedures will be involved in studying fungal hyphae as those used for soil pores. Thus, for a reliable visualization of a 10 μm diameter fungal hyphae, a resolution no bigger than 3 μm would be needed. And, as mentioned earlier, at present, such fine resolution scanning is only possible in small (<10 mm) soil samples, a size that is not big enough to have meaningful observations of fungal spread and interactions with the soil.

Successful fungi visualization via CT has been obtained in decomposing wood samples (Van den Bulcke et al., 2008, 2009), where high image resolution (0.7 μm) enabled visualization of fungi mycelia within wood cells (Fig. 10.3b).

Lilje et al. (2013) visualized fungi grown in mixtures of polystyrene beads, which were built to resemble soil and were X-ray translucent. Prior to scanning the fungi were fixed and stained to enhance visualization. Lammel et al. (2019) were successful in identifying hyphae on CT images of the soil samples after different dopants were applied to stain the organics and enhance its visualization. Another item to consider for CT studies that involve living fungi is that they can be more sensitive to radiation damage than other soil microorganisms (Jackson et al., 1967). Even though one formal assessment demonstrated no negative effects of radiation on fungal PLFA (Bouckaert et al., 2013), there are concerns that radiation can affect growth of some fungal species (Van den Bulcke et al., 2009).

A practical, even though low spatial resolution, approach to relate microbial data with CT information has been proposed by Kravchenko et al. (2013) and Wang et al. (2013). They added *Escherichia coli* to CT scanned macro-aggregates, which were then cut into small subsections and *E. coli*'s quantities in each subsection were determined using qPCR. Combining *E. coli* results with pore characteristics in the individual subsections enabled exploring the role of pores in defining where the *E. coli* might go upon entering soil macro-aggregates and whether/how much of it will leave the aggregates after the soil is subjected to saturated flow. Kravchenko, et al. (2014b) continued with this approach in relating bacterial community composition to pore characteristics. The CT scanned macro-aggregates were cut into subsections, where each subsection was subjected to 16S rRNA analysis. The benefit of this approach is that both physical and biological measurements can be obtained using the best techniques currently available; pore architecture via image analysis of fine resolution CT images and microbial abundance and community composition via standard well-tested biological techniques known to generate reliable results. The drawback is that the volume of the soil needed for the biological analyses must be relatively large. A minimal size of an individual soil subsection used to generate a single biological data point can hardly be less than 0.125 mm^3 ($500 \times 500 \times 500$ μm^3). Given that the size of an individual bacteria is ~1 μm and that distances at which the bacterial colonies are spatially correlated are ~100–500 μm (Nunan et al., 2002, 2003), such a sample, of course, provides only very coarse assessment of microbiological data. Such soil volume also contains multiple micro-environments with potentially very different features, resulting in only crude assessments of the relationships between biological and physical characteristics.

A different experimental approach enables a more direct focus on procuring inhabitants from specific soil micro-environments (i.e. pores of different sizes). The method is based on modifications of soil matric potentials for either extracting soil solution from pores of different sizes or for adding different substrates into pores of different sizes. In the first scenario the extracted solution can be subjected to biological and biochemical analyses (e.g., Bailey et al., 2017), while in the second indirect observations of the microbial activities can be conducted (e.g., Ruamps et al., 2011). While this approach is typically implemented by using pore information from water retention curves (Killham et al., 1993; Wright et al., 1995; Ruamps et al., 2013; Nunan et al., 2017), more detailed exploration of the pore characteristics can be obtained by CT scanning (Kravchenko et al., 2020). Yet, this method also has its own drawbacks, the most important of which is that it is impossible to get the contents of pores of a specific size without avoiding some degree of mixing and cross-contamination from pores of other sizes.

Very promising recent developments in efforts to obtain CT and biological data at comparable scales have used a method of resin impregnation of intact soil followed by analyses of thin sections procured from the soil surface. The method originated from soil micromorphology (Bullock et al., 1985) and also was employed in the microscopic analyses of soil microbial and pore characteristics (Nunan et al., 2001). Recent works have combined microbial data obtained from the surfaces of resin impregnated soil with 3-D CT data (Juyal et al., 2019, 2021). Microbial data in such analyses are obtained via fluorescent microscopy of the polished soil thin sections. The great advantage of this approach over the above-described methods is that the spatial resolutions in the physical and microbial data are very similar and locations of individual microorganisms or microbial colonies can be visualized and enumerated in relation to soil pores. However, the method suffers from two significant shortcomings. One is that preparation of resin impregnated thin sections is a very laborious, time-consuming, and costly process, and the other is that matching of the microscopy and CT images can be very difficult and imprecise. While the latter problem might be resolved by the latest progress in development of image analyses and registration tools (Schlüter et al., 2019a, 2019b), the tasks of resin impregnation and procuring the thin sections for the analyses are unlikely to become easier.

10.4 Indirect Use of CT to Observe the Outcomes of Microbial Activities

Since microorganism sizes are below the currently used resolutions in soil CT scanning, in addressing their roles in soil structure formation the CT is often used to view the outcomes of their activities, that is the changes in soil structure generated by microorganisms under different settings. It can be argued that such studies assess the microorganisms' influences only indirectly. However, the experiments that stimulate disparate microbial activities and where the influence on soil structure can be visualized via CT generate a lot of useful information.

In such studies CT is implemented as a tool to characterize soil structure and pore architecture, while specialized microbiological methods are used for microorganism characterization. Feeney et al. (2006) followed the CT analyses of the pore characteristics affected by plant roots and fungi by destructive fungal biomass measurements using ergosterol-based approach. The authors observed that soil porosity increased even in the bulk soil, which was not directly affected by plant roots—as long as the soil was accessible to fungi during plant growth. Studies that aim at comparisons among management practices often choose to work with sets of sub-samples where some are subjected to CT for pore characterization, while the other are used for chemical and biological measurements. For example, Rabbi et al.

(2016) used selected soil macro- and micro-aggregates collected from sites with contrasting land use, such as crop, pasture, woodlot, to characterize pore geometry via CT, while the other set of aggregates collected from the same sites was used to characterize diversity of microbial communities and organic decomposition rates. Harvey et al. (2020) used CT to examine pore characteristics of one subset of manufactured aggregates, while the other subset was populated with yeasts placed in different positions within the aggregates to explore protection of microorganisms within aggregate interiors. Juyal et al. (2018) built soil materials with contrasting bulk density levels and employed CT to procure in-detail quantification of the pore architectures of the obtained materials. Then inoculums of *Bacillus* sp. and *Pseudomonas* sp. were applied on one side of the sample and the abundance of the bacteria that reached the other side of the sample was assessed, enabling the authors to explore relationships between soil pore characteristics and bacteria movement through the soil.

Such studies have been instrumental in demonstrating the relationships between organic inputs, such as plant residues and roots, and soil pore formation driven by microbial decomposition. For example, presence of residues within soil aggregates stimulates pore formation; and pores follow a spatial gradient with more new pores forming at shorter distances to the residue (De Gryze et al., 2006). The nature of the added organic inputs, that is whether they are added as particulate organic carbon (C) or as dissolved organic C, results in contrasting microbial activities, which then lead to formation of contrasting pore architectures (Bucka et al., 2019).

Production and emission of N_2O, a potent greenhouse gas, is a function of microbial activity, including nitrification and denitrification, and can be strongly affected by soil pore characteristics. The activities of microorganisms involved in N_2O production and subsequent processing, for example, full denitrification of N_2O to N_2, are driven by the existence of micro-environmental conditions which are conducive to their growth. The presence of pores of tens-micron Ø range, their connectivity, and spatial distribution patterns affect travel times for gases and solutions from/to pores vs. surrounding soil matrix and water fill status of the pores. These characteristics are just a few of many key components that define hotspots of microbial activities in soil which can be studied using CT (Schlüter et al., 2019a, 2019b). X-ray CT has been demonstrated to be a very efficient tool for exploring pore effects in N_2O emission studies, as it enables relating properties of soil pore architecture to the emissions (Mangalassery et al., 2013, 2014; Rabot et al., 2015; Porre et al., 2016; Kravchenko et al., 2018b).

Despite the many complications and challenges facing the attempts of using CT in microbial studies, such attempts are most valuable, bringing new light into micro-environmental effects on soil microbial functioning. For example, the presence of large pores in macro-aggregates and their greater connectivity were found to be associated with greater C mineralization, but, surprisingly, with lower microbial diversity (Rabbi et al., 2016). It has been often suggested that micro-environments with low connectivity and smaller pores constitute a collection of non-connected niches, conducive to formation of high microbial diversity (e.g. Nunan et al., 2017). Studies combining CT with microbial analyses support this notion demonstrating that places with high pore connectivity and high microbial productivity are the ones where diversity might be reduced (Rabbi et al., 2016) and where copiotroph organisms might dominate (Kravchenko et al., 2014b). Lower bulk density, higher pore connectivity, and greater surfaces of the interfaces between solids and pores are the factors that facilitate bacterial transport through the soil, via pores and likely water films along pore boundaries (Juyal et al., 2021).

10.5 Effect of CT Scanning on Soil Organisms

When using CT in the studies aiming at exploring macro- and micro-organism activities within the scanned soil samples, the important question is whether X-rays have a negative effect on the organisms' wellbeing and functioning. Obviously, the presence of such effects would be a serious argument against implementation of CT in soil studies focusing on those soil functions that involve biological activities. After all, gamma-irradiation is a widely used tool for effective soil sterilization (McNamara et al., 2003). The doses used in gamma-irradiation soil sterilization studies are 10–20 kGy, where 10 kGy will eliminate actinomycetes, fungi and invertebrates and 20 kGy will be sufficient to eliminate most bacteria (McNamara et al., 2003). However, the radiation levels used currently in X-ray studies of plants and soils typically are <100 Gy (Zappala et al. (2013), that is, at least two orders of magnitude lower than those that are critically harmful for soil biota. Yet, even low doses might have an effect, which is important to be aware of and account for in experimental work and result interpretation. Here the studies that have looked at X-ray CT effects on specific macro- and micro- soil organisms are summarized in terms of indicators of soil biological activity, such as microbial biomass and soil respiration (Table 10.1).

The consensus from several independent works is that radiation doses delivered during CT soil scanning sessions have either minor or non-detectable effects on soil microbial biomass and microbial activity, as expressed via C mineralization. Effects on abundance of some of bacteria phyla or fungal species can be expected, but typically they do not translate into differences in the indicators of bacterial community structure and diversity indexes (see Fischer et al., 2013). Activities of extracellular enzymes is what appears to have experienced minor but consistent effects of CT scanning (Ganther et al., 2020). Yet the good news is that most of the scanning effects are temporal and disappear within days after scanning.

It also should be noted that the majority of the studies that explored the effect of CT scanning on microbial activity were conducted via benchtop scanners with very low radiation dozes (1–23 Gy). A study conducted at a synchrotron facility at an inherently higher radiation (1000 Gy) also did not report significant effects on bacterial community compositions (Kravchenko et al., 2014a, 2014b). But it should be noted that in that study the scanning was undertaken on air-dry samples, while the effect of radiation on soil organisms has been shown to be lower in dry than in moist soil. Yet, another synchrotron study that involved repeated scanning of moist soil

Table 10.1 Summary of the studies that explored the effect of CT scanning on soil microbial activity

Organism/indicator	Doze/ Duration	Scanner	Effect	Source
Microbial biomass	23 Gy over 6 repeated scans	Benchtop A Phoenix Nanotom X-ray CT scanner (GE sensing and Inspection technologies, GmbH, Wunstorf, Germany)	Slightly numerically higher microbial biomass in unscanned samples, no statistically significant differences.	Zappala et al. (2013)
C mineralization, PLFA analysis, enzyme activities	Single scan	Benchtop	No effect on C mineralization and PLFA, a slight temporal decrease in dehydrogenase activity	Bouckaert et al. (2013)
Bacterial community structure (16C rRNA), extracellular enzyme activities, microbial biomass	Single scan	Benchtop (phoenix vltomel x L 240, GE-Sensing & Inspection Technologies GmbH, Wunstorf, Germany)	Differences in microbial community structures, decrease in enzyme activities, e.g., beta-glucosidase, increase in microbial biomass a week after scanning.	Fischer et al. (2013)
Bacterial community structure (16S rRNA)	~1000 Gy single scan	Synchrotron, advanced photon source, Argonne	No effect on bray-Curtis dissimilarity clustering of bacterial communities. However, statistically significant differences between scanned and unscanned samples in selected bacterial groups.	Kravchenko et al. (2014b)
C mineralization, extracellular enzyme activity, microbial biomass, bacterial community structure (16S rRNA), archaea abundance	2.4 Gy	Benchtop ((HMX ST 225, Metris X-Tek, UK)	No effect on any of the studied microbial characteristics	Schmidt et al. (2015)
Microbial growth in response to glucose addition, microbial community composition, extracellular enzyme activities	0.81 Gy	Benchtop (X-TEK XTH 225, Nikon metrology)	No effect on microbial growth and bacterial community diversity indexes and community composition; decreased activity of leucine aminopeptidase and phosphomonoesterase, but not beta-glucosidase	Ganther et al. (2020)

samples did report that scanning somewhat affected cyanobacteria functioning, namely, it prevented a normal behaviour of migration to the surface after dry samples were rewetted (Couradeau et al., 2018).

10.6 CT and Soil Organic Matter Visualization

The organic materials within the soil are commonly classified into two broad types: particulate organic matter (POM) and non-particulate organic matter, where the latter will be referred to here as just soil organic matter (SOM). Visualization of these two types of material within intact soil samples via CT requires different approaches and strategies. Particulate organic matter consists of discernible fragments of plant and animal residues ranging in size from 0.053 to 2 mm (Cambardella & Elliott, 1992). Due to its size and attenuation properties, POM visualization on CT images is relatively straightforward, consisting of the steps similar to those needed to segment plant roots (e.g. Mooney et al., 2012; see Chap. 9). However, it should be noted that, unlike live/fresh plant roots, POM fragments imbedded into the soil matrix come from a wide range of origins and are at variable stages of decomposition. Thus, there is often a significant variability in the greyscale values of POM fragments due to differences in densities, chemical composition, potential incorporation of mineral grains, etc. For example, relatively fresh plant or animal residues can have darker grey values than those that are at a more advanced stage of decomposition, while biochar might have lighter greyscale values due to its higher density. The fragments of POM also lack connectivity and defined structure of plant root images, still further complicating their identification.

Typically, identification of POM is based on determining the range of greyscale values representative of POM in the studied images (e.g., De Gryze et al., 2006; Sleutel et al., 2008; Negassa et al., 2015). The greyscale images of POM are then used for thresholding, followed by a series of artefact removal steps, which are often needed to exclude image features that happen to have similar greyscale values as POM. One of such commonly encountered features are pore/solid boundaries subjected to partial volume effects; an artefact caused by some voxels being partially occupied by pores and partially by soil's solid material. A number of strategies, including filtering, erosion, denoising, separation by size, etc. can be implemented to remove such artefacts (Kravchenko, et al., 2014a; Piccoli et al., 2019). The usage of contrast/staining agents, chemicals that enhance the contrast between POM fragments and the surrounding soil on the images, can greatly improve POM determination (Peth et al., 2014; Lammel et al., 2019; Piccoli et al., 2019).

Visualization of SOM presents a much greater challenge than POM visualization. Due to the small size of non-particular SOM, it cannot be directly seen in CT images. Even though areas within an intact soil matrix with higher SOM often tend to have darker greyscale values due to their lower density and dominance of H and C atoms, using this information for quantitative assessments of SOM can be possible only in very specific circumstances. Quigley et al. (2018) demonstrated how this approach can generate valuable information to explore gradients of organic matter next to decomposing plant roots. But the use of this approach was less successful in other areas within the intact soil cores. Thus, using staining agents that react with SOM or its selected components, changing the attenuation characteristics is the only option for visualization of SOM in CT images. Staining with OsO_4 is a potentially useful approach that demonstrated substantial promise in applications with soil aggregates and small intact cores (Peth et al., 2014; Rawlins et al., 2016; Quigley et al., 2018; Arai et al., 2019; Zheng et al., 2020) (Fig. 10.4). Yet, the issue that plagues those SOM detection procedures that rely on the use of contrast agents is the difficulty in ensuring consistent accessibility of all SOM within the sample to the added agent. This problem can be less severe and thus less noticeable in well-mixed artificially prepared soil mixtures, which are commonly used in the studies that compare performances of different staining techniques (e.g., Van Loo et al., 2014; Lammel et al., 2019). How well can the staining agent reach the organics within intact soil samples is an important question; and the bigger the sample or the more heterogeneous it is in terms of its pore characteristics and solid matrix properties, the more questionable the uniformity of the staining becomes. Intact, heterogeneous soil samples are typically the main focus in the studies that use CT tools. Staining chemicals can be delivered to the soil in either liquid or gaseous form (Peth et al., 2014; Van Loo et al., 2014; Lammel et al., 2019), and staining that relies on gaseous diffusion of the contrast agent might lead to more uniform results. However, in the study that used OsO_4 vapours, the staining efficiency still decreased at distances >5 mm from the gas entry surface even after >7 days of exposure to OsO_4 vapours (Zheng et al., 2020).

A potential difficulty with implementing staining techniques that rely on adding the staining agent in the liquid form are a likely disturbance of the original soil structure both during initial liquid addition and during subsequent efforts of washing the non-reacted chemicals out of the soil (Van Loo et al., 2014; Arai et al., 2019; Lammel et al., 2019). The main reason for the use of staining agents in CT studies is explorations of the relationships between SOM and soil pore architecture. Thus, an intensive routine of saturation, drainage, and washing might not be suitable for many soils, especially those prone to swelling and shrinking. It can potentially significantly affect the initial soil pore structure, thus the initial positions of organic materials with respect to other elements of soil matrix, altering SOM-pore associations.

The next step after SOM visualization in CT images is obtaining numerical measures to characterize its quantities and spatial patterns, and at present the methods to undertake this are in a very early stage of development. Any fresh biological input into the soil consists of an extremely complex mixture of different organic compounds, where each of them can react differently with the added staining agents. Microbial decomposition of the fresh inputs greatly magnifies the complexity of the resultant products; and even within the same soil, the long-term differences in land use and management can lead to differences in SOM's chemical composition (Sleutel et al., 2007). Thus, it is unlikely that analytical quantifications based on considering specific chemical reactions between organic compounds and staining chemicals will ever become available in this context. Chemical reactions remain

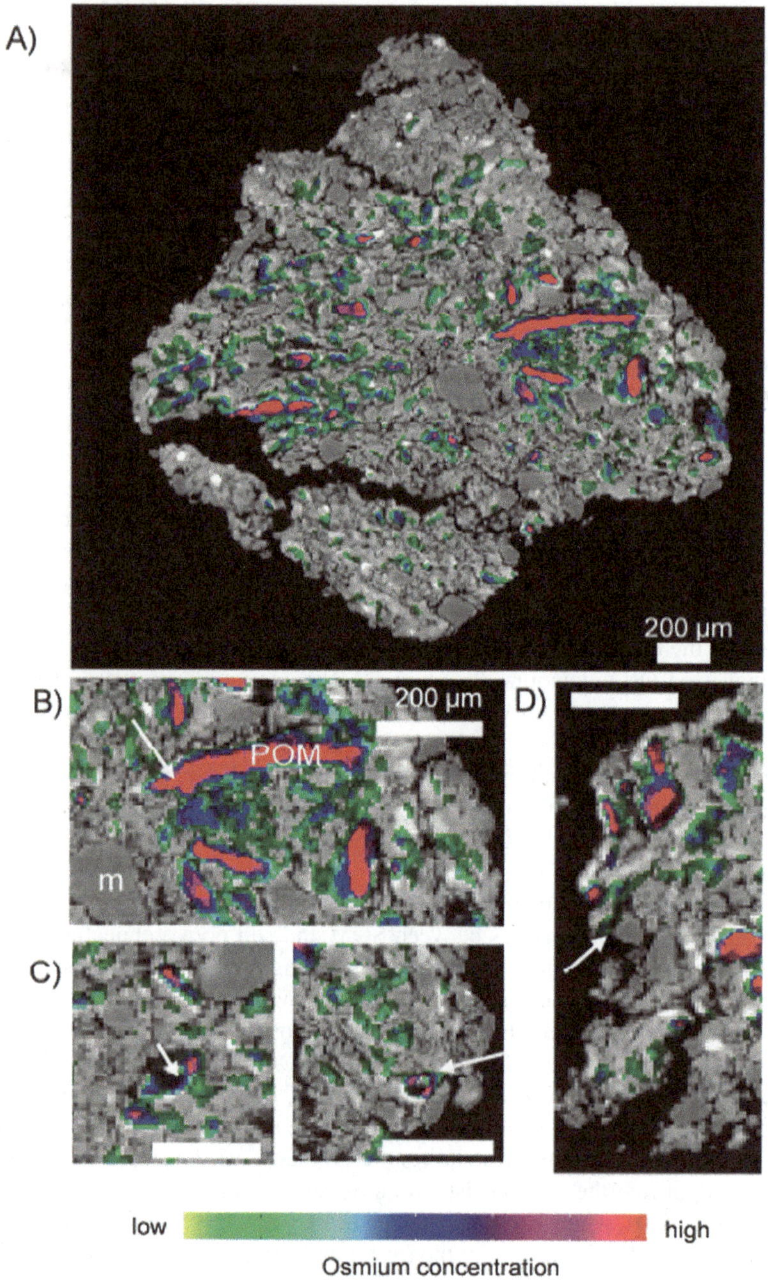

Fig. 10.4 Example of visualizing particulate (POM) and non-particulate (SOM) in an intact soil aggregate (**a**) and in its selected subsections (**b-d**) using staining with OsO_4 vapours and dual-energy CT scanning (from Peth et al., 2014)

poorly understood even for the staining of fresh biological tissues with OsO_4 vapours which is the technique that has been actively used by biologists for >70 years and continues to be a subject of active research (Bahr, 1954; Griffiths, 1974; Belazi et al., 2009). It is not clear what components of SOM and POM are reacting with staining agents, to what extent, and by which of several possible mechanisms (Lammel et al., 2019). Zheng et al. (2020) demonstrated that while plant root residues and soil with high organic matter content are well stained by OsO_4 vapours, biochar was not, and soils can vary widely in their biochar contents. Quantifications built on empirically derived relationships between measurements of soil C and the results of staining-assisted CT scanning appear to be the only possible route. Such efforts have shown initial promise in the studies with OsO_4 (Zheng et al., 2020) and I_2 vapours (Lammel et al., 2019).

An additional issue with using liquid contrast agents for quantitative purposes is that there might be not only chemical, but also hydrological drivers of their spatial distribution patterns. For example, fragments of plant residues within the soil matrix to a certain extent act as sponges that absorb moisture from the surrounding soil in quantities that depend both on their characteristics and on pores in the surrounding soil (Kravchenko et al., 2017, 2018a). Thus, when the contrast agent is added to a dry soil in a liquid form, the resultant high presence of the agent within a POM fragment can stem more from the liquid absorption by the POM, and less so from the organic matter reaction with the agent. On the other hand, applying the liquid contrast agent to a wet soil can lead to mixing of the added chemical with soil water decreasing quality of POM identification (Piccoli et al., 2019). From that point of view, gaseous staining agents might be more promising for quantification purposes.

10.7 Conclusions

Even though the primary focus of most X-ray CT applications in soil science concerns the physical structure of intact soil, given that that structure is both a habitat and a product of activities of soil biota, the benefits of CT use in soil biology are numerous and overarching. For soil meso-fauna, including earthworms and burrowing insects, CT offers capabilities of viewing the outcomes of their activities, the temporal dynamic in development of resultant soil architecture, and the locations and movements of the organisms themselves. For microorganisms, CT enables quantification of their influences on soil structure formation in space and time; as well as on what the resultant soil structure adds to a variety of soil processes, including soil C protection and sequestration, and greenhouse gas emissions. While the importance of physical structure for soil fauna has been always recognized, the traditional destructive methods of its classification (e.g. dry- and wet-aggregate sieving) precluded direct observations of soil-dwellers in their undisturbed environments. Despite its many technical limitations and difficulties in implementation, CT offers such capabilities. Its wider utilization in soil biology will lead to a greater appreciation of the soil as a unique habitat and a better understanding of the ecological principles that guide its functioning.

References

Arai, M., Uramoto, G. I., Asano, M., Uematsu, K., Uesugi, K., Takeuchi, A., Morono, Y., & Wagai, R. (2019). An improved method to identify osmium-stained organic matter within soil aggregate structure by electron microscopy and synchrotron X-ray micro-computed tomography. *Soil & Tillage Research, 191*, 275–281.

Auclerc, A., Capowiez, Y., Guerold, F., & Nahmani, J. (2013). Application of X-ray tomography to evaluate liming impact on earthworm burrowing activity in an acidic forest soil under laboratory conditions. *Geoderma, 202*, 45–50.

Bahr, G. F. (1954). Osmium tetroxide and ruthenium tetroxide and their reactions with biologically important substances-electron stains iii. *Experimental Cell Research, 7*, 457–479.

Bailey, V. L., Smith, A. P., Tfaily, M., Fansler, S. J., & Bond-Lamberty, B. (2017). Differences in soluble organic carbon chemistry in pore waters sampled from different pore size domains. *Soil Biology & Biochemistry, 107*, 133–143.

Bastardie, F., Capowiez, Y., de Dreuzy, J. R., & Cluzeau, D. (2003). X-ray tomographic and hydraulic characterization of burrowing by three earthworm species in repacked soil cores. *Applied Soil Ecology, 24*, 3–16.

Bastardie, F., Ruy, S., & Cluzeau, D. (2005). Assessment of earthworm contribution to soil hydrology: A laboratory method to measure water diffusion through burrow walls. *Biology and Fertility of Soils, 41*, 124–128.

Baveye, P. C., Otten, W., Kravchenko, A., Balseiro-Romero, M., Beckers, E., Chalhoub, M., Darnault, C., Eickhorst, T., Garnier, P., Hapca, S., Kiranyaz, S., Monga, O., Mueller, C. W., Nunan, N., Pot, V., Schluter, S., Schmidt, H., & Vogel, H. J. (2018). Emergent properties of microbial activity in heterogeneous soil microenvironments: Different research approaches are slowly converging. *Yet Major Challenges Remain. Frontiers in Microbiology, 9*.

Belazi, D., Sole-Domenech, S., Johansson, B., Schalling, M., & Sjovall, P. (2009). Chemical analysis of osmium tetroxide staining in adipose tissue using imaging ToF-SIMS. *Histochemistry and Cell Biology, 132*, 105–115.

Booth, S., Kurtz, B., de Heer, M. I., Mooney, S. J., & Sturrock, C. J. (2020). Tracking wireworm burrowing behaviour in soil over time using3-D X-ray computed tomography. *Pest Management Science, 76*, 2653–2662.

Bottinelli, N., Zhou, H., Capowiez, Y., Zhang, Z. B., Qiu, J., Jouquet, P., & Peng, X. H. (2017). Earthworm burrowing activity of two non-Lumbricidae earthworm species incubated in soils with contrasting organic carbon content (vertisol vs. Ultisol). *Biology and Fertility of Soils, 53*, 951–955.

Bouckaert, L., Van Loo, D., Ameloot, N., Buchan, D., Van Hoorebeke, L., & Sleutel, S. (2013). Compatibility of X-ray micro-computed tomography with soil biological experiments. *Soil Biology & Biochemistry, 56*, 10–12.

Bucka, F. B., Kolbl, A., Uteau, D., Peth, S., & Kogel-Knabne, I. (2019). Organic matter input determines structure development and aggregate formation in artificial soils. *Geoderma, 354*.

Bullock, P., Fedoroff, N., Jongerius, A., Stoops, G., & Tursina, T. (1985). *Handbook for soil thin section description*. Waine Research Publications.

Cambardella, C. A., & Elliott, E. T. (1992). Particulate soil organic-matter changes across a grassland cultivation sequence. *Soil Science Society of America Journal, 56*, 777–783.

Capowiez, Y., & Belzunces, L. (2001). Dynamic study of the burrowing behaviour of Aporrectodea nocturna and Allolobophora chlorotica: Interactions between earthworms and spatial avoidance of burrows. *Biology and Fertility of Soils, 33*, 310–316.

Capowiez, Y., Bottinelli, N., & Jouquet, P. (2014a). Quantitative estimates of burrow construction and destruction, by anecic and endogeic earthworms in repacked soil cores. *Applied Soil Ecology, 74*, 46–50.

Capowiez, Y., Bottinelli, N., Sammartino, S., Michel, E., & Jouquet, P. (2015). Morphological and functional characterisation of the burrow systems of six earthworm species (Lumbricidae). *Biology and Fertility of Soils, 51*, 869–877.

Capowiez, Y., Monestiez, P., & Belzunces, L. (2001). Burrow systems made by Aporrectodea nocturna and Allolobophora chlorotica in artificial cores: Morphological differences and effects of interspecific interactions. *Applied Soil Ecology, 16*, 109–120.

Capowiez, Y., Pierret, A., & Moran, C. J. (2003). Characterisation of the three-dimensional structure of earthworm burrow systems using image analysis and mathematical morphology. *Biology and Fertility of Soils, 38*, 301–310.

Capowiez, Y., Sammartino, S., & Michel, E. (2011). Using X-ray tomography to quantify earthworm bioturbation non-destructively in repacked soil cores. *Geoderma, 162*, 124–131.

Capowiez, Y., Sammartino, S., & Michel, E. (2014b). Burrow systems of endogeic earthworms: Effects of earthworm abundance and consequences for soil water infiltration. *Pedobiologia, 57*, 303–309.

Carrel, M., Beltran, M. A., Morales, V. L., Derlon, N., Morgenroth, E., Kaufmann, R., & Holzner, M. (2017). Biofilm imaging in porous media by laboratory X-ray tomography: Combining a non-destructive contrast agent with propagation-based phase-contrast imaging tools. *PLoS One, 12*.

Couradeau, E., Felde, V., Parkinson, D., Uteau, D., Rochet, A., Cuellar, C., Winegar, G., Peth, S., Northen, T. R., & Garcia-Pichel, F. (2018). In situ X-ray tomography imaging of soil water and cyanobacteria from biological soil crusts undergoing desiccation. *Frontiers in Environmental Science, 6*.

Curry, J. P., & Schmidt, O. (2007). The feeding ecology of earthworms–a review. *Pedobiologia, 50*, 463–477.

Darwin, C. R. (1881). *The formation of vegetable mould through the action of worms, with observations on their habits*. Murray.

Davit, Y., Iltis, G., Debenest, G., Veran-Tissoires, S., Wildenschild, D., Gerino, M., & Quintard, M. (2011). Imaging biofilm in porous media using X-ray computed microtomography. *Journal of Microscopy, 242*, 15–25.

De Gryze, S., Jassogne, L., Six, J., Bossuyt, H., Wevers, M., & Merck, X. R. (2006). Pore structure changes during decomposition of fresh residue: X-ray tomography analyses. *Geoderma, 134*, 82–96.

Feeney, D. S., Crawford, J. W., Daniell, T., Hallett, P. D., Nunan, N., Ritz, K., Rivers, M., & Young, I. M. (2006). Three-dimensional microorganization of the soil-root-microbe system. *Microbial Ecology, 52*, 151–158.

Fischer, D., Pagenkemper, S., Nellesen, J., Peth, S., Horn, R., & Schloter, M. (2013). Influence of non-invasive X-ray computed tomography (XRCT) on the microbial community structure and function in soil. *Journal of Microbiological Methods, 93*, 121–123.

Ganther, M., Yim, B., Ibrahim, Z., Bienert, M. D., Lippold, E., Maccario, L., Sorensen, S. J., Bienert, G. P., Vetterlein, D., Heintz-Buschart, A., Blagodatskaya, E., Smalla, K., & Tarkka, M. T. (2020). Compatibility of X-ray computed tomography with plant gene expression, rhizosphere bacterial communities and enzyme activities. *Journal of Experimental Botany, 71*, 5603–5614.

Garner, M. R. (1953). The preparation of latex casts of soil cavities for the study of tunneling activities of animals. *Science, 118*, 380–381.

Griffiths, W. (1974). Osmium tetraoxide and its applications. *Platinum Metals Review, 18*, 94–96.

Harvey, H. J., Wildman, R. D., Mooney, S. J., & Avery, S. V. (2020). Soil aggregates by design: Manufactured aggregates with defined microbial composition for interrogating microbial activities in soil microhabitats. *Soil Biology & Biochemistry, 148*.

Iltis, G. C., Armstrong, R. T., Jansik, D. P., Wood, B. D., & Wildenschild, D. (2011). Imaging biofilm architecture within porous media using synchrotron-based X-ray computed microtomography. *Water Resources Research, 47*.

Jackson, N. E., Corey, J. C., Frederick, L. R., & Picken, J. C. (1967). Gamma irradiation and the microbial population of soils at two water contents. *Soil Science Society of America Journal, 31*, 491–494.

Jegou, D., Brunotte, J., Rogasik, H., Capowiez, Y., Diestel, H., Schrader, S., & Cluzeau, D. (2002). Impact of soil compaction on earthworm burrow systems using X-ray computed tomography: Preliminary study. *European Journal of Soil Biology, 38*, 329–336.

Jegou, D., Hallaire, V., Cluzeau, D., & Trehen, P. (1999). Characterization of the burrow system of the earthworms Lumbricus terrestris and Aporrectodea giardi using X-ray computed tomography and image analysis. *Biology and Fertility of Soils, 29*, 314–318.

Joschko, M., Graff, O., Muller, P. C., Kotzke, K., Lindner, P., Pretschner, D. P., & Larink, O. (1991). A nondestructive method for the morphological assessment of earthworm burrow systems in 3 dimensions by X-ray computed-tomography. *Biology and Fertility of Soils, 11*, 88–92.

Joschko, M., Muller, P. C., Kotzke, K., Dohring, W., & Larink, O. (1993). Earthworm burrow system-development assessed by means of X-ray computed-tomography. *Geoderma, 56*, 209–221.

Juyal, A., Eickhorst, T., Falconer, R., Baveye, P. C., Spiers, A., & Otten, W. (2018). Control of pore geometry in soil microcosms and its effect on the growth and spread of pseudomonas and bacillus sp. *Frontiers in Environmental Science, 6*.

Juyal, A., Otten, W., Baveye, P. C., & Eickhorst, T. (2021). Influence of soil structure on the spread of Pseudomonas fluorescens in soil at microscale. *European Journal of Soil Science, 72*, 141–153.

Juyal, A., Otten, W., Falconer, R., Hapca, S., Schmidt, H., Baveye, P. C., & Eickhorst, T. (2019). Combination of techniques to quantify the distribution of bacteria in their soil microhabitats at different spatial scales. *Geoderma, 334*, 165–174.

Kautz, T., Amelung, W., Ewert, F., Gaiser, T., Horn, R., Jahn, R., Javaux, M., Kemna, A., Kuzyakov, Y., Munch, J. C., Patzold, S., Peth, S., Scherer, H. W., Schloter, M., Schneider, H., Vanderborght, J., Vetterlein, D., Walter, A., Wiesenberg, G. L. B., & Kopke, U. (2013). Nutrient acquisition from arable subsoils in temperate climates: A review. *Soil Biology & Biochemistry, 57*, 1003–1022.

Keiluweit, M., Wanzek, T., Kleber, M., Nico, P., & Fendorf, S. (2017). Anaerobic microsites have an unaccounted role in soil carbon stabilization. *Nature Communications, 8*.

Killham, K., Amato, M., & Ladd, J. N. (1993). Effect of substrate location in soil and soil pore-water regime on carbon turnover. *Soil Biology & Biochemistry, 25*, 57–62.

Koestel, J., Dathe, A., Skaggs, T. H., Klakegg, O., Ahmad, M. A., Babko, M., Giménez, D., Farkas, C., Nemes, A., & Jarvis, N. (2018). Estimating the permeability of naturally structured soil from percolation theory and pore space characteristics imaged by X-ray. *Water Resources Research, 54*, 9255–9263.

Koestel, J., & Schluter, S. (2019). Quantification of the structure evolution in a garden soil over the course of two years. *Geoderma, 338*, 597–609.

Kravchenko, A., Chun, H. C., Mazer, M., Wang, W., Rose, J. B., Smucker, A., & Rivers, M. (2013). Relationships between intra-aggregate pore structures and distributions of Escherichia coli within soil macro-aggregates. *Applied Soil Ecology, 63*, 134–142.

Kravchenko, A., Falconer, R. E., Grinev, D., & Otten, W. (2011). Fungal colonization in soils with different management histories: Modeling growth in three-dimensional pore volumes. *Ecological Applications, 21*, 1202–1210.

Kravchenko, A. N., Fry, J. E., & Guber, A. K. (2018a). Water absorption capacity of soil-incorporated plant leaves can affect N2O emissions and soil inorganic N concentrations. *Soil Biology & Biochemistry, 121*, 113–119.

Kravchenko, A. N., Guber, A. K., Gunina, A., Dippold, M., & Kuzyakov, Y. (2020). Pore-scale view of microbial turnover: Combining 14C imaging, μCT, and zymography after adding soluble carbon to soil pores of specific sizes. *European Journal of Soil Science*. https://doi.org/10.1111/ejss.13001

Kravchenko, A. N., Guber, A. K., Quigley, M. Y., Koestel, J., Gandhi, H., & Ostrom, N. E. (2018b). X-ray computed tomography to predict soil N2O production via bacterial denitrification and N2O emission from soils in contrasting bioenergy cropping systems. *Global Change Biology Bioenergy*, 1–17.

Kravchenko, A. N., Negassa, W., Guber, A. K., & Schmidt, S. (2014a). New approach to measure soil particulate organic matter in intact samples using X-ray computed microtomography. *Soil Science Society of America Journal, 78*, 1177–1185.

Kravchenko, A. N., Negassa, W. C., Guber, A. K., Hildebrandt, B., Marsh, T. L., & Rivers, M. L. (2014b). Intra-aggregate pore structure influences phylogenetic composition of bacterial Community in Macroaggregates. *Soil Science Society of America Journal, 78*, 1924–1939.

Kravchenko, A. N., Toosi, E. R., Guber, A. K., Ostrom, N. E., Yu, J., Azeem, K., Rivers, M. L., & Robertson, G. P. (2017). Hotspots of soil N2O emission enhanced through water absorption by plant residue. *Nature Geoscience, 10*, 496–500.

Lamande, M., Schjonning, P., Dal Ferro, N., & Morari, F. (2020). Soil pore system evaluated from gas measurements andCTimages: A conceptual study using artificial, natural and 3-D-printed soil cores. *European Journal of Soil Science.*

Lammel, D. R., Arlt, T., Manke, I., & Rillig, M. C. (2019). Testing contrast agents to improve micro computerized tomography (mu CT) for spatial location of organic matter and biological material in soil. *Frontiers in Environmental Science, 7.*

Le Bayon, R.-C., Guenat, C., Schlaepfer, R., Fischer, F., Luiset, A., Schomburg, A., & Turberg, P. (2020). Use of X-ray microcomputed tomography for characterizing earthworm-derived belowground soil aggregates. *European Journal of Soil Science.*

Le Mer, G., Jouquet, P., Capowiez, Y., Maeght, J., Tran, T. M., Doan, T. T., & Bottinelli, N. (2021). Age matters: Dynamics of earthworm casts and burrows produced by the anecic Amynthas khami and their effects on soil water infiltration. *Geoderma, 382.*

Lee, K. E., & Foster, R. C. (1991). Soil fauna and soil structure. *Austr J Soil Res., 29*, 745–775.

Leue, M., Uteau-Puschmann, D., Peth, S., Nellesen, J., Kodesova, R., & Gerke, H. H. (2019). Separation of soil macropore types in three-dimensional X-ray computed tomography images based on pore geometry characteristics. *Vadose Zone Journal, 18.*

Ligthart, T. N., Peek, G., & Taber, E. J. (1993). A method for the 3-dimensional mapping of earthworm burrow systems. *Geoderma, 57*, 129–141.

Lilje, O., Lilje, E., Marano, A. V., & Gleason, F. H. (2013). Three dimensional quantification of biological samples using micro-computer aided tomography (microCT). *Journal of Microbiological Methods, 92*, 33–41.

Mangalassery, S., Sjogersten, S., Sparkes, D. L., Sturrock, C. J., Craigon, J., & Mooney, S. J. (2014). To what extent can zero tillage lead to a reduction in greenhouse gas emissions from temperate soils? *Scientific Reports, 4.*

Mangalassery, S., Sjogersten, S., Sparkes, D. L., Sturrock, C. J., & Mooney, S. J. (2013). The effect of soil aggregate size on pore structure and its consequence on emission of greenhouse gases. *Soil & Tillage Research, 132*, 39–46.

McNamara, N. P., Black, H. I. J., Beresford, N. A., & Parekh, N. R. (2003). Effects of acute gamma irradiation on chemical, physical and biological properties of soils. *Applied Soil Ecology, 24*(2), 117–132. https://doi.org/10.1016/S0929-1393(03)00073-8

Meurer, K., Barron, J., Chenu, C., Coucheney, E., Fielding, M., Hallett, P., Herrmann, A. M., Keller, T., Koestel, J., Larsbo, M., Lewan, E., Or, D., Parsons, D., Parvin, N., Taylor, A., Vereecken, H., & Jarvis, N. (2020). A framework for modelling soil structure dynamics induced by biological activity. *Global Change Biology, 26*, 5382–5403.

Mombo, S., Laplanche, C., Besson, P., Sammartino, S., Schreck, E., Dumat, C., & Capowiez, Y. (2018). Metal soil pollution differentially affects both the behaviour and exposure of A-caliginosa and L-terrestris: A mesocosm study. *Biology and Fertility of Soils, 54*, 319–328.

Mooney, S. J., Pridmore, T. P., Helliwell, J., & Bennett, M. J. (2012). Developing X-ray computed tomography to non-invasively image 3-D root systems architecture in soil. *Plant and Soil, 352*, 1–22.

Negassa, W., Guber, A. K., Kravchenko, A. N., Marsh, T. L., Hildebrandt, B., & Rivers, M. L. (2015). Properties of soil pore space regulate pathways of plant residue decomposition and community structure of associated bacteria. *PLoS One*.

Nunan, N., Leloup, J., Ruamps, L. S., Pouteau, V., & Chenu, C. (2017). Effects of habitat constraints on soil microbial community function. *Scientific Reports, 7*, 4280.

Nunan, N., Ritz, K., Crabb, D., Harris, K., Wu, K. J., Crawford, J. W., & Young, I. M. (2001). Quantification of the in situ distribution of soil bacteria by large-scale imaging of thin sections of undisturbed soil. *FEMS Microbiology Ecology, 37*, 67–77.

Nunan, N., Ritz, K., Rivers, M., Feeney, D. S., & Young, I. M. (2006). Investigating microbial micro-habitat structure using X-ray computed tomography. *Geoderma, 133*, 398–407.

Nunan, N., Wu, K., Young, I. M., Crawford, J. W., & Ritz, K. (2002). In situ spatial patterns of soil bacterial populations, mapped at multiple scales, in an arable soil. *Microbial Ecology, 44*, 296–305.

Nunan, N., Wu, K. J., Young, I. M., Crawford, J. W., & Ritz, K. (2003). Spatial distribution of bacterial communities and their relationships with the micro-architecture of soil. *FEMS Microbiology Ecology, 44*, 203–215.

Oades, J. M. (1993). The role of biology in the formation, stabilization and degradation of soil structure. *Geoderma, 56*, 377–400.

Or, D., Smets, B. F., Wraith, J. M., Dechesne, A., & Friedman, S. P. (2007). Physical constraints affecting bacterial habitats and activity in unsaturated porous media–a review. *Advances in Water Resources, 30*, 1505–1527.

Otten, W., Pajor, R., Schmidt, S., Baveye, P. C., Hague, R., & Falconer, R. E. (2012). Combining X-ray CT and 3-D printing technology to produce microcosms with replicable, complex pore geometries. *Soil Biology & Biochemistry, 51*, 53–55.

Pagenkemper, S. K., Athmann, M., Uteau, D., Kautz, T., Peth, S., & Horn, R. (2015). The effect of earthworm activity on soil bioporosity–investigated with X-ray computed tomography and endoscopy. *Soil & Tillage Research, 146*, 79–88.

Pelosi, C., Grandeau, G., & Capowiez, Y. (2017). Temporal dynamics of earthworm-related macroporosity in tilled and non-tilled cropping systems. *Geoderma, 289*, 169–177.

Peszynska, M., Trykozko, A., Iltis, G., Schlueter, S., & Wildenschild, D. (2016). Biofilm growth in porous media: Experiments, computational modeling at the porescale, and upscaling. *Advances in Water Resources, 95*, 288–301.

Peth, S., Chenu, C., Leblond, N., Mordhorst, A., Garnier, P., Nunan, N., Pot, V., Ogurreck, M., & Beckmann, F. (2014). Localization of soil organic matter in soil aggregates using synchrotron-based X-ray microtomography. *Soil Biology & Biochemistry, 78*, 189–194.

Piccoli, I., Dal Ferro, N., Delmas, P. J., Squartini, A., & Morari, F. (2019). Contrast-enhanced repacked soil cores as a proxy for soil organic matter spatial arrangement. *Soil Research, 57*, 535–545.

Porre, R. J., van Groenigen, J. W., De Deyn, G. B., de Goede, R. G. M., & Lubbers, I. M. (2016). Exploring the relationship between soil mesofauna, soil structure and N2O emissions. *Soil Biology & Biochemistry, 96*, 55–64.

Quigley, M. Y., Rivers, M. L., & Kravchenko, A. N. (2018). Patterns and sources of spatial heterogeneity in soil matrix from contrasting long term management practices. *Frontiers in Environmental Science, 6*.

Rabbi, S. M. F., Daniel, H., Lockwood, P. V., Macdonald, C., Pereg, L., Tighe, M., Wilson, B. R., & Young, I. M. (2016). Physical soil architectural traits are functionally linked to carbon decomposition and bacterial diversity. *Scientific Reports, 6*.

Rabot, E., Lacoste, M., Henault, C., & Cousin, I. (2015). Using X-ray computed tomography to describe the dynamics of nitrous oxide emissions during soil drying. *Vadose Zone Journal, 14*.

Rawlins, B. J. W., Reinhard, C., Atwood, R. C., Houston, A., Lark, R. M., & Rudolph, S. (2016). Three-dimensional soil organic matter distribution, accessibility and microbial respiration in macroaggregates using osmium staining and synchrotron X-ray computed tomography. *The Soil, 2*, 659–671.

Rogaar, H., & Boswinkel, J. A. (1978). Some soil morphological effects of earthworm activity–field data and X-ray radiography. *Netherlands Journal of Agricultural Science, 26*, 145–160.

Ruamps, L. S., Nunan, N., & Chenu, C. (2011). Microbial biogeography at the soil pore scale. *Soil Biology & Biochemistry, 43*, 280–286.

Ruamps, L. S., Nunan, N., Pouteau, V., Leloup, J., Raynaud, X., Roy, V., & Chenu, C. (2013). Regulation of soil organic C mineralisation at the pore scale. *FEMS Microbiology Ecology, 86*, 26–35.

Schaetzl, R. J., & Anderson, S. (2005). *Soils: Genesis and geomorphology*. Cambridge University Press.

Schlüter, S., Zawallich, J., Vogel, H.-J., & Doersch, P. (2019a). Physical constraints for respiration in microbial hotspots in soil and their importance for denitrification. *Biogeosciences, 16*, 3665–3678.

Schlüter, S., Eickhorst, T., & Müller, C. W. (2019b). Correlative imaging reveals holistic view of soil microenvironments. *Environmental Science & Technology, 53*, 829–837.

Schmidt, H., Vetterlein, D., Kohne, J. M., & Eickhorst, T. (2015). Negligible effect of X-ray mu-CT scanning on archaea and bacteria in an agricultural soil. *Soil Biology & Biochemistry, 84*, 21–27.

Schrader, S., Rogasik, H., Onasch, I., & Jegou, D. (2007). Assessment of soil structural differentiation around earthworm burrows by means of X-ray computed tomography and scanning electron microscopy. *Geoderma, 137*, 378–387.

Sleutel, S., Cnudde, V., Masschaele, B., Vlassenbroek, J., Dierick, M., Van Hoorebeke, L., Jacobs, P., & De Neve, S. (2008). Comparison of different nano- and micro-focus X-ray computed tomography set-ups for the visualization of the soil microstructure and soil organic matter. *Computers & Geosciences, 34*, 931–938.

Sleutel, S., Kader, M. A., Leinweber, P., D'Haene, K., & De Neve, S. (2007). Tillage management alters surface soil organic matter composition: A pyrolysis mass spectroscopy study. *Soil Science Society of America Journal, 71*, 1620–1628.

Taina, I. A., Heck, R. J., & Elliot, T. R. (2008). Application of X-ray computed tomography to soil science: A literature review. *Canadian Journal of Soil Science, 88*, 1–20.

Van den Bulcke, J., Boone, M., Van Acker, J., & Van Hoorebeke, L. (2009). Three-dimensional X-ray imaging and analysis of fungi on and in Wood. *Microscopy and Microanalysis, 15*, 395–402.

Van den Bulcke, J., Masschaele, B., Dierick, M., Van Acker, J., Stevens, M., & Van Hoorebeke, L. (2008). Three-dimensional imaging and analysis of infested coated wood with X-ray submicron CT. *International Biodeterioration & Biodegradation, 61*, 278–286.

Van Loo, D., Bouckaert, L., Leroux, O., Pauwels, E., Dierick, M., Van Hoorebeke, L., Cnudde, V., De Neve, S., & Sleutel, S. (2014). Contrast agents for soil investigation with X-ray computed tomography. *Geoderma, 213*, 485–491.

Vogel, H. J., Weller, U., & Schluter, S. (2010). Quantification of soil structure based on Minkowski functions. *Computers & Geosciences, 36*, 1236–1245.

Wang, W., Kravchenko, A. N., Johnson, T., Srinivasan, S., Ananyeva, K. A., Smucker, A. J. M., Rose, J. B., & Rivers, M. L. (2013). Intra-aggregate pore structures and Escherichia coli distribution by water flow within and movement out of soil macroaggregates. *Vadose Zone Journal, 12*.

Warner, G. S., Nieber, J. L., Moore, I. D., & Geise, R. A. (1989). Characterizing macropores in soil by computed-tomography. *Soil Science Society of America Journal, 53*, 653–660.

Wendt, K. H., & Larink, O. J. P. (1990). Computer-aided analysis of earthworm burrows from sectioned soil columns. *Pedobiologia, 34*, 221–225.

Whalen, J. K., Han, L. W., & Dutilleul, P. (2015). Burrow refilling behavior of Aporrectodea turgida (Eisen) and Lumbricus terrestris L. as revealed by X-ray computed tomography scanning: Graphical and quantitative analyses. *Canadian Journal of Soil Science, 95*, 231–235.
Wright, D. A., Killham, K., Glover, L. A., & Prosser, J. I. (1995). Role of pore-size location in determining bacterial-activity during predation by protozoa in soil. *Applied and Environmental Microbiology, 61*, 3537–3543.
Zappala, S., Helliwell, J. R., Tracy, S. R., Mairhofer, S., Sturrock, C. J., Pridmore, T., Bennett, M., & Mooney, S. J. (2013). Effects of X-ray dose on rhizosphere studies using X-ray computed tomography. *PLoS One, 8*.
Zhang, Z. B., Liu, K. L., Zhou, H., Lin, H., Li, D. M., & Peng, X. H. (2018). Three dimensional characteristics of biopores and non-biopores in the subsoil respond differently to land use and fertilization. *Plant and Soil, 428*, 453–467.
Zhang, Z. B., Peng, X., Zhou, H., Lin, H., & Sun, H. (2015). Characterizing preferential flow in cracked paddy soils using computed tomography and breakthrough curve. *Soil & Tillage Research, 146*, 53–65.
Zheng, H., Kim, K., Kravchenko, A., Rivers, M., & Guber, A. (2020). Testing Os staining approach for visualizing soil organic matter patterns in intact samples via X-ray dual-energy tomography scanning. *Environmental Science & Technology, 54*, 8980–8989.

Chapter 11
Integrating X-ray CT Data into Models

Xavier Portell, Valerie Pot, Ali Ebrahimi, Olivier Monga, and Tiina Roose

11.1 Introduction

X-ray Computed Tomography (X-ray CT) offers important 4-D (i.e., 3-D scanning over time) structural information of the soil architecture. This imaging tool provides access to the 3-D morphological properties of the soil pore space such as the 3-D connectivity of pores that are essential to the understanding of water, solute, and gas transport processes. Other morphological properties such as pore-size distribution,

Supplementary Information The online version contains supplementary material available at [https://doi.org/10.1007/978-3-031-12176-0_11].

X. Portell (✉)
School of Water, Energy and Environment, Cranfield University, Bedfordshire, UK

Departamento de Ciencias, IS-FOOD, Universidad Pública de Navarra, Pamplona, Spain
e-mail: xavier.portell@unavarra.es

V. Pot
Université Paris-Saclay, INRAE, AgroParisTech, UMR ECOSYS, Palaiseau, France
e-mail: Valerie.Pot@inrae.fr

A. Ebrahimi
Department of Civil and Environmental Engineering, Massachusetts Institute of Technology, Cambridge, MA, USA
e-mail: alieb@mit.edu

O. Monga
Sorbonne Université, IRD, UMMISCO, Bondy, France
e-mail: olivier.monga@ird.fr

T. Roose
Bioengineering Sciences Research Group, Department of Mechanical Engineering, School of Engineering, Faculty of Engineering and Physical Sciences, University of Southampton, Southampton, UK
e-mail: t.roose@soton.ac.uk

S. Jon Mooney et al. (eds.), *X-ray Imaging of the Soil Porous Architecture*,
https://doi.org/10.1007/978-3-031-12176-0_11

specific surface area, or spatial heterogeneity of soil can be obtained from the X-ray CT images. Many studies have used this technique to better understand the evolution of macroscopic soil physical properties such as structural stability and relate it to spatial descriptors of soil pore space morphology when the soil undergoes wetting/drying cycles (e.g., Diel et al., 2019) or when it is subjected to different agricultural practices (e.g., Papadopoulos et al., 2009; Dal Ferro et al., 2013; Caplan et al., 2017). Non-equilibrium transfer processes, such as preferential transport, have also been related to the quantification of macropores in X-ray CT images (e.g., Larsbo et al., 2014; Katuwal et al., 2015; Soto-Gómez et al., 2018). In addition, X-ray CT data have proved particularly useful for reconstructing the skeletons of biopore networks, such as those burrowed by earthworms (Capowiez et al., 1998), and for monitoring their temporal dynamics (Joschko et al., 1993) (see Chap. 10). The role of air-filled soil pores and in particular their connectivity in 3-D in the transport of microbial-generated gaseous products (N_2O, CO_2) have been hypothesized (Rabot et al., 2015; Porre et al., 2016). X-ray CT data have also provided new knowledge about the 3-D architecture of root systems (e.g., Helliwell et al., 2013) and their impact on the 3-D soil architecture (see Chap. 9). For instance, root hairs were shown to modify the pore-size distribution and connectivity in the rhizosphere (e.g., Keyes et al., 2013; Koebernick et al., 2017, 2019). X-ray CT measurements have also allowed imaging aerenchymatous roots and the gas bubbles entrapped in the soil of rice paddies to explain transport of CO_2 and O_2 between roots and the atmosphere (Kirk et al., 2019). The dynamics of the spatial dispersion of soil microorganisms could be related to the 3-D description of the pore space obtained by X-ray CT (Juyal et al., 2020). The role of some pore-size classes could also be linked with soil carbon storage (Kravchenko et al., 2020) (see Chap. 10).

Many experimental studies have highlighted the link between soil functioning and structural information of soil architecture extracted from X-ray CT. This is often performed by looking for correlations between measured soil functions and morphological properties of the pore space. From this vantage point, modelling offers an interesting and complementary methodological tool to these studies. In addition to their predictive character, models open the way to the testing of hypotheses about the role of the complex soil architecture on soil biological and physical processes, by carrying out structurally realistic *in-silico* experiments that can now be set up thanks to X-ray CT imaging. Development of spatially explicit pore-scale modelling in the 1990s, encompassed with a steep increase in desktop processors computing power and development of parallel computing algorithms, has made the routine use of X-ray CT images in modelling common. Early examples of the integration of X-ray CT data into spatially explicit models are found in the rock physics domain (Spanne et al., 1994; Hazlett, 1995). The inclusion of X-ray CT data into models focusing on soils occurred at a later stage (Falconer et al., 2005; Monga et al., 2008; Keyes et al., 2013). Consequently, the rock physics domain accounts for a large number of studies dealing with water flow, multiphase flow, and reactive transport (e.g., Blunt et al., 2013; Bultreys et al., 2016; de Paulo et al., 2020), whilst the soil domain has been the object of a lower number of studies, possibly because of its importance not being widely recognized and/or funding being more abundant for the problems dealing with oil and gas movement in the rocks. Synthetic porous media or packing of sands have also been imaged and processed into pore-scale models to study fundamental physical and biological processes in granular media (e.g., Schaap et al., 2007; Sukop et al., 2008; Peszynska et al., 2016). Such approaches are a valuable step forward to account for some of the soil complexity, but outputs obtained should be taken cautiously because the studied system can behave fundamentally differently from the real soil situation.

The main approach in using X-ray CT information in image-based models is the direct integration of the 3-D organization of the soil phases in the numerical grid of pore-scale models. Soil processes such as water flow, multiphase flow, diffusion, or reactive transport are subsequently computed in the 3-D pore space using Computational Fluid Dynamics (CFD) tools such as Finite-Element, Finite-Volume, or lattice-Boltzmann approach. In an alternative indirect approach, X-ray CT data is used to obtain key image-based structural qualities to parameterize soil features that are needed as input parameters into macroscopic soil models that ignore the explicit 3-D description of soil. One example of such a morphometric statistical approach is the extraction of critical pore diameters to be used in macroscopic analytical relations to predict saturated hydraulic conductivity. The structural morphometric parameters obtained from X-ray CT can also be used to generate an equivalent 3-D pore space, for instance extracting the Euler number (a number used for quantifying the connectivity of the pore space) to parameterize a pore network model (PNM). The latter approach is also classified as indirect. Soil structure can also be generated from explicit soil images by using more sophisticated statistical methods such as the Markov-chain Monte-Carlo method as performed by Wu et al. (2004). Although the images used in this latter work were 2-D thin sections, the method can be applied to X-ray CT data (Lin et al., 2020).

This chapter focuses on soil modelling approaches that make use of X-ray CT. Other porous media, such as carbonate rocks, sandstone, etc., have been thoroughly addressed in other reviews (e.g., Golparvar et al., 2021) to which readers are referred to. First, we focus on the practical aspects of the integration of X-ray CT data of the main soil phases into spatially explicit models highlighting the main approaches that can be used for such integration. We then provide an overview of how X-ray CT data has been integrated in soil modelling research distinguishing between work on a direct or an indirect use of X-ray CT.

11.2 Direct Integration of X-ray CT Data-Derived Soil Phases into Image-Based Models

In this section, we focus on the integration of X-ray CT data of the main soil phases, namely pore space (sect. 11.2.1), air and water (sect. 11.2.2), organic matter (sect. 11.2.3), and roots (sect. 11.2.4), highlighting the main modelling strategies that can be adopted.

A number of preliminary steps must be carried out before the integration of the soil phases in image-based models can be undertaken. These processes allow for the acquisition of a greyscale representation of the soil structural information that is usually segmented to produce binary images, i.e., black and white. Following this, the different soil phases are then labelled by different colours identifying solid particles and non-solid particles (pores), which are normally used by image-based models despite providing less signal information than greyscale images. Readers interested in segmentation and post-treatment processes are referred to Chaps. 5 and 6 along with the reviews of Iassonov et al. (2009) and Schlüter et al. (2014).

11.2.1 Integration of X-ray CT Data of Pore Space Geometry

Different strategies can be adopted to integrate the structural information of pore space geometry provided by the segmented X-ray CT images into image-based models. We will distinguish four strategies differing in the degree of simplification of the soil structural information and the immediacy of their use: voxel-based, finite-element meshing, morphological, and irregular PNM approaches. The main characteristics of these strategies are summarized in Fig. 11.1.

The most straightforward strategy is to make use of all the structural information elements provided by the image to reconstruct the numerical grid. Depending on the CFD tools used, two approaches can be distinguished, the so-called 'voxel-based' and 'finite-element' modelling techniques. Voxel-based models directly map the voxels of the segmented X-ray CT image to the model's numerical grid, as is typically done when using lattice-Boltzmann models. Depending on the desired degree of accuracy of the soil processes, the lattice-Boltzmann grid can be further refined (e.g., Sukop et al., 2008). Finite-Element solvers require the reconstruction of the numerical grid using more or less sophisticated meshing software (e.g., iso2mesh, Tran et al., 2020).

Another strategy is to reduce the number of computational elements of the segmented X-ray CT image while keeping key structural information, as performed by the so-called 'morphological' and 'irregular Pore Networks' models. In these cases, the integration of X-ray CT data requires a number of operations specific for each modelling approach. The main steps required are reported in Fig. 11.2. Morphological models approximate the complex pore space morphology with a set of simplified geometrical primitives such as balls (e.g., Hilpert & Miller, 2001; Vogel et al., 2005; Lehmann et al., 2006; Monga et al., 2008), ellipsoids or generalized cylinders (Ngom et al., 2012; Kemgue et al., 2019), located along the median axes of the pore space. This set of maximal balls recover the skeleton of the pore space. These balls also function as the nodes of an adjacency graph, which considers the neighbourhood information of each geometric primitive (e.g., balls) where the soil processes are locally computed (Monga et al., 2007). They can also be used to construct generalized cylinders or ellipsoids (Ngom et al., 2012; Kemgue et al., 2019). Like morphological models, irregular PNMs construct networks of nodes

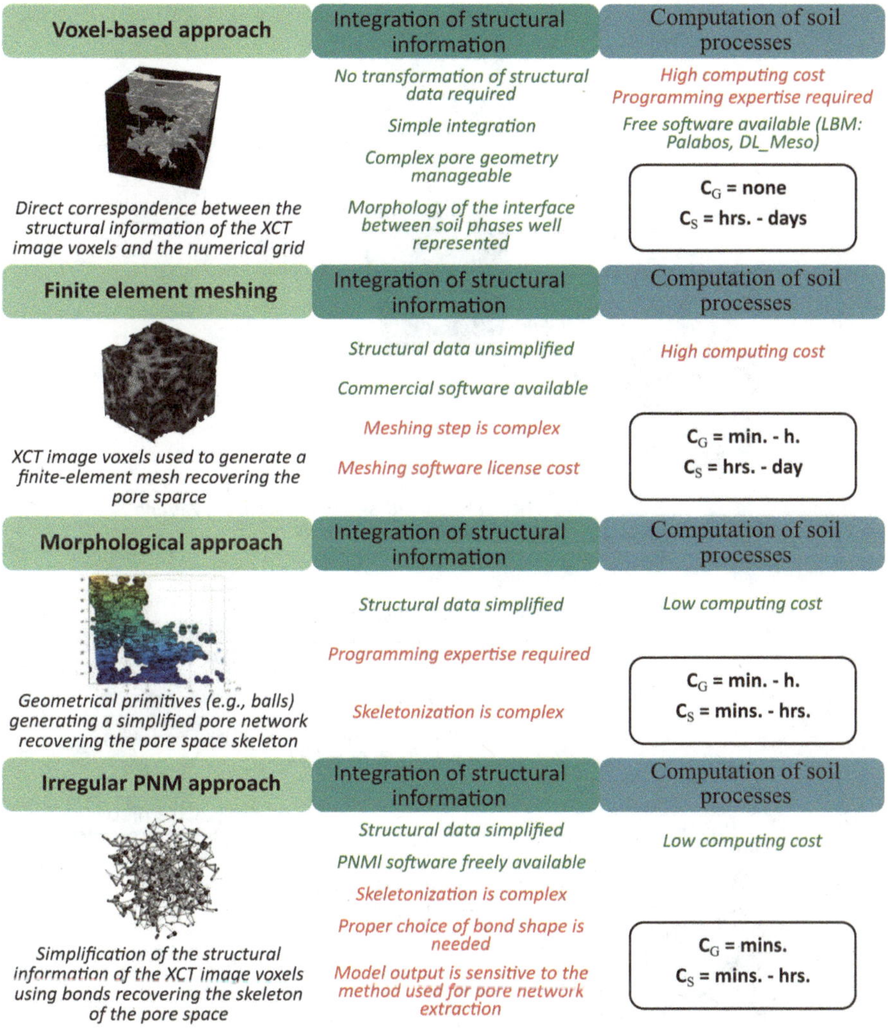

Fig. 11.1 Overview of four main methods used to integrate X-ray CT 3-D structural data in image-based models. In the figure, points that can be seen as advantages are in green while disadvantages are depicted in red. PNMs stands for pore network models. In the figure, C_G is an indication of the computing time required to *generate* the numerical grid and C_S is an indication of the computing time required to *simulate* diffusion of a chemical species through the network for 10 days. An X-ray CT image of 512^3 voxels and an average desktop computer (e.g., AMD Ryzen 7 or Intel i7 processor, and 16/32 GB of RAM) have been assumed on providing these indicative numbers

located along the median axes of the imaged pore space (e.g., Ebrahimi et al., 2013). Irregular PNMs use the same operations than the morphological approach to extract clusters of maximal balls. Then these clusters of maximal balls are converted to pore-throat network (Fig. 11.2). In both approaches, the size of the given geometrical

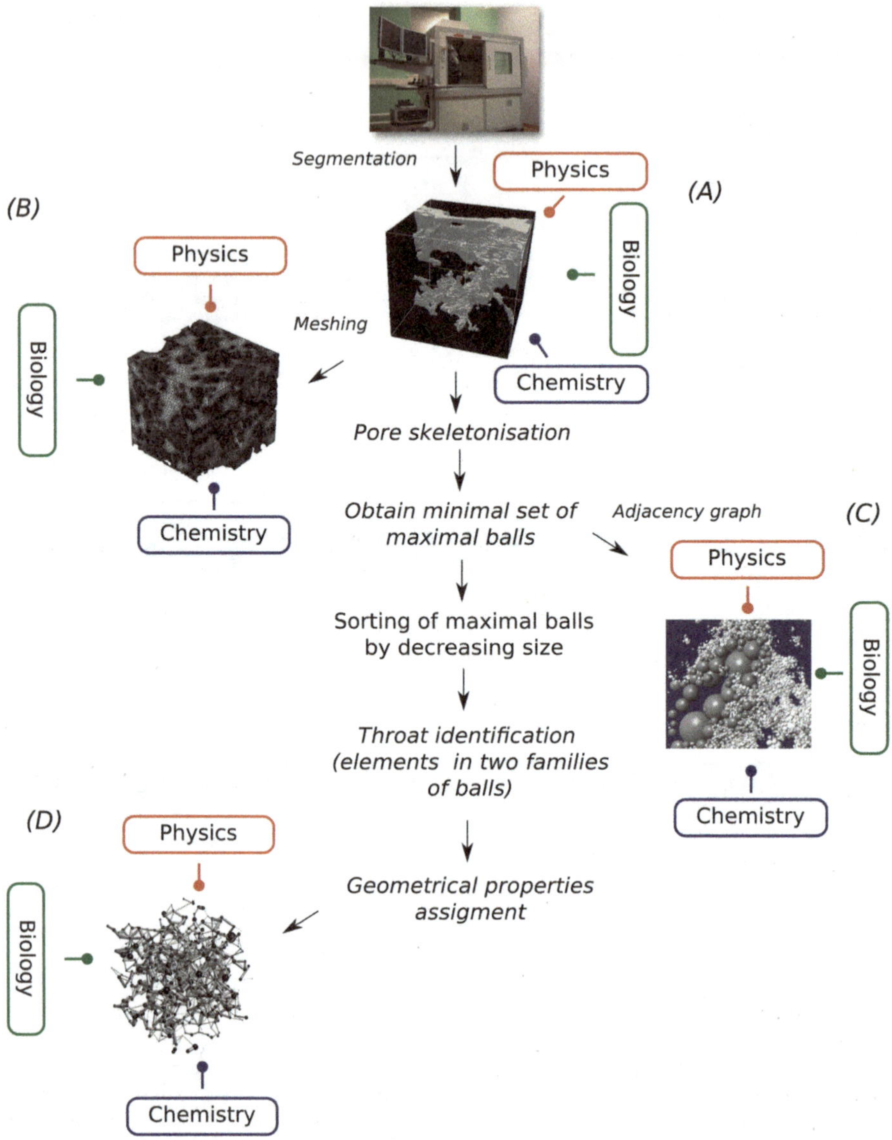

Fig. 11.2 Overview of the operations required to integrate X-ray CT images of the soil architecture into image-based models. For illustrative purpose, we sketch a binary segmented image showing pore and solid space. Note that other soil phases may be added (e.g., water phase, solid organic matter). (**a**) The segmented X-ray CT image is used directly by voxel-based approaches. Note that only the pore phase is represented in the figure. (**b**) Finite-element approaches use a mesh approaching where the segmented pore space is located. (**c**) The morphological approach requires the skeletonization of the segmented pore space image, which is used as a base to construct a minimal set of maximal balls that approach the soil pore space. Once identified, the relationships and connectivity of the maximal balls are written in an adjacency graph, which is valuated by attaching to each ball a set of attributes (e.g., air or water, organic matter mass). (**d**) In addition to the steps described by the morphological approach, irregular pore networking models extract clusters of maximal balls that are converted to pore-throat networks. This requires sorting the maximal balls by

primitives (in morphological models) and the bonds (in irregular PNMs) are locally computed from the measured area of inscribed circles or balls within the pores (e.g., Perez-Reche et al., 2012). PNMs classically use cylindrical or angular bonds (throats). There are standalone open software tools (http://openpnm.org, Gostick et al., 2016; http://porespy.org, Gostick et al., 2019) that allow constructions of these irregular pore networks.

In PNMs, the choice of the bond shape can have important consequences on the model behaviour. Microscopic soil images revealed that natural soil pores consist of heterogeneous and angular cross-sections rather than cylindrical shapes (Blank & Fosberg, 1989; Hamamoto et al., 2016). Quantifying such angular form of pores is not only crucial for a realistic representation of soil microscale structure, but it also has a significant impact on water configuration and soil hydraulic conductivity (Raoof & Hassanizadeh, 2012). For instance, using fast tomography imaging in simplified porous media made of packing of glass beads and in sands, Hoogland et al. (2016) characterized corner flow and conductivity dynamics during soil drainage process. As pores are drained or dried, a fraction of water remained in the pores' angular corner (Tuller et al., 1999). Many studies have shown that the assumption of cylindrical pore cross-section that only allows for single-phase occupancy was insufficient to represent water configuration and hydraulic continuity (Zhou et al., 2000; Tuller & Or, 2001). On the other hand, angular pores allow retaining two or more fluids simultaneously and thus better represent liquid configuration in soil. Joekar Niasar et al. (2009) developed a geometry-based approach for pore network modelling to characterize and produce various ranges of pore cross-sections to describe natural pore geometry, including irregular hyperbolic triangles and regular hyperbolic polygons. The pore geometry and angularity are also essential factors controlling colloid and microbial transport in unsaturated soil (Bradford & Torkzaban, 2008; Saiers & Lenhart, 2003). For instance, Ebrahimi and Or (2014) showed that pore angularity impacts effective water film thickness necessary for bacterial motility in unsaturated soil. In addition, soil macropores and their orientation are known to significantly impact hydraulic conductivity by allowing the rapid flow of water in the soil. Micropores are created by worm holes, plant roots, soil aggregation, and soil wetting/drying. Recent advances using X-ray CT imaging allowed for characterization and quantification of such soil pores at higher resolution compared to traditional methods that used dye tracing, spectral image analysis, and sectioning soil into thin layers (Katuwal et al., 2015; Li et al., 2016; Hu et al., 2020). It is worth mentioning that as with all imaging and image-based techniques, there is a constant temptation to include more and more detail on different physical processes into models. Novel technologies provide a detailed view of soil pore structure

Fig. 11.2 (continued) decreasing sizes, with larger balls ranked as ancestors and small ones as daughters, the identification of common element between families (bonds between pores), and the assignment of geometrical properties to the bonds identified (e.g., angularity, surface area, shape factors). Further details of the substeps required by the morphological and irregular pore network approaches can be found in Table S11.1

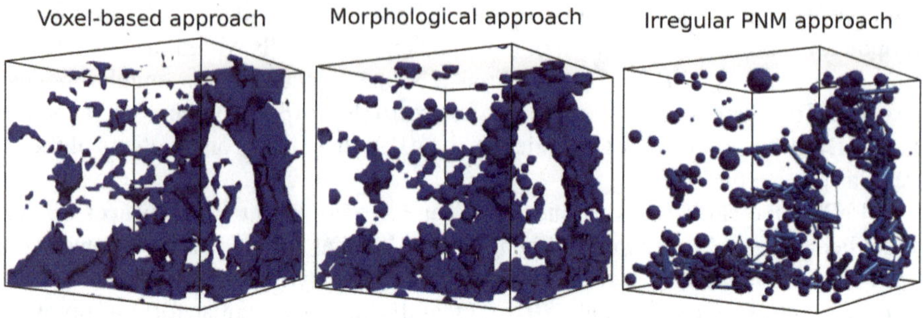

Fig. 11.3 3-D view of a pore geometry integrated in the voxel-based, morphological, and PNM approaches. In the morphological approach, balls are used as geometric primitives. In the irregular PNM approach, cylindrical bonds are used. The size of the ROI is 56x56x56 voxels. The voxel resolution is 9.77 μm (Peth et al., 2014)

however, the challenge remains in finding the right balance in terms of the number of measurable details that need to be captured in the models and observations to effectively describe soil function while reducing the inherent complexity of soil. Not all details matter but direct simulation alone or in combination with homogenization approaches (discussed later) is needed to determine which features matter and when; this process allows for systematic classification of processes/phenomena into important and unimportant functional traits.

The advantage of morphological models and irregular PNMs is that they provide a close approximation of the X-ray CT image-based soil pore space as they retain the spatial heterogeneity of the clustering of pores and connectivity in addition to the pore-size distribution (Blunt, 2001; Xiong et al., 2016; Perez-Reche et al., 2012). These methods compact the data information provided by the X-ray CT images and represent an interesting alternative to estimate soil processes in larger-scale simulations than can do the voxel-based models.

As an illustrative example, we have used a set of synchrotron X-ray CT images of soil aggregates from Peth et al. (2014) to show how pore space is integrated in three selected image-based model approaches: a lattice-Boltzmann Method (LBM), a morphological model, and an irregular PNM. Before segmenting the image, we reduced the noise of the image using a non-local filter mean and passed an unsharp mask to enhance contrast between phases (Schlüter et al., 2014). We then applied a local segmentation using a watershed algorithm (Marker-Controlled Watershed in MorpholibJ plugin of Fiji/ImageJ, Legland et al., 2016) to obtain a binary image, with white voxels corresponding to the solid phase and black voxels corresponding to the pore phase. Post-processing of the binary image was further performed with a majority filter to remove segmentation noise. The segmented image was then integrated in the LBM which is a voxel-based model by making a direct correspondence between each X-ray CT voxels and the nodes of the model numerical grid. The same segmented image was then integrated in the morphological model of Monga et al. (2014) and the irregular PNM of Ebrahimi et al. (2013). Figure 11.3 displays the same region of interest (ROI) of the image to better highlight the differences in

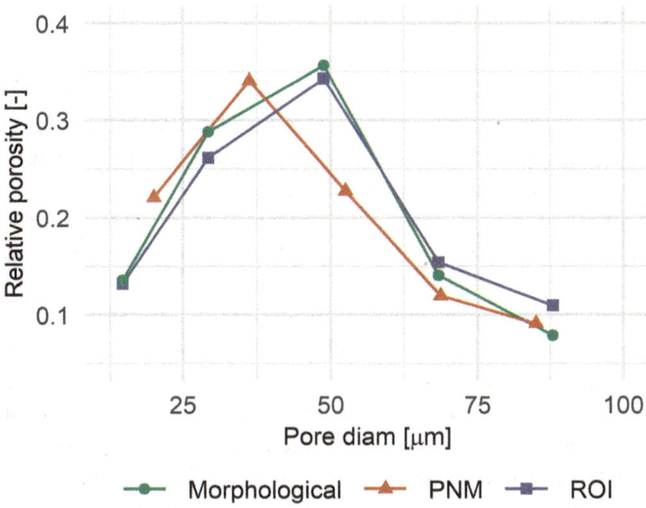

Fig. 11.4 Pore-size distribution of the ROI (56×56×56 voxels) of Peth et al. (2014) image, obtained with a voxel-based (purple), morphological (green), and the irregular PNM (orange) depicted in Fig. 11.3

pore space recovering by the three approaches. The porosity obtained with the three approaches was 0.1233, 0.1054, and 0.0986 for the voxel-based, morphological, and irregular PNM approaches, respectively. Comparison of the pore-size distribution obtained with the three methods can be seen in Fig. 11.4. A priori, underestimations of the pore space volume recovered by the irregular PNM and morphological model can be expected due to the approximations these methods assume. Nonetheless, for this image, these approximations did not translate into a substantial loss of the pore-size classes of the pore volume. The morphological approach, which consists of a minimal set of maximal balls that barely overlap, reproduced the pore-size distribution of the ROI computed with the morphological opening released in the MorpholibJ plugin of Fiji/ImageJ (Legland et al., 2016) well. The slight underestimation of pore volume in PNM is associated with the throat assignment step in which the neighbouring maximal balls to a pore are converted to a throat with a constant size. The throat size is chosen based on the average size of maximal balls connecting two pores that tends to overrepresent small balls.

11.2.2 Integration of X-ray CT Data of Air–Water Distribution

The fact that attenuation of X-rays by water is comparable to that of organic matter hinders distinguishing these two phases in soil X-ray CT images. Imaging protocols are not always straightforward (e.g., Rezanezhad et al. (2010) did not distinguish water from organic matrix of peat soils). In the case of peat soils, distinguishing water from organic matrix relies on the equilibration time between peat and water, which results in some solutes being incorporated to the water surrounding peat, blurring the distinction between phases. If water in the saturated peat is replaced with deionized water, contrast is enhanced and distinction of the two phases is possible (Iain Young, personal communication). High quality images and a resolution targeted to the visualization of air–water interfaces in well resolved pores are thus required. For instance, air-filled and water-filled pores were successfully segmented in X-ray CT images of soil columns (e.g., Mooney, 2002; Daly et al., 2015). Imaging the spatial distribution of water and air phases in the soil pore space can be more easily measured using synchrotron radiation X-ray CT (e.g., Carminati et al., 2008; Pot et al., 2015; Daly et al., 2017). Another option is to stain water by heavy contrast agents. Li and Tang (2019) used an iodine-based contrast medium to visualize water films in glass beads and sand, but they could not observe them in a clayey soil. Our previous experiences have shown that the drawback of using contrast agents, particularly in structured clayey soils, is the massive change in electrolyte concentration in the soil solution brought by adding contrast agent solution (such as KI), which could lead to strong soil dispersion (Pot et al., 2011). Scotson et al. (2019) investigated the use of gold nanoparticles for X-ray CT contrast enhancement, which might well provide a promising avenue, especially for soil biological processes, as gold is inert.

The water distribution and movement can be more easily measured using neutron imaging. Neutrons interact strongly with light elements such as hydrogen allowing for better distinction of soil water at the expense of a poorer representation of soil minerals. Combining X-ray and Neutron CT images using data registration techniques allows exploiting this complementarity and obtaining the solid, air, and water phase distributions experimentally (Clark et al., 2019). This process requires de-resolving the higher resolution image (usually X-ray CT) to match the lower resolution data, which can be done, for instance using Fiji/ImageJ. Despite that in principle, the whole range of soil water content can be resolved, the accuracy is largely influenced by the spatial and temporal resolution achievable by the tomographic method (Tötzke et al., 2021). For instance, Moradi et al. (2011) reported rhizosphere volumetric water content profiles of a sandy soil sieved to <2 mm that ranged from 0.05 to 0.4, approximately.

To date, only a few image-based models have integrated X-ray CT imaging of water and air distributions directly (Keyes et al., 2013; Daly et al., 2015, 2016). Rather, these distributions have been computed using drainage and imbibition algorithms that calculate the typical capillary pressure head that allows air or water

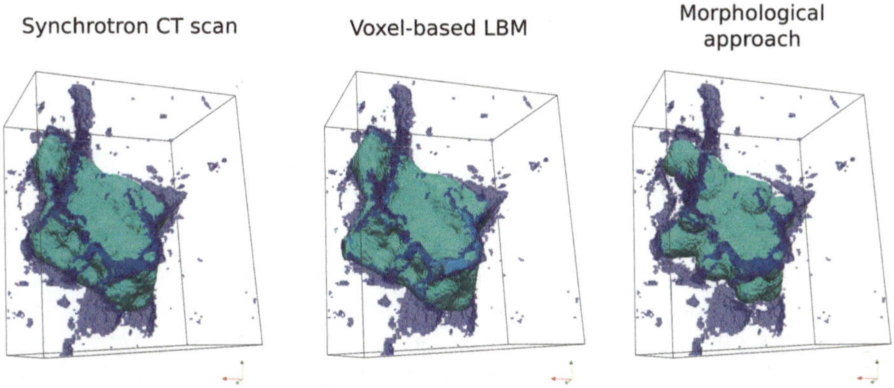

Fig. 11.5 Images of air–water interfaces in a disturbed soil sample, as measured by synchrotron X-ray CT imaging and as produced by a voxel-based approach and a morphological model (Pot et al., 2015). Cyan is the iso-contour of the air–water interface and dark blue is the liquid phase

invasion in the pore network. The sequence of pore invasion by air or water is determined by pore size and pore connectivity only and ignoring the viscous effects and gravity. In the case of drainage, the network of pores is initially assumed to be water filled. Then pores connected to the network entry or already invaded neighbouring pores are drained in a sequence of decreasing pore size. Each pore is evaluated for air invasion based on its size and experienced pressure head according to the Young–Laplace equation. This is adopted in modelling frameworks using morphological models and PNMs (e.g., Carminati et al., 2008; Monga et al., 2014; Ebrahimi & Or, 2015; Pot et al., 2015; Hoogland et al., 2016; Kemgue et al., 2019). Another option is to apply interfacial (van der Waals) and capillary forces to the simulated fluid to generate a phase separation between regions of high fluid density and low fluid density, respectively, attributed to water and air. This approach is adopted in the voxel-based LBM (e.g., Falconer et al., 2012; Genty & Pot, 2013) and in finite-element solvers such as the commercial software COMSOL Multiphysics (Cooper et al., 2017, 2018). LBM solvers with free licenses for research such as PALABOS (Latt et al., 2021) and DL_MESO (Seaton et al., 2013) are available. When interfacial and capillary forces are computed there is a remarkable agreement of the position of the air–water menisci to that measured in X-ray CT images (Pot et al., 2015). Good agreement can also be obtained using morphological models and PNMs (Carminati et al., 2008; Pot et al., 2015) even if the air–water interface presents as an abstracted geometry of the real pore space using Euclidean geometries with defined angularities (Hoogland et al., 2016).

The success of numerical methods to compute water and air distributions in pore space geometries extracted from X-ray CT images and the difficulty in obtaining X-ray CT imaging of water and air distribution in complex soil samples explain the wide adoption of these methods. As an example, Fig. 11.5 displays a comparison between synchrotron X-ray CT imaging and air–water interfaces computed using the LBM and the morphological approach. The mean absolute error between simulations and measurements of the position of the menisci were 5% and 18% for the LBM and morphological approach, respectively. As can be seen in the picture, performance of the methods is remarkable, especially for the LBM approach.

11.2.3 Integration of X-ray CT Data of Solid Organic Matter Distribution

The similar attenuation of X-rays by water and solid organic matter also hinders the imaging of soil organic matter in X-ray CT images (see Chap. 10). However, large fragments of organic matter, the particulate organic matter (POM), can be rather easily visually identified due to their recognizable shape, as was shown when comparing X-ray CT with micromorphological study of organo-mineral assemblages in thin sections (Elyeznasni et al., 2012). Multi-modal segmentation tools can be used to correctly segment POM and avoid false coatings of organic matter around large pores, due to partial volume effect (e.g., Elyeznasni et al., 2012; Schlüter et al., 2014). However, the identification of small organic matter embedded in the soil matrix remains a challenge and needs the adoption of organic matter staining (Peth et al., 2014; Van Loo et al., 2014; Quigley et al., 2018; Piccoli et al., 2019a; Zheng et al., 2020; Maenhout et al., 2021). Despite that some measured X-ray CT organic matter distributions are available in the literature, this information has not been used in modelling studies yet. Instead, scenarios of organic matter placements in the segmented X-ray CT images have been used (e.g., Falconer et al., 2015; Portell et al., 2018).

To initiate integration of realistic 3-D X-ray CT images of spatial distribution of organic matter, we take the example of the soil image obtained by Peth et al. (2014) that we partially displayed in Fig. 11.3 and extracted the structural information required by image-based modelling (Fig. 11.6). Peth et al. (2014) irreversibly fixed osmium to the organic compounds of the soil. To locate the stained organic matter, the soil sample was scanned at a synchrotron at two photon energies above and below the absorption edge of X-ray by osmium. To obtain Fig. 11.6, we first applied a noise reduction pre-processing on both images, using a median 3-D filter and passed an unsharp mask to enhance contrast between phases (Schlüter et al., 2014). Following the procedure in Peth et al. (2014), we subtracted these images and obtained a 3-D image where the highest grey levels correspond to the highest concentrations of osmium associated with large fragments of organic matter (POM) and lower grey levels correspond to organic matter coatings of pore walls or organic matter residing in smaller pores within the soil matrix (Peth et al., 2014). A global segmentation using three thresholds whose values were initially chosen using Otsu and C-means segmentation algorithms was carried out. This resulted in a multi-modal image where the three types of organic matter identified in Peth et al. (2014), associated to different levels of osmium concentration, were distinguished. A majority filter was passed on the resulting image to remove noise associated with

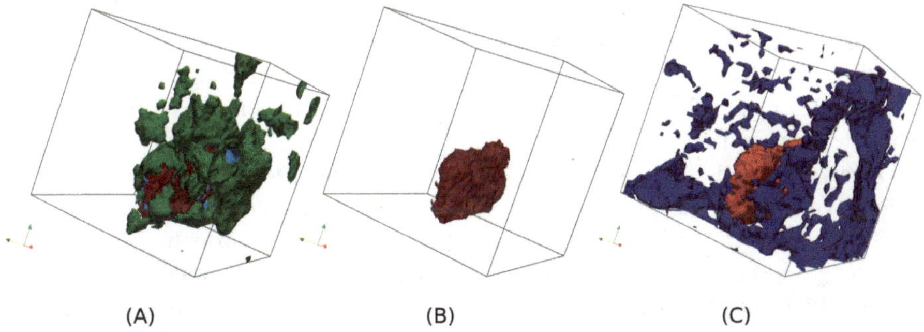

Fig. 11.6 3-D views of soil fragment showing osmium-stained soil organic matter. (a) Total organic matter with high (red), medium (cyan), and low (green) relative osmium concentration, (b) Piece of organic matter showing a high relative osmium concentration, (c) Pore phase (dark blue) and the same organic matter stained with a high relative osmium concentration. The size of the ROI is 56×56×56 voxels. The voxel resolution is 9.77 μm (Peth et al., 2014)

the segmentation step. To integrate this structural information on organic matter in an image-based model, we merged this information with the former binary image containing information on the pore and solid phase as discussed in sect. 11.2.1. To do so we simply added both images to get a multi-modal image. This structural information can now be directly integrated in an image-based model, e.g., the lattice-Boltzmann model of Pot et al. (2010), where different reactivity of the solid–pore interface is included.

11.2.4 Integration of X-ray CT Data of Roots

Despite the recent popularity for rhizosphere related research, few rhizosphere image-based modelling approaches are available to date, probably due to the rhizosphere complexity (both in terms of processes and soil phases involved) and to the challenges faced by the structural imaging of plant–soil interactions such as the identification of soil phases and difficulty of imaging fluxes (see Roose et al., 2016). Distinction among root and other soil phases is increasingly being performed yet it remains challenging, suffering from the limitations such as partial volume effects already highlighted previously. For instance, the pioneering work of Keyes et al. (2013) used Synchrotron X-ray CT and successfully identified root hairs, soil, pore space, and root surface regions. Also, along with the characterization of mucilage and root hairs, and water sorption of chickpea roots, Rabbi et al. (2018) computed the radial and axial water permeability using lattice-Boltzmann modelling based on X-ray CT images of sieved and repacked soil. Given that they are usually filled with water, roots can also be identified using neutrons. For instance, Zarebanadkouki et al. (2012, 2014) used neutron radiography to visualize the transport of deuterated water and developed a method to reconstruct flow into roots. Segmentation of the reconstructed images is complex, often involving semiautomatic methods (Keyes et al., 2013; Daly et al., 2016). Open-source semiautomated tools to follow root growth such as RooTrak (Mairhofer et al., 2016) or the Fiji/ImageJ plugin Smart Root (Lobet et al., 2011) as well as commercial tools such as WinRHIZO are available (see Chap. 9).

Once segmented, the root structural information is used to set up the processes of interest. To date, the most studied processes are nutrient uptake (mainly P), uptake of water in saturated and unsaturated conditions, and root growth. Unlike root architecture-based approaches, which use volumetric reactions/uptake terms, spatially explicit image-based approaches use the root structural information to define the boundary conditions suitable to the process of interest (Ruiz et al., 2020b). Keyes et al. (2013) used Michaelis–Menten nutrient uptake conditions to simulate P uptake by root and root hairs. McKay Fletcher et al. (2019) also assumed a Michaelis–Menten nutrient uptake boundary condition on all active roots but added a constant citrate exudation rate per surface on root tips to study organic-acid enhanced P uptake. Daly et al. (2018) assumed a water uptake rate proportional to the radial conductivity of water through root epidermis and the pressure drop between root internal xylem pressure and the water soil potential. The internal xylem pressure adopted was 1 MPa during the day and 0 MPa at night. For an in-depth explanation of the mathematical expressions adopted by current approaches, readers are referred to the reviews of Ruiz et al., (2020b) and Daly et al. (2016). So far, assumptions about the specific uptake/exudation regions need to be tackled using expert knowledge due to the lack of structurally explicit data. Root (or root hair) growth is currently being studied using time-dependent boundary conditions. A common technique to do so is assuming a given geometry and activating the appropriate boundary condition according to a (measured) growth rate (e.g., McKay Fletcher et al., 2020). In this approach, the developed root segment is used, but the uptake/exudation properties are activated at the appropriate time after having started the simulation.

11.3 Overview of Image-Based Modelling Works

11.3.1 Modelling of Soil Physical Processes

11.3.1.1 Saturated Water Conditions

The macroscopic properties of transport in soils depend on the 3-D organization of the pores and their size (Dullien, 1992). In that regard, the direct integration into transport models of X-ray CT images accurately describing soil cavities and their connectivity is a unique opportunity to explain soil hydrodynamic properties. Several works have numerically calculated 3-D water velocity fields in X-ray CT images of soil samples (e.g., Daly et al., 2015; Dal Ferro et al., 2015; Pot et al., 2020; Zhang et al., 2021a). To carry out image-based modelling, the images are segmented into binary information representing the soil pores and the solid phase (see Sect. 11.2.1). Mathematically, rigorous up-scaling for fluid flow computation, where Darcy's law is derived from Stokes' equations solved at the microscopic scale (Daly & Roose, 2014), enables the calculation of the hydraulic conductivity and soil water retention curve, based on a representative element volume (REV)(see Chap.3). The periodicity of the soil architecture of the REV is also required. However, the search for an image size that is representative of the physical properties of the soil such as porosity and/or hydraulic conductivity has been rarely carried out (Daly et al., 2015; Gharedaghloo et al., 2018; Gerke & Karsanina, 2020). In other cases, the aim is to understand how flow paths emerge in the complex pore space (e.g., Pot et al., 2020) and how the macroscopic property of interest (calculated at the ROI scale), such as permeability, emerges from the 3-D water velocity fields calculated at the microscopic scale (pore organization within the ROI, e.g., Menon et al., 2011; Deurer et al., 2012; Khan et al., 2012; Zhang et al., 2021b).

In almost all the studies above, the voxels of the segmented image were directly mapped to grid nodes and the Stokes' water flow was computed in the complex pore geometry using either the Lattice-Boltzmann approach (Menon et al., 2011, 2015; Hyväluoma et al., 2012; Zhou et al., 2018; Pot et al., 2020; Zhang et al., 2021b), finite-element solvers (e.g., Daly et al., 2015, 2017; Gerke et al., 2018; Gerke & Karsanina, 2020) or a finite-volume solver (Gackiewicz et al., 2019). Other approaches such as smoothed particle hydrodynamics (Dal Ferro et al., 2015) have also been used. Gharedaghloo et al. (2018) assumed an irregular PNM model. The network of pores extracted from segmented X-ray CT images consisted of cylindrical throats and volumeless nodes (see Sect. 11.2.1) and the water flow through the throats was calculated by solving a system of linear water balance equations (Valvatne & Blunt, 2004). To carry out the flow calculation, periodic conditions at the edge of the images are applied and a pressure gradient or force field (gravity) imposed. Once the steady state is reached, the 3-D water velocity field, which is the velocity at each node of the computational grid, is saved. Visualization of these 3-D fields reveals preferential flow paths. As a matter of example, visualization and characterization of flow pathways evidenced that pores of similar size contributed unequally to flow resistance in biological crusts (Menon et al., 2011). This work also provided qualitative insights on soil surface seal functioning by showing how only a few flow channels penetrated through the surface seal and revealed strong lateral flows below these seal structures (Hyväluoma et al., 2012). Permeability (or saturated hydraulic conductivity) is then calculated by integrating the 3-D fluid velocity fields over the ROI. In that respect, image-based modelling outputs have highlighted the major role that large conductive pores have on water flow (e.g., Dal Ferro et al., 2015). They have also confirmed that: (i) variability in permeability is largely affected by size and by a few key pore throats (e.g., Gharedaghloo et al., 2018; Pot et al., 2020; Zhang et al., 2021a) and (ii) that porosity is not the sole morphological parameter explaining such permeability, suggesting that pore-size distribution and the pore network connectivity and tortuosity must be considered as well (e.g., Zhou et al., 2018).

The outputs of image-based models have also been compared to X-ray CT-derived pore space morphological parameters such as porosity, pore-size distribution, connectivity (Euler number), tortuosity, and critical diameter of the percolating pores. Many of these parameters can already be calculated with existing open-source free image analysis software (e.g., Fiji/ImageJ, QuantIm), while others, including tortuosity, remain more complicated to obtain (e.g., Gommes et al., 2009; Roque & Costa, 2020). Tortuosity can be estimated from spectral dimension (e.g., Crawford et al., 1993) or by computed 3-D fluid velocity fields or solute diffusion fields in 3-D images (e.g., Koponen et al., 1996; Berg & Held, 2016; Zhang et al., 2021b). Correlations between these morphological parameters and outputs of the image-based models are commonly investigated in the literature. For instance, prediction of a reduction of flow rate when biological crusts and repacked soil aggregates are submitted to compaction was related not only to the reduction in total porosity but also to a modified pore-size distribution (Menon et al., 2011, 2015).

In an alternative indirect approach, X-ray CT images are first analysed to deliver morphological parameters such as the ones described above. These parameters are subsequently used in macroscopic empirical models that predict water content, hydraulic conductivity, and solute diffusivity. Examples of these empirical models are the well-known empirical Kozeny–Carman equation and other relations based on percolation theory and critical path analysis (Katz & Thompson, 1986) which relate the saturated hydraulic conductivity with specific surface area, tortuosity, and a characteristic length of the pore size. As found in works dealing with porous rocks (e.g., Nishiyama & Yokoyama, 2017), this indirect integration of 3-D X-ray CT data of soils has allowed to demonstrate that the critical path diameter is a better parameter to be used in the Kozeny–Carman equation as compared to the mean hydraulic radius (Koestel et al., 2018; Pot et al., 2020). Luo and Lin (2009) used X-ray CT imaging to show that fractal dimension alone was insufficient to characterize soil macropore structure and flow patterns including a lacunarity function that reflects size distribution of macropores as a diagnostic value to better describe soil structural properties. Packages included in existing image analysis software can also be used to compute diffusion processes. The QuantIm software (Vogel, 2008) solves Fick's law of diffusion in 3-D with a finite difference approach. Using this package, Deurer et al. (2012) calculated the apparent diffusion coefficient of macropore networks around soil aggregates. They extracted small ROIs from segmented X-ray CT images of soil cores sampled under tree-rows and wheel-tracks and computed the diffusive flux within them. The size of the ROI was chosen to encompass a sufficiently large number of soil aggregates (almost 100). They attributed the absence of significant differences in the gas diffusion coefficient between ROIs to the similarity in the volume of pores larger than 1 mm and to the fact that they were well connected. Lattice-Boltzmann model (Zhou et al., 2019) and PNM (Gharedaghloo et al., 2018) were also used to compute solute transport properties. According to theory of transport in porous media, Zhou et al. (2019) predicted an increase of the solute dispersion coefficient in soil macroaggregates amended with biochar resulting from an increased saturated hydraulic conductivity. For peat soils, the diffusion coefficient decreased with depth driven by an increased tortuosity whereas dispersivity did not vary considerably (Gharedaghloo et al., 2018).

11.3.1.2 Unsaturated Water Conditions

Soil water movement has been modelled classically using Richards' equation, which consists of a mass conservation combined with two equilibrium measurements for the partially saturated hydraulic conductivity and soil suction characteristic that relates the level of relative soil saturation to the soil water pore pressure. As such, both measurements can take a long time and hence the need for intelligent estimation of these values based on various morphological and textural information. In addition, both hydraulic conductivity and suction characteristic exhibit hysteresis, i.e., the shape of the curve as a function of saturation is different depending on if the measurement is done whilst drying or wetting the sample. Currently, this compounds the problem for measurements as to capture the hysteresis one would need to be doing numerous long wetting and drying experiments (weeks and months) very accurately so the locations of the Haine's jumps are accurately captured for hysteresis; not to mention the extra sets to capture scanning curves between the wetting and drying at the intermediate points. Nonetheless, as suggested by Berg et al. (2013), who were able to directly image Haine's jumps in sandstone samples using X-ray CT, future developments may allow to speed up the process.

Currently, Haine's jumps can be predicted using modelling. The mathematical foundations for this that were used on X-ray CT images were established by Daly and Roose (2015). Cooper et al. (2018) expanded the previous work of Daly and Roose (2015) and Cooper et al. (2017) to include the effect of plant exudates on the water release curve (and hence Haine's jumps) and the hydraulic conductivity. We next discuss how this was undertaken by giving a flavour of mathematical analysis involved.

We have already discussed above the ways X-ray CT measurements can be used to capture some of the statistics for estimating saturated hydraulic conductivity. An essentially similar approach can be employed to capture the partially saturated hydraulic conductivity curves if one scans the soil samples whilst wetting and drying and segments out the water–mineral and water–air interfaces. However, this is a very intensive image analysis endeavour. Rather, it would be much better to be able to predict computationally in combination with soil pore space imaging how the wetting and drying influence partially saturated hydraulic conductivity and soil suction characteristics. The fundamental science/mathematics concerning this was established by Daly and Roose (2014, 2015) and implemented directly on images by Cooper et al. (2017) and expanded to include root exudates that change the water wetting and drying properties in the pore space by Cooper et al. (2018). All this work was possible due to the rigorous application of the homogenization theory. Homogenization theory is a perturbation approach which assumes that a composite material, i.e., soil in this case, is comprised of many small things. This means that there is a microscale x and macroscale y and that the ratio of the two $x/y = \varepsilon$ is small. The next step in homogenization assumes that the solution depends on each of these independently and upon application of the perturbation in ε, a sequence of analytic mathematical problems can be found that determine what microscale features matter at the macroscopic scale. For example, after Daly and Roose (2014, 2015) and Cooper et al. (2017, 2018), this approach gave a computational analytic approach to calculating partially the saturated hydraulic conductivity tensor and soil suction characteristic, including wetting and drying Haine's jumps, based on the images of the pore space.

Constrained by the intensive image analysis tasks involved, very few works have attempted to use successive X-ray CT scans of soil samples at different water contents. Using an indirect integration of X-ray CT data, Rezanezhad et al. (2009, 2010) proposed a modified Kozeny–Carman equation taking into account the soil air-filled porosity. The parameters of the air-filled fraction were extracted from X-ray CT images of peat soils scanned at different water contents. Reasonably good agreement was obtained between the predicted conductivity and the experimentally measured one. The increasing number, volume, and tortuosity of the air-filled pores imaged were key parameters in these organic soils to explain the decrease of the unsaturated hydraulic conductivity. Fast scanning X-ray CT data of soil columns subjected to rainfall infiltration were also used to extract two key parameters to be used in a 1-D macroscopic preferential water flow model coupling a soil-matrix capillary flow and a gravity-driven macropore flow: the percolating macroporosity and the characteristic size of the porous medium, which was defined as the average half-distance between percolating macropores (Lissy et al., 2020). The latter parameter controls the lateral flow from macropores to soil matrix and was calculated using a Voronoï diagram of the X-ray CT image. Fixing these parameter values in the model strongly improved the mechanistic understanding of water exchanges between macropores and soil matrix. In particular, as it could be expected, the model was able to generate large water exchanges between macropores and the soil matrix for initially drier soil conditions.

Indirect integration of X-ray CT data has also been used to compute the partially saturated hydraulic conductivity and the soil suction characteristic. Knowledge of the pore-size distribution was delivered by 3-D X-ray CT imaging and the water release curve was calculated from it using the Young–Laplace equation (Beckers et al., 2014). Computation of the soil hydraulic function was subsequently carried out using the macroscopic numerical unimodal (van Genuchten, 1980) and dual porosity (Durner, 1994) models calibrated with the water release curve predicted from the pore-size distribution. The addition of supplementary X-ray CT-image-based data in the water retention curve allowed an improved calibration of the dual porosity model and enhanced predictions of the near-saturated hydraulic function.

Pore-scale imaging can also be used to characterize fractal geometry of pore space and water distribution with implications to determine gas transport and permeability of porous media (Yu & Liu, 2004; Liu et al., 2019; Piccoli et al., 2019b; San José Martínez et al., 2010). For instance, Liu et al. (2019) used X-ray CT imaging to show that the water and gas distributions of quartzitic sands are fractal. Using an indirect approach, they extracted the area and tortuosity fractal dimensions from X-ray CT images in order to develop a fractal theory-based relative permeability model to describe water and gas distributions in pores. Others have developed a number of fractal based models to describe soil unsaturated hydraulic conductivity using a fractal model for the water retention curve and the pore-size distributions (Tyler & Wheatcraft, 1990; Xu, 2004; Alfaro Soto et al., 2017; Ghanbarian & Hunt, 2017).

X-ray CT image-based information has also been indirectly integrated into 3-D macroscopic numerical models of water and solute transport. Unlike a direct integration, a computation mesh covering the whole domain, not only the pore space, is generated in this approach. Values of the parameters of the water release curve and hydraulic conductivity function are locally calculated as a function of either the type of material (denser regions defined as high grey values of the X-ray CT image and less dense regions defined as low grey values, Kasteel et al., 2000) or as a function of the local porosity obtained after segmentation of the X-ray CT images (Dohnal et al., 2013). These studies confirmed that solute spreading simulated at the column scale was sensitive to the difference in the local saturated hydraulic conductivity but not to the local dispersivities (Kasteel et al., 2000). Additionally, Dohnal et al. (2013) complemented X-ray CT data with 3-D data of water content obtained by Magnetic Resonance Imaging (MRI) to determine maps of entrapped air that was added to the 3-D macroscopic model as no-flow regions. Their modelling results supported the assumption that the observed decrease in outflow rate during repeated infiltration experiments could be caused by entrapped air in large pores. Moreover, the spatial distribution of the entrapped air had a greater impact than its volume.

Comparing predictions of an image-based model relying on accurate representations of 3-D air–water interfaces to long established semi-empirical equations from experimental determinations was used to better understand solute transport in unsaturated conditions (Genty & Pot, 2014). At the same time, this allowed testing the robustness of the direct image-based models used. Genty and Pot (2014) integrate X-ray CT pore geometries in an image-based model that computes the phase field equation (described in Sect. 11.2.2) obtaining explicit air–water interfaces at different water contents and at equilibrium. The air–water distributions were subsequently used by a second image-based model of solute transport. The 3-D solute diffusion fields at specific water contents produced by this second model allowed the estimation of the effective diffusion coefficient at the scale of the ROI through an inversion of a macroscopic analytical solution of the Fick's law. A good agreement with the macroscopic semi-empirical equation based on Archie's law, which relates the effective solute diffusion coefficient and the water saturation, confirmed the classic value of the exponent of Archie's law used for soils (Genty & Pot, 2014). Hu et al. (2018) also used knowledge of equilibrated air and water distributions, this time calculated by a morphological image-based approach, and directly integrated this 3-D structural information in a simplified free-surface model considering static equilibrium to predict unsaturated water flow.

The final approach to simulating the unsaturated fluid flow in soils found in the literature is based on an indirect use of X-ray CT data to generate simpler spatially explicit models where the soil processes of interest are subsequently simulated. This was first done by Vogel and Roth (2001), who integrated soil pore-size distribution and topology data on the pore connectivity obtained from a 3-D reconstruction of serial thin sections of soil to produce a large number of equivalent pore networks, using a regular PNM. These authors demonstrated that this structural information was sufficient to predict the soil suction characteristic and hydraulic properties. Later, Köhne et al. (2011) applied the procedure described by Vogel and Roth (2001) to obtain an equivalent pore network based on X-ray CT images of two arable soils. It is worth noting that these works are following an indirect approach to the integration of X-ray CT data. Unlike irregular PNMs, regular PNMs are generated using regular assemblies of nodes interconnected by bonds of simplified geometry (e.g., cylinders). The length, radius, and number of bonds per nodes are parameters of the PNM. The modelled architecture reproduces an equivalent pore network of similar physical properties of given soil samples such as the pore geometry, size distribution, and degree of connectivity, which are extracted after image analysis of the target X-ray CT data.

11.3.2 Modelling of Biological Processes

11.3.2.1 Modelling of Microbial Activity

Image-based models are progressively moving towards the inclusion of soil biological activity, including fungal (e.g., Falconer et al., 2007), roots (e.g., Keyes et al., 2013), and bacterial (e.g., Vogel et al., 2015) processes, as it has been reported in recent reviews (Roose et al., 2016; König et al., 2020; Golparvar et al., 2021). Image-based modelling of soil microbial activity has essentially emphasized the role of soil architecture in establishing diverse spatial accessibility of organic substrates to decomposers that explain the spatial distribution and growth of fungi (Falconer et al., 2007, 2012, 2015; Pajor et al., 2010; Kravchenko et al., 2011; Cazelles et al., 2013), the bacterial diversity (Portell et al., 2018), and the microbial functions such as the decomposition of organic matter and CO_2 emissions (Ngom et al., 2011; Vogel et al., 2015, 2018; Monga et al., 2008, 2014). For instance, X-ray CT images of undisturbed sandy loam soil were directly integrated in an image-based model of fungal growth dynamics by Falconer et al. (2012). These authors simulated fungal biomass spreading in an X-ray CT pore space directly integrated in the numerical grid of the model. Fungal mycelia biomass was conceptualized by three biomass fractions: insulated (i.e., old inactive biomass), non-insulated biomass (i.e., active hyphal tips), and mobile (i.e., internal translocating resource) biomasses. Insulated biomass is immobile while non-insulated biomass and the internal resource spread through a diffusion process constrained by the pore space and existing fungal biomass subdomains, respectively. The system of partial differential equations was solved using the classical Crank-Nicholson scheme. Local fungal spread was modulated by a diffusion coefficient that is a function of the water content present in the numerical grid.

The water content of the grid elements was computed by an LBM approach (as described in Sect. 11.2.2) implemented in the open-source PALABOS (Latt et al., 2021). Falconer et al. (2012) predicted a significant decrease in fungal growth by increased water saturation conditions. Water-filled pores alter the connectivity of the air phase, which slows down fungal colonization in line with the model assumptions. Nonetheless, the explicit integration of the 3-D distribution of air-water phases in the model suggested that it is the spatial location of water-filled pores rather than the water content itself that is responsible of this loss of connectivity and, therefore, that the bulk water content of the soil samples is insufficient to predict fungal growth and spatial expansion.

The explicit soil architecture imaged with X-ray CT can also be used to study survival and coexistence of different bacterial species in soil. Survival and coexistence of different bacterial species in simplified porous media have been shown to be controlled by the connectivity of the pores, the water saturation conditions, and the respective locations of the nutrients and bacteria (e.g., Long & Or, 2005; Ebrahimi & Or, 2015). As an example, Portell et al. (2018) investigated the role of soil architecture and location of organic matter in the coexistence of three bacterial strains showing competitive, generalist, and poorly competitive profiles. To do so, X-ray CT data of an undisturbed loamy soil was directly integrated in the numerical grid. The methodology coupled a lattice-Boltzmann model, accounting for the diffusion of dissolved organic carbon through the water phase (Genty & Pot, 2013, 2014), and an individual-based model, accounting for the activity of soil microorganisms. The approach also considers explicitly the distribution of organic matter hydrolysing over time following a first-order kinetic of constant rate. Hydrolysed organic carbon is released into the water phase neighbouring the particulate carbon, if there is any. The water–air distribution in the pore space was computed for three saturation levels (0.25, 0.5, and 1) using an LBM approach (see Sect. 11.2.2) described and parameterized in the literature (Genty & Pot, 2013; Pot et al., 2015). The lack of multimodal data covering the organic matter distribution (see Sect. 11.2.3) and bacteria position was tackled here using simulation scenarios assuming a fixed bulk amount of organic matter initialized as solid fragments. Bacteria were positioned randomly on the interface between the pore and solid face, performing repetitions to account for the uncertainty of this distribution. Portell et al. (2018) predicted large effects of the spatial distribution of organic matter on the strain's distribution but a largely unaffected total microbial biomass evolution. The poorly competitive strain was found to grow to the same extent of the other strains when located near large pieces of organic matter. Nonetheless, the presence of more competitive strains in the same microhabitat prevents the growth of the least competitive strain. These outputs are in agreement with predictions performed using simplified porous media that found that survival and coexistence of different bacterial species are controlled by the connectivity of the pores, the water saturation conditions, and the respective locations of the nutrients and bacteria (e.g., Long & Or, 2005; Ebrahimi & Or, 2015).

Integration of X-ray CT structural information into image-based models allowed the investigation of the role of differentiated soil porous architectures of a silt loam soil either subject to conventional tillage or under a grassland on organic matter decomposition (Ngom et al., 2011). In this study, the pore space of the segmented X-ray CT data was approached by means of a morphological approach based on spheres (see Sect. 11.2.1). The water retention curve of the ROI was simulated using the Young–Laplace law with a zero-contact angle between water and soil (Sect. 11.2.2). The biological sub-model described in Monga et al. (2008) was then used to simulate organic matter decomposition. This model simulates the growth, respiration, mortality, and dormancy of densities of microorganisms placed in its constituents' maximal balls. To mimic extracellular enzymatic decomposition, the model assumes that soil organic matter within a ball can be decomposed by microorganisms placed in all balls connected through the water-filled path. Organic matter decomposition is modelled assuming an offer/demand approach comparing the maximal decomposable masses (offer) and the maximal microorganisms' needs (demand). When the offer is less than the demand, the former is distributed according to the geodesic distance between the resources and the microorganisms using a polynomial repartition law. Simulation scenarios conducted by Ngom et al. (2011) used a temporal extent of 100 days and distributed organic matter and bacteria in pores of different sizes by placing organic matter in pores smaller than 20 μm and immobile bacteria in larger, more aerated pores. Ngom et al. (2011) suggested that small, isolated pores largely decreased organic matter availability to microbes, hindering mineralization in the conventional tillage treatment, whereas more connected pore space facilitated mineralization of most of the organic matter on the grassland treatment.

In a recent study, the impact of fertilization on the health of microbial communities was tackled using a direct finite-element based approach. This study (Ruiz et al., 2020a) examined how the soil structure and soil moisture content influence nitrogen movements around an ammonium nitrate fertilizer pellet and the subsequent impact on microorganisms' activity. Nitrogen species accounted for in the model are ammonium, nitrate, and dissolved organic nitrogen, which were subjected to the processes of N immobilization, ammonification, nitrification as well as diffusion in the water phase. Samples were imaged with X-ray CT from where the soil phase, air phase, and the fertilizer pellet were segmented. The water distribution in the segmented pore space was computed using a two-phase water–air model (see Sect. 11.2.2). Release of nitrogen to the water phase is modelled using appropriate boundary conditions on the surface of the pellet. Boundary conditions around the soil phase were used to model the sorption and desorption of chemical species as well as chemical transformations mediated by microorganisms. The microbial population size responsible for these transformations was assumed to be fixed for the time length of the simulation. The reduction in the microbial activity due to cytotoxic concentrations of fertilizer was calibrated using preliminary experiments. Ruiz et al. (2020a) found that diffusion and concentration of nitrogen in the water phase is critically dependent upon soil moisture and chemical species considered and provides predictions for the maximum concentrations of ammonium and nitrate around the pellet under dry conditions.

In addition to the direct approach, X-ray CT data has also been indirectly used in modelling studies to understand the consequences of structural heterogeneities in soil microbial diversity and functioning. For instance, experimental studies have shown that low pore connectivity and heterogeneous water configuration in soil increase microbial diversity (Carson et al., 2010) and interaction (Tecon et al., 2018). Using statistical estimates of pore sizes and distances between interstitial sites obtained from existing X-ray CT images of soil samples, Resat et al. (2012) produced simplified soil aggregate structures with random, continuously-connected pores. They studied the dynamics of microorganisms using two distinct carbon obtention strategies: microorganisms with membrane-associated hydrolase and microorganisms releasing extracellular hydrolases. The authors found that coexistence of both strategies produced higher cellulose utilization efficiency and a reduced stochasticity. They also concluded that bacterial growth dynamics was the result of a balance between the substrate degradation kinetics, the dynamics of enzymes produced by the microorganisms, and the diffusive processes taking place in the modelled pore space. Their modelling work also supports previous evidence that bacteria tended to grow preferentially near the carbon source.

Burrowing activity of soil fauna makes the soil architecture a highly dynamic variable. A realistic consideration of the modification of soil architecture by its biological components is needed as it impacts not only soil microbial functions, but soil physical functions as well. The latter were investigated by Bastardie et al. (2002) who built simplified 3-D networks of earthworm burrows in which they simulated saturated water flow. The networks were made of a series of connected cylinders, which represented segments of earthworm burrow systems. Using an indirect approach, they extracted statistical estimates of length and orientation of these segments from X-ray CT images of a repacked soil core colonized by a single earthworm (Capowiez et al., 1998) to generate a large number of equivalent porous media. Earthworm trajectories were computed by predefined rules of worm movement obtained from experimental observations in 2-D terraria (Capowiez & Belzunces, 2001). Therefore, like regular PNMs, physical properties of these equivalent porous media such as length and orientation of the burrow segments are reproduced but their actual spatial arrangement is disregarded. Bastardie et al. (2002) showed parameters such as the volume of pore space and the interconnectedness of burrows were not relevant to explain the permeability of the porous media. Instead, the connectivity of the burrow systems (i.e., the number of independent paths) was found to be a more relevant parameter. So far, actual representation of temporal modification of soil architecture by earthworms has been undertaken using repeated X-ray CT scanning at a low temporal resolution, mainly to highlight modification caused by the treatment applied (Pagenkemper et al., 2015). Nonetheless, this information has not yet been directly used in a model. Instead, Blanchart et al. (2009) generated fractal soil geometry to investigate the role of earthworm burrowing including ingestion and production of casts in compacting and uncompacting soil using an equivalent porous medium. Along the same line, it is worth reporting modelling studies simulating temporal modification of soil architecture by soil microorganisms in which a simplified arrangement of solid particles on a regular numerical grid has been used by some authors (Crawford et al., 2012;

Ray et al., 2017) to tackle the role of the production of gluing agents by microorganisms in gathering soil particles.

It is worth highlighting the role that image-based modelling of soil microbial activity can play in generating mechanistic knowledge of biologically driven processes in complex environments such as soil. One of the approaches used in that respect is the realization of sensitivity analysis where the soil architecture is explicitly taken into account using X-ray CT data (Cazelles et al., 2013; Vogel et al., 2015, 2018). Water–air distributions in the imaged pore space are then usually computed using image-based modelling for different water contents (see Sect. 11.2.2). Varied organic matter distributions (see Sect. 11.2.3) and positions of microorganisms are additionally considered for conducting simulation scenarios. Far from just being a mathematical exercise, such analysis provides valuable insights into how much soil functions and processes are affected by different soil architectures (provided by X-ray CT) and by changes on further abiotic (amount and distribution of organic matter, moisture content, etc.) and biotic (distribution of microorganisms, growth parameters, etc.) inputs. This allows testing of hypotheses about the role of soil architecture on soil microbial activity by unravelling, for instance, the relative importance of the biotic and abiotic components. As an example, analysing fungal physiological processes, soil architecture, and initial carbon distribution, Cazelles et al. (2013) found the spatial extension of hyphal tips and the fungal biomass yield were highly affected by the fungal physiological parameters associated with the growth and recycling processes. Also, Vogel et al. (2018) found that under low accessibility of organic matter to soil bacteria, conditioned by the local structural heterogeneities in the bacteria micro-habitats, the decomposition of organic matter was no longer dependent on the physiological parameters. A general point related with this kind of analysis is that the model assumptions and limitations need to be properly communicated and understood so the results can be appropriately interpreted.

11.3.2.2 Modelling of Root Processes

The prospective of steering rhizosphere management to boost plant productivity while reducing agrochemicals has attracted much attention of the scientific community, including the development of many rhizosphere models. Nonetheless, the number of image-based models existing to date is still very limited (Ruiz et al., 2020b). If we consider models where the soil architecture is considered explicitly, the list of approaches is even more reduced. Keyes et al. (2013) used Synchrotron X-ray CT and developed a finite-element image-based model of phosphate uptake by wheat roots and root hairs and implemented it in COMSOL. The volume mesh used by the model includes root hairs, soil, root surface, and water defined separately. Different grey levels were used to identify the meshed elements thanks to the high quality of the synchrotron light. Potential heterogeneity on the P transporters along the root was not considered due to the lack of data and uptake properties were assumed to be the same over the surface. These authors found that the root surface was responsible for a slightly superior P uptake (1%) than the root hairs, despite their closest proximity to the soil surfaces, where most of the phosphate was located. In a follow-up study, Daly et al. (2016) expanded the previous model to include both bulk and rhizosphere soil. The soil near the root (i.e., rhizosphere soil) was explicitly described through the integration of X-ray CT data and surrounded by a large/infinite bulk soil (i.e., away from the root). The latter was achieved with the use of a boundary condition obtained by means of homogenization theory. Root-hair growth was mimicked by considering the structure and a time-dependent uptake boundary condition. The study found that the key criterion for convergence of the P uptake was the size of the rhizosphere region considered.

Further work on P uptake by roots focuses on complexifying the P uptake mechanism but also simplifying the structural description of the soil phase. McKay Fletcher et al. (2020) added the exudation of citrate, which can enhance P uptake and study the effect of root system architecture on organic-acid enhanced P uptake of 12-day-old wheat plants. Time-lapse X-ray CT was used to define the root architectures. Boundary conditions were placed as dictated by the measured X-ray CT structures and growth simulated by activating the appropriate conditions mimicking the X-ray CT root growth measured. The soil volume was identified in the X-ray CT images, but the soil architecture was disregarded and a homogenous soil, including a parameter to account for an approximate volumetric water content, adopted. McKay Fletcher et al. (2019) further evolved the approach to investigate how changes in root system architectures of barley (*Hordeum vulgare*) affect P uptake from a fertilizer pellet. The study elegantly combines modelling, X-ray CT, Scanning Electron Microscopy with Electron Dispersive X-ray Spectroscopy (SEM-EDS), and soil thin sections. As in McKay Fletcher et al. (2020), X-ray CT identified solid, water, and pores were assumed as a single homogeneous volume and modelled using a single finite-element mesh. Appropriate boundary conditions are then placed on the boundaries of the root and fertilizer pellet, which were also identifiable in the X-ray CT images.

Daly et al. (2018) used time resolved X-ray CT and modelling to test the accuracy of spatially averaging the water uptake of young wheat plants grown in pots of 50 mm diameter x 80 mm height. Roots, solid, and pore space were identified in the X-ray CT images. The water–air interfaces were computed by implementing Richards' equation for partially saturated flow parameterized with the water release characteristics and the saturation-dependent hydraulic conductivity measured. A finite-element mesh describing the root surface, the soil architecture, and the air phase was created and model outputs for a number of root systems were compared to the averaged model of Roose and Fowler (2004) to account for the effect of explicitly including the root architecture. Comparison of these approaches for a number of root system architectures and two soil types found less than 2% divergence. Nonetheless, the simplified model could not capture the heterogeneous water distribution seen in the image-based approach. Since Daly et al. (2018) performed a number of image treatments averaging many soil particles, it has been suggested that this may actually be masking some nuances at the pore scale not captured in this study (Ruiz et al., 2020b). A more comprehensive review about the use of image-based modelling to describe water dynamics in the rhizosphere is provided by Daly et al. (2017).

Landl et al. (2019) simulated the impact of naturally created biopores originated from decayed plant roots and earthworm activity on root growth, soil water flow, and root water uptake. In this work, soil properties are defined at the nodes of regular cubic grids, each having a hydraulic state and soil properties. A soil element (volume) was subsequently defined by the properties of the 8 surrounding nodes. They assumed two regular cubic grids, a coarse grid (1 cm) used to compute water flow and a finer one (0.5 cm) to compute growth development affected by soil local conditions. Biopores were modelled as a soil material with a very low penetration resistance to the roots and assumed to be air-filled, thus, not contributing to the water flow. An X-ray CT image stack (231 μm thick) was initially used to set up the position of the biopores directly. A downgrading procedure involving an initial skeletonization of the X-ray CT data was used to assign biopore voxels to the fine grid. The method was found to produce an increased bioporosity while keeping biopore length densities fairly constant. Model parameters were calibrated via the optimization of the root architecture parameters so that simulated and experimentally observed root length distributions were in agreement with each other. Simulation scenarios with the calibrated model showed biopores sustainedly mitigated transpiration deficits by permitting roots to take up water from deeper soil layers. This was found also when a reduced root water uptake in biopores due to limited root–soil contact was assumed.

Assuming an indirect integration approach, X-ray CT data of root architectures growing in soil can be used to parameterize existing root system models. Root system models are approaches simulating the structure, topology, and 3-D placements of roots. Root growth is simulated assuming a set of rules regarding elongation, branching, and death and a discrete approach defining individual segments of root that may have attributes such as radius, age, or hydraulic properties associated is usually adopted (Schnepf et al., 2020). A set of benchmarks, including reference calibration data sets, have recently been proposed by Schnepf et al. (2020), which suggest that methodologies may converge in the near future. As a matter of example of the methodology, Landl et al. (2021) coupled a 3-D root architecture with a rhizodeposition model that accounts for the transport of rhizodeposition. The coupled model is used to investigate spatio-temporal distribution patterns of citrate and mucilage in soil. Root system architectures of *Vicia faba* and *Zea mays* were extracted from X-ray CT images and saved as Root System Markup Language (Lobet et al., 2015). An in-house computer code was then used to extract the model parameters of the root architecture model (e.g., internodal distance, root radius, branching angle, maximal root length). Soil architecture was not explicitly accounted for in this approach. After running a number of simulation scenarios, Landl et al. (2021) found that: (i) rhizodeposition hotspot volumes around roots reached maximum values at intermediate root growth rates, (ii) root branching led to an increased volume of rhizodeposition hotspot zones due to the rhizosphere of individual roots overlapping, and (iii) that the rhizodeposition hotspot values were smaller around the *Zea mays* root system as compared to the one predicted by *Vicia faba*.

11.4 Conclusions

Whether direct or indirect, the combination of X-ray CT and image-based modelling is a powerful tool to understand the consequences of the soil structural heterogeneity on soil functions. Indirect modelling methods allow consideration of this structural heterogeneity in an implicit manner. Direct approaches explicitly describe structural heterogeneity. The power of the direct approach is that it allows focus on particular aspects of soil, which often are difficult, or simply impossible, to manipulate experimentally. This strength comes at the cost of requiring higher computing resources which, in practice, limit the volume of soil simulated. These detailed microscale description of soil processes can then be extrapolated to the macroscale using mathematical homogenization.

Despite providing new insights on soil functioning, the robustness of image-based modelling frameworks relies on the quality of the X-ray CT data, as well as the mathematical and computational tools available. Caution must be made because computation tools may introduce their own errors into the problem being tackled, which can depend on the particular process or components being considered. Another critique often made regarding spatially explicit image-based models, mainly those focused on the soil microscale, is the unfeasibility of 'validating' the model outputs. This is due to the considerable heterogeneity found in soils and to our lack of knowledge of the soil functioning at this scale, which often requires using inverse modelling techniques and educated guesses. In simpler cases of validating image-based models simulating saturated water flow, some discrepancies between the predicted saturated hydraulic conductivity calculated in the X-ray CT images and experimental measures have been reported (Elliot et al., 2010; Dal Ferro et al., 2015; Koestel et al., 2018). This relatively poor performance has been suggested to be related to an inadequate imaging resolution that hinders the visualization of smaller pores in the X-ray CT images that still contribute to the water flow (Dal Ferro et al., 2015; Koestel et al., 2018). The first point to make here is that, in this context, models are first a tool to test and develop alternative hypotheses that can be subsequently tested experimentally. The second point is that when the purpose of modelling is prediction, not being able to obtain a specific experimental measure does not necessarily mean that the model cannot be validated. Confidence in model outputs can be obtained using multiple experimental data or knowledge data measured at different scales and/or for related scenarios (e.g., Grimm et al., 2005). Being able to reproduce situations and outputs for which the model was not explicitly calibrated or designed for suggests that the mechanisms implemented are robust. A closer exchange and collaboration among experimentalist and modellers would revert in the identification of experimental approaches and measurable (indirect) metrics allowing validating current modelling approaches.

A limitation of image-based modelling is inherent to trade-off between sample size and image resolution. The larger the soil sample, the higher the unresolved porosity that we are missing. For some processes such as water flow, this may not be a big deal but for others, such as the ones related to the diffusion of solutes (Zhang et al., 2016) and the biological activity of bacteria and viruses, it may not be acceptable. Decreasing sample size complicates experimental manipulation and may require increasing the number of repetitions. The unresolved porosity issue could be, in principle, partially tackled using greyscale images instead of segmented X-ray CT (Baveye et al., 2017). Reviewing the current work, it is self-evident that very few studies attempt to use this greyscale information which, in our view, should be increasingly explored in the near future. Further opportunities to tackle the resolution limit can also rely on the combination of methods. The representation of reactive species transport in soils, challenging at the time, offers examples of this method combination. In reactive transport, the bottleneck is often associated with the low resolution of X-ray CT images to estimate pore surface area. Some studies have suggested to use a sorption correction function which accounts for the deviation between the complex pore surface area measured in the X-ray CT images and the simpler one generated in an equivalent pore network (Köhne et al., 2011). Such correction functions allow for incorporating the information on the pore surface area into pore network models. Using this strategy, Köhne et al. (2011) improved the transport prediction of a reactive species through intact soil samples. Overall, further studies are required to systematically combine submicron and nanometre scale roughness of soil pore surfaces for accurate prediction of solute and colloid transport. Nanoscale surface heterogeneities have been shown to both increase and decrease colloid attachment in porous surfaces resulted from different influences of roughness under energy barrier absent or present, respectively (Shen et al., 2011; Rasmuson et al., 2017; Torkzaban & Bradford, 2016). As a matter of example, Lin et al. (2021) used a PNM and showed that nanoscale roughness alters the sticking efficiency and colloid attachment non-monotonically with tendency to increase with roughness.

The chapter has also highlighted that the simultaneous use of multi-modal image-based description of the soil is, with few exceptions, still to be translated into modelling exercises. This is the case for imaged organic carbon but, to a certain extent, also for soil microorganisms, although methods are being developed (Baveye et al., 2018). The most common method to date seems to rely on thin slicing of resin impregnated soil cores combined with a specific (FISH) or unspecific (calcofluor white) staining method to visualize microorganisms. These 2-D maps are then aligned with the 3-D structural X-ray CT data following complex alignment procedures (Hapca et al., 2015; Schlüter et al., 2019).

Acknowledgements We kindly thank Prof Stephan Peth for providing the images of the ROI used to compare modelling methods and Dr. Miriam Patricia Ortega-Ramirez for her help in segmenting the image. XP is a María Zambrano Fellow at the Public University of Navarra (UPNA) and acknowledges funding from the European Union—NextGenerationEU through the Spanish program 'Ayuda para la Recualificación del Sistema Universitario Español'. AE acknowledges funding from Swiss National Science Foundation: Grants P2EZP2 175128 and P400PB_186751. TR was funded by ERC Consolidator grant 646809 DIMR.

References

Alfaro Soto, M. A., Chang, H. K., & van Genuchten, M. T. (2017). Fractal-based models for the unsaturated soil hydraulic functions. *Geoderma, 306*, 144–151. https://doi.org/10.1016/j.geoderma.2017.07.019

Bastardie, F., Cannavacciuolo, M., Capowiez, Y., de Dreuzy, J.-R., Bellido, A., & Cluzeau, D. (2002). A new simulation for modeling the topology of earthworm burrow systems and their effects on macropore flow in experimental soils. *Biology and Fertility of Soils, 36*, 161–169. https://doi.org/10.1007/s00374-002-0514-0

Baveye, P. C., Pot, V., & Garnier, P. (2017). Accounting for sub-resolution pores in models of water and solute transport in soils based on computed tomography images: Are we there yet? *Journal of Hydrology, 555*, 253–256. https://doi.org/10.1016/j.jhydrol.2017.10.021

Baveye, P. C., Otten, W., Kravchenko, A., Balseiro-Romero, M., Beckers, É., Chalhoub, M., Darnault, C., Eickhorst, T., Garnier, P., Hapca, S., Kiranyaz, S., Monga, O., Mueller, C. W., Nunan, N., Pot, V., Schlüter, S., Schmidt, H., & Vogel, H.-J. (2018). Emergent properties of microbial activity in heterogeneous soil microenvironments: Different research approaches are slowly converging, yet major challenges remain. *Frontiers in Microbiology, 9*, 1929. https://doi.org/10.3389/fmicb.2018.01929

Beckers, E., Plougonven, E., Roisin, C., Hapca, S., Léonard, A., & Degré, A. (2014). X-ray microtomography: A porosity-based thresholding method to improve soil pore network characterization? *Geoderma, 219–220*, 145–154. https://doi.org/10.1016/j.geoderma.2014.01.004

Berg, C. F., & Held, R. (2016). Fundamental transport property relations in porous media incorporating detailed pore structure description. *Transp Porous Med, 112*, 467–487. https://doi.org/10.1007/s11242-016-0661-7

Berg, S., Ott, H., Klapp, S. A., Schwing, A., Neiteler, R., Brussee, N., Makurat, A., Leu, L., Enzmann, F., Schwarz, J.-O., Kersten, M., Irvine, S., & Stampanoni, M. (2013). Real-time 3-D imaging of Haines jumps in porous media flow. *Proceedings of the National Academy of Sciences, 110*, 3755–3759. https://doi.org/10.1073/pnas.1221373110

Blanchart, E., Marilleau, N., Chotte, J.-L., Drogoul, A., Perrier, E., & Cambier, C. (2009). SWORM: An agent-based model to simulate the effect of earthworms on soil structure. *European Journal of Soil Science, 60*, 13–21. https://doi.org/10.1111/j.1365-2389.2008.01091.x

Blank, R. R., & Fosberg, M. A. (1989). Cultivated and adjacent virgin soils in northcentral South Dakota: I. chemical and physical comparisons. *Soil Science Society of America Journal, 53*, 1484–1490. https://doi.org/10.2136/sssaj1989.03615995005300050031x

Blunt, M. J. (2001). Flow in porous media—Pore-network models and multiphase flow. *Curr Opinion Colloid Interf Sci, 6*, 197–207. https://doi.org/10.1016/S1359-0294(01)00084-X

Blunt, M. J., Bijeljic, B., Dong, H., Gharbi, O., Iglauer, S., Mostaghimi, P., Paluszny, A., & Pentland, C. (2013). Pore-scale imaging and modelling. *Advances in Water Resources, 51*, 197–216. https://doi.org/10.1016/j.advwatres.2012.03.003

Bradford, S. A., & Torkzaban, S. (2008). Colloid transport and retention in unsaturated porous media: A review of Interface-, collector-, and pore-scale processes and models. *Vadose Zone Journal, 7*, 667–681. https://doi.org/10.2136/vzj2007.0092

Bultreys, T., De Boever, W., & Cnudde, V. (2016). Imaging and image-based fluid transport modeling at the pore scale in geological materials: A practical introduction to the current state-of-the-art. *Earth-Science Reviews, 155*, 93–128. https://doi.org/10.1016/j.earscirev.2016.02.001

Caplan, J. S., Giménez, D., Subroy, V., Heck, R. J., Prior, S. A., Runion, G. B., & Torbert, H. A. (2017). Nitrogen-mediated effects of elevated CO2 on intra-aggregate soil pore structure. *Glob Change Biol, 23*, 1585–1597. https://doi.org/10.1111/gcb.13496

Capowiez, Y., Pierret, A., Daniel, O., Monestiez, P., & Kretzschmar, A. (1998). 3-D skeleton reconstructions of natural earthworm burrow systems using CAT scan images of soil cores. *Biology and Fertility of Soils, 27*, 51–59. https://doi.org/10.1007/s003740050399

Capowiez, Y., & Belzunces, L. (2001). Dynamic study of the burrowing behavior of *Apporrectodea nocturna* and *Allolobophora chlorotica*: Interactions between earthworm and spatial avoidance of burrows. *Biology and Fertility of Soils, 33*, 310–316. https://doi.org/10.1007/s003740000327

Carminati, A., Kaestner, A., Lehmann, P., & Flühler, H. (2008). Unsaturated water flow across soil aggregate contacts. *Advances in Water Resources, 31*(2008), 1221–1232. https://doi.org/10.1016/j.advwatres.2008.01.008

Carson, J. K., Gonzalez-Quiñones, V., Murphy, D. V., Hinz, C., Shaw, J. A., & Gleeson, D. B. (2010). Low pore connectivity increases bacterial diversity in soil. *Applied and Environmental Microbiology, 76*, 3936–3942. https://doi.org/10.1128/AEM.03085-09

Cazelles, K., Otten, W., Baveye, P. C., & Falconer, R. E. (2013). Soil fungal dynamics: Parameterisation and sensitivity analysis of modelled physiological processes, soil architecture and carbon distribution. *Ecological Modelling, 248*, 165–173. https://doi.org/10.1016/j.ecolmodel.2012.08.008

Clark, T., Burca, G., Boardman, R., & Blumensath, T. (2019). Correlative X-ray and neutron tomography of root systems using cadmium fiducial markers. *Journal of Microscopy, 277*, 170–178. https://doi.org/10.1111/jmi.12831

Cooper, L. J., Daly, K. R., Hallett, P. D., Naveed, M., Koebernick, N., Bengough, A. G., George, T. S., & Roose, T. (2017). Fluid flow in porous media using image-based modelling to parametrize Richards' equation. *Proc.R.Soc.A, 473*, 20170178. https://doi.org/10.1098/rspa.2017.0178

Cooper, L. J., Daly, K. R., Hallett, P. D., Koebernick, N., George, T. S., & Roose, T. (2018). The effect of root exudates on rhizosphere water dynamics. *Proc. R. Soc. A, 474*, 20180149. https://doi.org/10.1098/rspa.2018.0149

Crawford, J. W., Deacon, L., Grinev, D., Harris, J. A., Ritz, K., Singh, B. K., & Young, I. (2012). Microbial diversity affects self-organization of the soil–microbe system with consequences for function. *J. R. Soc. Interface, 9*, 1302–1310. https://doi.org/10.1098/rsif.2011.0679

Crawford, J. W., Ritz, K., & Young, I. (1993). Quantification of fungal morphology, gaseous transport and microbial dynamics in soil: An integrated framework utilising fractal geometry. *Geoderma, 56*, 157–172. https://doi.org/10.1016/0016-7061(93)90107-V

Dal Ferro, N., Charrier, P., & Morari, F. (2013). Dual-scale X-ray CT assessment of soil structure in a long-term fertilization experiment. *Geoderma, 204*, 84–93. https://doi.org/10.1016/j.geoderma.2013.04.012

Dal Ferro, N., Strozzi, A. G., Duwig, C., Delmas, P., Charrier, P., & Morari, F. (2015). Application of smoothed particle hydrodynamics (SPH) and pore morphologic model to predict saturated water conductivity from X-ray CT imaging in a silty loam Cambisol. *Geoderma, 255–256*, 27–34. https://doi.org/10.1016/j.geoderma.2015.04.019

Daly, K. R., & Roose, T. (2014). Multiscale modelling of hydraulic conductivity in vuggy porous media. *Proc. R. Soc. A, 470*, 20130383. https://doi.org/10.1098/rspa.2013.0383

Daly, K. R., & Roose, T. (2015). Homogenization of two fluid flow in porous media. *Proc. R. Soc. A, 471*, 20140564. https://doi.org/10.1098/rspa.2014.0564

Daly, K. R., Mooney, S. J., Bennett, M. J., Crout, N. M., Roose, T., & Tracy, S. R. (2015). Assessing the influence of the rhizosphere on soil hydraulic properties using X-ray computed tomography and numerical modelling. *Journal of Experimental Botany, 66*, 2305–2314. https://doi.org/10.1093/jxb/eru509

Daly, K. R., Keyes, S. D., Masum, S., & Roose, T. (2016). Image based modeling of nutrient movement in and around the rhizosphere. *Journal of Experimental Botany, 67*, 1059–1070. https://doi.org/10.1093/jxb/erv544

Daly, K. R., Cooper, L. J., Koebernick, N., Evaristo, J., Keyes, S. D., van Veelen, A., & Roose, T. (2017). Modelling water dynamics in the rhizosphere. *Rhizosphere, 4*, 139–151. https://doi.org/10.1016/j.rhisph.2017.10.004

Daly, K. R., Tracy, S. R., Crout, N. M. J., Mairhofer, S., Pridmore, T., Mooney, S. J., & Roose, T. (2018). Quantification of root water uptake in soil using X-ray Computed Tomography and image based modelling. *Plant, Cell and Environment, 41*, 121–133. https://doi.org/10.1111/pce.12983

Deurer, M., Müller, K., Kim, I., Huh, K. Y., Young, I., Jun, G. I., & Clothier, B. E. (2012). Can minor compaction increase soil carbon sequestration? A case study in a soil under a wheel-track in an orchard. *Geoderma, 183-184*, 74–79. https://doi.org/10.1016/j.geoderma.2012.02.013

de Paulo, F. L., Surmas, R., Nelis Tonietto, S., Antunes Pereira da Silva, M., & Pires Peçanha, R. (2020). Modeling reactive flow on carbonates with realistic porosity and permeability fields. *Advances in Water Resources, 139*(2020), 103564. https://doi.org/10.1016/j.advwatres.2020.103564

Diel, J., Vogel, H.-J., & Schlüter, S. (2019). Impact of wetting and drying cycles on soil structure dynamics. *Geoderma, 345*, 63–71. https://doi.org/10.1016/j.geoderma.2019.03.018

Dohnal, M., Jelinkova, V., Snehota, M., Dusek, J., & Brezina, J. (2013). Tree-dimensional numerical analysis of water flow affected by entrapped air: Application of noninvasive imaging techniques. *Vadose Zone Journal, 12*, 1–12. https://doi.org/10.2136/vzj2012.0078

Dullien, F. A. L. (1992). Pore structure. In *Porous media*. Elsevier.

Durner, W. (1994). Hydraulic conductivity estimation for soils with heterogeneous pore structure. *Water Resources Resesearch, 30*, 211–233. https://doi.org/10.1029/93WR02676

Ebrahimi, A., Jamshidi, S., Iglauer, S., & Boozarjomehry, R. B. (2013). Genetic algorithm-based pore network extraction from micro-computed tomography images. *Chemical Engineering Science, 92*, 157–166. https://doi.org/10.1016/j.ces.2013.01.045

Ebrahimi, A., & Or, D. (2014). Microbial dispersal in unsaturated porous media: Characteristics of motile bacterial cell motions in unsaturated angular pore networks. *Water Resources Research, 50*, 7406–7429. https://doi.org/10.1002/2014WR015897

Ebrahimi, A., & Or, D. (2015). Hydration and diffusion processes shape microbial community organization and function in model soil aggregates. *Water Resources Research, 51*, 9804–9827. https://doi.org/10.1002/2015WR017565

Elliot, T. R., Reynolds, W. D., & Heck, R. J. (2010). Use of existing pore models and X-ray computed tomography to predict saturated soil hydraulic conductivity. *Geoderma, 156*, 133–142. https://doi.org/10.1016/j.geoderma.2010.02.010

Elyeznasni, N., Sellami, F., Pot, V., Benoit, P., Vieublé-Gonod, L., Young, I., & Peth, S. (2012). Exploration of soil micromorphology to identify coarse-sized OM assemblages in X-ray CT images of undisturbed cultivated soil cores. *Geoderma, 179-180*(38), 45. https://doi.org/10.1016/j.geoderma.2012.02.023

Falconer, R. E., Bown, J. K., White, N. A., & Crawford, J. W. (2005). Biomass recycling and the origin of phenotype in fungal mycelia. *Proceedings of the Royal Society B, 272*, 1727–1734. https://doi.org/10.1098/rspb.2005.3150

Falconer, R. E., Brown, J. L., White, N. A., & Crawford, J. W. (2007). Biomass recycling: A key to efficient foraging by fungal colonies. *Oikos, 116*, 1558–1568. https://doi.org/10.1111/j.2007.0030-1299.15885.x

Falconer, R., Houston, A. N., Otten, W., & Baveye, P. C. (2012). Emergent behavior of soil fungal dynamics: Influence of soil architecture and water distribution. *Soil Science, 177*, 111–119. https://doi.org/10.1097/SS.0b013e318241133a

Falconer, R. E., Battaia, G., Schmidt, S., Baveye, P., Chenu, C., & Otten, W. (2015). Microscale heterogeneity explains experimental variability and non-linearity in soil organic matter mineralisation. *PLoS One, 10*, e0123774. https://doi.org/10.1371/journal.pone.0123774

Gackiewicz, B., Lamorski, K., & Sławiński, C. (2019). Saturated water conductivity estimation based on X-ray CT images–evaluation of the impact of thresholding errors. *International Agrophysics, 33*, 49–60. https://doi.org/10.31545/intagr/104376

Genty, A., & Pot, V. (2013). Numerical simulation of 3-D liquid-gas distribution in porous media by a two-phase TRT lattice Boltzmann method. *Transport in Porous Media, 96*, 271–294. https://doi.org/10.1007/s11242-012-0087-9

Genty, A., & Pot, V. (2014). Numerical calculation of effective diffusion in unsaturated porous media by the TRT lattice Boltzmann method. *Transp Porous Med, 105*, 391–410. https://doi.org/10.1007/s11242-014-0374-8

Gerke, K. M., Vasilyev, R. V., Khirevich, S., Collins, D., Karsanina, M. V., Sizonenko, T. O., Korost, D. V., Lamontagne, S., & Mallants, D. (2018). Finite-difference method stokes solver (FDMSS) for 3-D pore geometries: Software development, validation and case studies. *Computers & Geosciences, 114*, 41–58. https://doi.org/10.1016/j.cageo.2018.01.005

Gerke, K. M., & Karsanina, M. V. (2020). How pore structure non-stationarity compromises flow properties representativity (REV) for soil samples: Pore-scale modelling and stationarity analysis. *European Journal of Soil Science, 72*, 1–19. https://doi.org/10.1111/ejss.13055

Ghanbarian, B., & Hunt, A. G. (2017). Improving unsaturated hydraulic conductivity estimation in soils via percolation theory. *Geoderma, 303*, 9–18. https://doi.org/10.1016/j.geoderma.2017.05.004

Gharedaghloo, B., Price, J. S., Rezanezhad, F., & Quinton, W. L. (2018). Evaluating the hydraulic and transport properties of peat soil using pore network modeling and X-ray micro computed tomography. *Journal of Hydrology, 561*, 494–508. https://doi.org/10.1016/j.jhydrol.2018.04.007

Golparvar, A., Kästner, M., & Thullner, M. (2021). Pore-scale modeling of microbial activity: What we have and what we need. *Vadose Zone Journal, 20*, e20087. https://doi.org/10.1002/vzj2.20087

Gommes, C. J., Bons, A.-J., Blacher, S., Dunsmuir, J. H., & Tsou, A. H. (2009). Practical methods for measuring the tortuosity of porous materials from binary or grey-tone tomographic reconstructions. *AICHE Journal, 55*(8), 2000–2012. https://doi.org/10.1002/aic.11812

Gostick, J., Aghighi, M., Hinebaugh, J., Tranter, T., Hoeh, M. A., Day, H., Spellacy, B., Sharqawy, M. H., Bazylak, A., Burns, A., & Lehnert, W. (2016). OpenPNM: A pore network modeling package. *Computing in Science & Engineering, 18*(4), 60–74. https://doi.org/10.1109/MCSE.2016.49

Gostick, J., Khan, Z. A., Tranter, T. G., Kok, M. D. R., Agnaou, M., Sadeghi, M. A., & Jervis, R. (2019). PoreSpy: A python toolkit for quantitative analysis of porous media images. *Journal of Open Source Software, 4*(37), 1296. https://doi.org/10.21105/joss.01296

Grimm, V., Revilla, E., Berger, U., Jeltsch, F., Wolf, M., Mooij, W. M., Railsback, S. F., Thulke, H.-H., Weiner, J., Wiegand, T., & De Angelis, D. L. (2005). Pattern-oriented modeling of agent-based complex systems: Lessons from ecology. *Science, 310*, 987–991. https://doi.org/10.1126/science.1116681

Hamamoto, S., Moldrup, P., Kawamoto, K., Sakaki, T., Nishimura, T., & Komatsu, T. (2016). Pore network structure linked by X-ray CT to particle characteristics and transport parameters. *Soils and Foundations, 56*(4), 676–690. https://doi.org/10.1016/j.sandf.2016.07.008

Hapca, S. M., Baveye, P. C., Wilson, C., Lark, R. M., & Otten, W. (2015). Three-dimensional mapping of soil chemical characteristics at micrometric scale by combining 2-D SEM-EDX data and 3-D X-Ray CT images. *PLoS One, 10*(9), e0137205. https://doi.org/10.1371/journal.pone.0137205

Hazlett, R. D. (1995). Simulation of capillary-dominated displacements in microtomographic images of reservoir rocks. *Transport in Porous Media, 20*, 21–35. https://doi.org/10.1007/BF00616924

Helliwell, J. R., Sturrock, C. J., Greyling, K. M., Tracy, S. R., Flavel, R. J., Young, I. M., Whalley, W. R., & Mooney, S. J. (2013). Applications of X-ray computed tomography for examining biophysical interactions and structural development in soil systems: A review. *European Journal of Soil Science, 64*, 279–297. https://doi.org/10.1111/ejss.12028

Hilpert, M., & Miller, C. T. (2001). Pore-morphology-based simulation of drainage in totally wetting porous media. *Advances in Water Resources, 24*(3–4), 243–255. https://doi.org/10.1016/S0309-1708(00)00056-7

Hoogland, F., Lehmann, P., Mokso, R., & Or, D. (2016). Drainage mechanisms in porous media: From piston-like invasion to formation of corner flow networks. *Water Resources Research, 52*, 8413–8436. https://doi.org/10.1002/2016WR019299

Hu, W., Liu, G., & Zhang, X. (2018). A pore-scale model for simulating water flow in unsaturated soil. *Microfluidics and Nanofluidics, 22*, 71. https://doi.org/10.1007/s10404-018-2090-0

Hu, X., Li, X.-Y., Li, Z.-C., Gao, Z., Wu, X.-C., Wang, P., Lyu, Y.-L., & Liu, L.-Y. (2020). Linking 3-D soil macropores and root architecture to near saturated hydraulic conductivity of typical meadow soil types in the Qinghai Lake watershed, northeastern Qinghai-Tibet plateau. *Catena, 185*, 104287. https://doi.org/10.1016/j.catena.2019.104287

Hyväluoma, J., Thapaliya, M., Alaraudanjoki, J., Sirén, T., Mattila, K., Timonen, J., & Turtola, E. (2012). Using microtomography, image analysis and flow simulations to characterize soil surface seals. *Computational Geosciences, 48*, 93–101. https://doi.org/10.1016/j.cageo.2012.05.009

Iassonov, P., Gebrenegus, T., & Tuller, M. (2009). Segmentation of X-ray computed tomography images of porous materials: A crucial step for characterization and quantitative analysis of pore structures. *Water Resources Research, 45*, W09415. https://doi.org/10.1029/2009WR008087

Joekar Niasar, V., Hassanizadeh, S. M., Pyrak-Nolte, L. J., & Berentsen, C. (2009). Simulating drainage and imbibition experiments in a high-porosity micromodel using an unstructured pore network model. *Water Resources Research, 45*, W02430. https://doi.org/10.1029/2007WR006641

Joschko, M., Müller, P. C., Kotzke, K., Döhring, W., & Larink, O. (1993). Earthworm burrow system development assessed by means of X-ray computed tomography. *Geoderma, 56*(1–4), 209–221. https://doi.org/10.1016/0016-7061(93)90111-W

Juyal, A., Otten, W., Baveye, P. C., & Eickhorst, T. (2020). Influence of soil structure on the spread of Pseudomonas fluorescens in soil at microscale. *European Journal of Soil Science, 72*, 141–153. https://doi.org/10.1111/ejss.12975

Katuwal, S., Norgaard, T., Moldrup, P., Lamandé, M., Wildenschild, D., & de Jonge, L. W. (2015). Linking air and water transport in intact soils to macropore characteristics inferred from X-ray computed tomography. *Geoderma, 237–238*, 9–20. https://doi.org/10.1016/j.geoderma.2014.08.006

Kasteel, R., Vogel, H.-J., & Roth, K. (2000). From local hydraulic properties to effective transport in soil. *European Journal of Soil Science, 51*, 81–91. https://doi.org/10.1046/j.1365-2389.2000.00282.x

Katz, A. J., & Thompson, A. H. (1986). Quantitative prediction of permeability in porous rock. *Physical Review B, 34*(11), 8179–8181. https://doi.org/10.1103/PhysRevB.34.8179

Kemgue, A. T., Monga, O., Moto, S., Pot, V., Garnier, P., Baveye, P. C., & Bouras, A. (2019). From spheres to ellipsoids: Speeding up considerably the morphological modeling of pore space and water retention in soils. *Computers and Geosciences, 123*, 20–37. https://doi.org/10.1016/j.cageo.2018.11.006

Keyes, S. D., Daly, K. R., Gostling, N. J., Jones, D. L., Talboys, P., Pinzer, B. R., Boardman, R., Sinclair, I., Marchant, A., & Roose, T. (2013). High-resolution synchrotron imaging of wheat root hairs growing in soil and image based modelling of phosphate uptake. *New Phytologist, 198*, 1023–1029. https://doi.org/10.1111/nph.12294

Khan, F., Enzmann, F., Kersten, M., Wiegmann, A., & Steiner, K. (2012). 3-D simulation of the permeability tensor in a soil aggregate on basis of nanotomography imaging and LBE solver. *Journal of Soils and Sediments, 12*, 86–96. https://doi.org/10.1007/s11368-011-0435-3

Kirk, G. J. D., Boghi, A., Affholder, M.-C., Keyes, S. D., Heppell, J., & Roose, T. (2019). Soil carbon dioxide venting through rice roots. *Plant, Cell & Environment, 42*, 3197–3207. https://doi.org/10.1111/pce.13638

Koebernick, N., Daly, K. R., Keyes, S. D., George, T. S., Brown, L. K., Raffan, A., Cooper, L. J., Naveed, M., Bengough, A. G., Sinclair, I., Hallett, P. D., & Roose, T. (2017). High-resolution synchrotron imaging shows that root hairs influence rhizosphere soil structure formation. *The New Phytologist, 216*, 124–135. https://doi.org/10.1111/nph.14705

Koebernick, N., Daly, K. R., Keyes, S. D., Bengough, A. G., Brown, L. K., Cooper, L., George, T. S., Hallett, P. D., Naveed, M., Raffan, A., & Roose, T. (2019). Imaging microstructure of the barley rhizosphere: Particle packing and root hair influences. *New Phytologist, 221*(4), 1878–1889. https://doi.org/10.1111/nph.15516

Koestel, J., Dathe, A., Skaggs, T. H., Klakegg, O., Ahmad, M. A., Babko, M., Giménez, D., Farkas, C., Nemes, A., & Jarvis, N. (2018). Estimating the permeability of naturally structured soil from percolation theory and pore space characteristics imaged by X-Ray. *Water Resources Research, 54*, 9255–9263. https://doi.org/10.1029/2018WR023609

Köhne, M., Schlüter, S., & Vogel, H.-J. (2011). Predicting solute transport in structured soil using pore network models. *Vadose Zone Journal, 10*, 1082–1096. https://doi.org/10.2136/vzj2010.0158

König, S., Vogel, H.-J., Harms, H., & Worrich, A. (2020). Physical, chemical and biological effects on soil bacterial dynamics in microscale models. *Frontiers in Ecology and Evolution, 8*, 53. https://doi.org/10.3389/fevo.2020.00053

Koponen, A., Kataja, M., & Timonen, T. (1996). Tortuous flow in porous media. *Physical Review E, 54*, 406–410. https://doi.org/10.1103/PhysRevE.54.406

Kravchenko, A., Falconer, R. E., Grinev, D., & Otten, W. (2011). Fungal colonization in soils with different management histories: Modeling growth in three-dimensional pore volumes. *Ecological Applications, 21*, 1202–1210. https://doi.org/10.2307/23022990

Kravchenko, A., Guber, A., Gunina, A., Dippold, M., & Kuzyakov, Y. (2020). Pore-scale view of microbial turnover: Combining 14C imaging, μCT and zymography after adding soluble carbon to soil pores of specific sizes. *European Journal of Soil Science, 72*, 593–607. https://doi.org/10.1111/ejss.13001

Larsbo, M., Koestel, J., & Jarvis, N. (2014). Relations between macropore network characteristics and the degree of preferential solute transport. *Hydrology and Earth System Sciences, 18*, 5255–5269. https://doi.org/10.5194/hess-18-5255-2014

Landl, M., Haupenthal, A., Leitner, D., Kroener, E., Vetterlein, D., Bol, R., Vereecken, H., Vanderborght, J., & Schnepf, A. (2021). Simulating rhizodeposition patterns around growing and exuding root systems. *bioRxiv, 2021.02.25.432851*. https://doi.org/10.1101/2021.02.25.432851

Landl, M., Schnepf, A., Uteau, D., Peth, S., Athmann, M., Kautz, T., Perkons, U., Vereecken, H., & Vanderborght, J. (2019). Modeling the impact of biopores on root growth and root water uptake. *Vadose Zone Journal, 18*(1–20), 180196. https://doi.org/10.2136/vzj2018.11.0196

Latt, J., Malaspinas, O., Kontaxakis, D., Parmigiani, A., Lagrava, D., Brogi, F., Ben Belgacem, M., Thorimbert, Y., Leclaire, S., Li, S., Marson, F., Lemus, J., Kotsalos, C., Conradin, R., Coreixas, C., Petkantchin, R., Raynaud, F., Beny, J., & Chopard, B. (2021). Palabos: Parallel Lattice Boltzmann Solver. *Computers & Mathematics with Applications, 81*, 334–350. https://doi.org/10.1016/j.camwa.2020.03.022

Legland, D., Arganda-Carreras, I., & Andrey, P. (2016). MorphoLibJ: Integrated library and plugins for mathematical morphology with ImageJ. *Bioinformatics, 32*, 3532–3534. https://doi.org/10.1093/bioinformatics/btw413

Lehmann, P., Wyss, P., Flisch, A., Lehmann, E., Vontobel, P., Krafczyk, M., Kaestner, A., Beckmann, F., Gygi, A., & Flühler, H. (2006). Tomographical imaging and mathematical description of porous media used for the prediction of fluid distribution. *Vadose Zone Journal, 5*, 80–97. https://doi.org/10.2136/vzj2004.0177

Li, T. C., Shao, M. A., & Jia, Y. H. (2016). Application of X-ray tomography to quantify macropore characteristics of loess soil under two perennial plants. *European Journal of Soil Science, 67*, 266–275. https://doi.org/10.1111/ejss.12330

Li, Z.-S., & Tang, L.-S. (2019). Using synchrotron-based X-Ray microcomputed tomography to characterize water distribution in compacted soils. *Advances in Materials Science and Engineering, 2019*, 7147283.

Lin, D., Hu, L., Bradford, S. A., Zhang, X., & Lo, I. M. C. (2021). Simulation of colloid transport and retention using a pore-network model with roughness and chemical heterogeneity on pore surfaces. *Water Resources Research, 57*, e2020WR028571. https://doi.org/10.1029/2020WR028571

Lin, W., Li, X., Yang, Z., Xiong, S., Luo, Y., & Zhao, X. (2020). Modeling of 3-D rock porous media by combining X-Ray CT and Markov chain Monte Carlo. *ASME Journal Energy Resource Technolology, 142*(1), 013001. https://doi.org/10.1115/1.4045461

Lissy, A.-S., Sammartino, S., & Ruy, S. (2020). Can structure data obtained from CT images substitute for parameters of a preferential flow model? *Geoderma, 380*, 114643. https://doi.org/10.1016/j.geoderma.2020.114643

Liu, L., Dai, S., Ning, F., Cai, J., Liu, C., & Wu, N. (2019). Fractal characteristics of unsaturated sands–implications to relative permeability in hydrate-bearing sediments. *Journal of Natural Gas Science and Engineering, 66*, 11–17. https://doi.org/10.1016/j.jngse.2019.03.019

Long, T., & Or, D. (2005). Aquatic habitats and diffusion constraints affecting microbial coexistence in unsaturated porous media. *Water Resources Research, 41*, W08408. https://doi.org/10.1029/2004WR003796

Lobet, G., Pagès, L., & Draye, X. (2011). A novel image analysis toolbox enabling quantitative analysis of root system architecture. *Plant Physiology, 157*, 29–39. https://doi.org/10.1104/pp.111.179895

Lobet, G., Pound, M. P., Diener, J., Pradal, C., Draye, X., Godin, C., Javaux, M., Leitner, D., Meunier, F., Nacry, P., Pridmore, T. P., & Schnepf, A. (2015). Root system markup language: Toward a unified root architecture description language. *Plant Physiology, 167*, 617–627. https://doi.org/10.1104/pp.114.253625

Luo, L., & Lin, H. (2009). Lacunarity and fractal analyses of soil macropores and preferential transport using micro-X-Ray computed tomography. *Vadose Zone Journal, 8*, 233–241. https://doi.org/10.2136/vzj2008.0010

McKay Fletcher, D. M., Keyes, S. D., Daly, K. R., van Veelen, A., & Roose, T. (2019). A multi image-based approach for modelling plant-fertiliser interaction. *Rhizosphere, 10*, 100152. https://doi.org/10.1016/j.rhisph.2019.100152

McKay Fletcher, D. M., Ruiz, S., Dias, T., Petroselli, C., & Roose, T. (2020). Linking root structure to functionality: The impact of root system architecture on citrate-enhanced phosphate uptake. *The New Phytologist, 227*, 376–391. https://doi.org/10.1111/nph.16554

Mairhofer, S., Johnson, J., Sturrock, C. J., et al. (2016). Visual tracking for the recovery of multiple interacting plant root systems from X-ray μCT images. *Machine Vision and Applications, 27*, 721–734. https://doi.org/10.1007/s00138-015-0733-7

Maenhout, P., De Neve, S., Wragg, J., et al. (2021). Chemical staining of particulate organic matter for improved contrast in soil X-ray CT images. *Scientific Reports, 11*, 370. https://doi.org/10.1038/s41598-020-79681-5

Menon, M., Yuan, Q., Jia, X., Dougill, A. J., Hoon, S. R., Thomas, A. D., & Williams, R. A. (2011). Assessment of physical and hydrological properties of biological soil crusts using X-ray microtomography and modeling. *Journal of Hydrology, 397*, 47–54. https://doi.org/10.1016/j.jhydrol.2010.11.021

Menon, M., Jia, X., Lair, G. J., Faraj, P. H., & Blaud, A. (2015). Analysing the impact of compaction of soil aggregates using X-ray microtomography and water flow simulations. *Soil and Tillage Research, 150*, 147–157. https://doi.org/10.1016/j.still.2015.02.004

Monga, O., Bousso, M., Garnier, P., & Pot, V. (2008). 3-D geometric structures and biological activity: Application to microbial soil organic matter decomposition in pore space. *Ecological Modelling, 216*, 291–302. https://doi.org/10.1016/j.ecolmodel.2008.04.015

Monga, O., Ngom, N. F., & Delerue, J. F. (2007). Representing geometric structures in 3-D tomography soil images: Application to pore space modelling. *Computers and Geosciences, 33*, 1140–1161. https://doi.org/10.1016/j.cageo.2006.11.015

Monga, O., Garnier, P., Pot, V., Coucheney, E., Nunan, N., Otten, W., & Chenu, C. (2014). Simulating microbial degradation of organic matter in a simple porous system using the 3-D diffusion-based model MOSAIC. *Biogeosciences, 11*, 2201–2209. https://doi.org/10.5194/bg-11-2201-2014

Mooney, S. (2002). Three-dimensional visualization and quantification of soil macroporosity and water flow patterns using computed tomography. *Soil Use and Management, 18*, 142–151. https://doi.org/10.1111/j.1475-2743.2002.tb00232.x

Moradi, A. B., Carminati, A., Vetterlein, D., Vontobel, P., Lehmann, E., Weller, U., Hopmans, J. W., Vogel, H. J., & Oswald, S. E. (2011). Three-dimensional visualization and quantification of water content in the rhizosphere. *The New Phytologist, 192*(3), 653–663. https://doi.org/10.1111/j.1469-8137.2011.03826.x

Ngom, N. F., Garnier, P., Monga, O., & Peth, S. (2011). Extraction of three-dimensional soil pore space from microtomography images using a geometrical approach. *Geoderma, 163*, 127–134. https://doi.org/10.1016/j.geoderma.2011.04.013

Ngom, N. F., Monga, O., Mohamed, M. M. O., & Garnier, P. (2012). 3-D shape extraction segmentation and representation of soil microstructures using generalized cylinders. *Computers & Geosciences, 39*, 50–63. https://doi.org/10.1016/j.cageo.2011.06.010

Nishiyama, N., & Yokoyama, T. (2017). Permeability of porous media: Role of the critical pore size. *Journal of Geophysical Research–Solid Earth, 122*, 6955–6971. https://doi.org/10.1002/2016JB013793

Pagenkemper, S. K., Athmann, M., Uteau, D., Kautz, T., Peth, S., & Horn, R. (2015). The effect of earthworm activity on soil bioporosity–investigated with X-ray computed tomography and endoscopy. *Soil and Tillage Research, 146*(Part A), 79–88. https://doi.org/10.1016/j.still.2014.05.007

Pajor, R., Falconer, R., Hapca, S., & Otten, W. (2010). Modelling and quantifying the effect of heterogeneity in soil physical conditions on fungal growth. *Biogeosciences, 7*, 3477–3501. https://doi.org/10.5194/bg-7-3731-2010

Papadopoulos, A., Bird, N. R. A., Whitmore, A. P., & Mooney, S. J. (2009). Investigating the effects of organic and conventional management on soil aggregate stability using X-ray computed tomography. *European Journal of Soil Science, 60*, 360–368. https://doi.org/10.1111/j.1365-2389.2009.01126.x

Perez-Reche, F. J., Taraskin, S. N., Otten, W., Viana, M. P., Costa, L. F., & Gilligan, C. A. (2012). Prominent effect of soil network heterogeneity on microbial invasion. *Physical Review Letters, 109*, 098102. https://doi.org/10.1103/PhysRevLett.109.098102

Peszynska, M., Trykozko, A., Iltis, G., Schlüter, S., & Wildenschild, W. (2016). Biofilm growth in porous media: Experiments, computational modeling at the porescale, and upscaling. *Advances in Water Resources, 95*, 288–301. https://doi.org/10.1016/j.advwatres.205.07.008

Peth, S., Chenu, C., Leblond, N., Mordhorst, A., Garnier, P., Nunan, N., Pot, V., Ogurreck, M., & Beckmann, F. (2014). Localization of soil organic matter in soil aggregates using synchrotron-based X-ray microtomography. *Soil Biology & Biochemistry, 78*, 189e194. https://doi.org/10.1016/j.soilbio.2014.07.024

Piccoli, I., Dal Ferro, N., Delmas, P. J., Squartini, A., & Morari, F. (2019a). Contrast-enhanced repacked soil cores as a proxy for soil organic matter spatial arrangement. *Soil Research, 57*, 535–545. https://doi.org/10.1071/SR18191

Piccoli, I., Schjønning, P., Lamandé, M., Zanini, F., & Morari, F. (2019b). Coupling gas transport measurements and X-ray tomography scans for multiscale analysis in silty soils. *Geoderma, 338*, 576–584. https://doi.org/10.1016/j.geoderma.2018.09.029

Porre, R. J., van Groenigen, J. W., De Deyn, G. B., de Goede, R. G. M., & Lubbers, I. M. (2016). Exploring the relationship between soil mesofauna, soil structure and N2O emissions. *Soil Biology and Biochemistry, 96*, 55–64. https://doi.org/10.1016/j.soilbio.2016.01.018

Portell, X., Pot, V., Garnier, P., Otten, W., & Baveye, P. C. (2018). Microscale heterogeneity of the spatial distribution of organic matter can promote bacterial biodiversity in soils: Insights from computer simulations. *Frontiers in Microbiology, 9*, 1583. https://doi.org/10.3389/fmicb.2018.01583

Pot, V., Hammou, H., Elyeznasni, N., Ginburg, I. (2010). *Role of soil heterogeneities onto pesticide fate: A pore-scale study with lattice Boltzmann.* 1st International conference and exploratory workshop on soil architecture and physico-chemical functions, CESAR, Tjele, Danemark (2010-11-30–2010-12-02) pp. 307–310.

Pot V, Peth S, Garnier P, Genty A, Monga O, Vieublé-Gonod L, Pinheiro M, Ogurreck M, Beckmann F (2011, June–July 26). *Visualisation and modelling of water-air interface in soil porosity: impact on microbial degradation.* Poster presented at the Soil Interfaces in a Changing World 6th ISMOM 2011 Conference.

Pot, V., Peth, S., Monga, O., Vogel, L. E., Genty, A., Garnier, P., Vieublé-Gonod, L., Ogurreck, M., Beckmann, F., & Baveye, P. C. (2015). Three-dimensional distribution of water and air in soil pores: Comparison of two-phase two-relaxation-times lattice-Boltzmann and morphological model outputs with synchrotron X-ray computed tomography data. *Advances in Water Resources, 84*, 87–102. https://doi.org/10.1016/j.advwatres.2015.08.006

Pot, V., Zhong, X., & Baveye, P. C. (2020). Effect of resolution, reconstruction settings, and segmentation methods on the numerical calculation of saturated soil hydraulic conductivity from 3-D computed tomography images. *Geoderma, 362*, 114089. https://doi.org/10.1016/j.geoderma.2019.114089

Quigley, M. Y., Rivers, M. L., & Kravchenko, A. N. (2018). Patterns and sources of spatial heterogeneity in soil matrix from contrasting Long term management practices. *Frontiers in Environmental Science, 6*, 28. https://doi.org/10.3389/fenvs.2018.00028

Rabbi, S. M. F., Tighe, M. K., Flavel, R. J., Kaiser, B. N., Guppy, C. N., Zhang, X., & Young, I. M. (2018). Plant roots redesign the rhizosphere to alter the three-dimensional physical architecture and water dynamics. *New Phytologist, 219*, 542–550. https://doi.org/10.1111/nph.15213

Rabot, E., Cousin, I., & Hénault, C. A. (2015). Modeling approach of the relationship between nitrous oxide fluxes from soils and the water-filled pore space. *Biogeochemistry, 122*, 395–408. https://doi.org/10.1007/s10533-014-0048-1

Raoof, A., & Hassanizadeh, S. M. (2012). A new formulation for pore-network modeling of two-phase flow. *Water Resource Research, 48*, W01514. https://doi.org/10.1029/2010WR010180

Rasmuson, A., Pazmino, E., Assemi, S., & Johnson, W. P. (2017). Contribution of Nano- to microscale roughness to heterogeneity: Closing the gap between unfavorable and favorable colloid attachment conditions. *Environmental Science & Technology, 51*, 2151–2160. https://doi.org/10.1021/acs.est.6b05911

Ray, N., Rupp, A., & Prechtel, A. (2017). Discrete-continuum multiscale model for transport, biomass development and solid restructuring in porous media. *Advances in Water Resources, 107*, 393–404. https://doi.org/10.1016/j.advwatres.2017.04.001

Resat, H., Bailey, V., McCue, L. A., & Konpka, A. (2012). Modeling microbial dynamics in heterogeneous environments: Growth on soil carbon sources. *Microbial Ecology, 63*, 883–897. https://doi.org/10.1007/s00248-011-9965-x

Rezanezhad, F., Quinton, W. L., Price, J. S., Elrick, D., Elliot, T. R., & Heck, R. J. (2009). Examining the effect of pore size distribution and shape on flow through unsaturated peat using computed tomography. *Hydrology and Earth System Sciences, 13*, 1993–2002. https://doi.org/10.5194/hess-13-1993-2009

Rezanezhad, F., Quinton, W. L., Price, J. S., Elliot, T. R., Elrick, D., & Shook, K. R. (2010). Influence of pore size and geometry on peat unsaturated hydraulic conductivity computed from 3-D computed tomography image analysis. *Hydrological Processes, 24*, 2983–2994. https://doi.org/10.1002/hyp.7709

Roose, T., & Fowler, A. C. (2004). A model for water uptake by plant roots. *Journal of Theoretical Biology, 228*, 155–171. https://doi.org/10.1016/j.jtbi.2003.12.012

Roose, T., Keyes, S., Daly, K., Carminati, A., Otten, W., Vetterlein, D., & Peth, S. (2016). Challenges in imaging and predictive modeling of rhizosphere processes. *Plant and Soil, 407*, 9–38. https://doi.org/10.1007/s11104-016-2872-7

Roque, W. L., & Costa, R. R. A. (2020). A plugin for computing the pore/grain network tortuosity of a porous medium from 2-D/3-D MicroCT image. *Applied Computing and Geosciences., 5*, 100019. https://doi.org/10.1016/j.acags.2020.100019

Ruiz, S. A., McKay Fletcher, D. M., Boghi, A., Williams, K. A., Duncan, S. J., Scotson, C. P., Petroselli, C., Dias, T. G. S., Chadwick, D. R., Jones, D. L., & Roose, T. (2020a). Image-based quantification of soil microbial dead zones induced by nitrogen fertilization. *Science Total Environment, 727*, 138197. https://doi.org/10.1016/j.scitotenv.2020.138197

Ruiz, S. A., Mckay Fletcher, D., Williams, K., Roose, T. (2020b). Review of plant-soil modelling: Root growth, nutrient and water transport/uptake, and mechanics. *Annual Plant Reviews*.

San José Martínez, F., Martín, M. A., Caniego, F. J., Tuller, M., Guber, A., Pachepsky, Y., & García-Gutiérrez, C. (2010). Multifractal analysis of discretized X-ray CT images for the characterization of soil macropore structures. *Geoderma, 156*, 32–42. https://doi.org/10.1016/j.geoderma.2010.01.004

Saiers, J. E., & Lenhart, J. J. (2003). Colloid mobilization and transport within unsaturated porous media under transient-flow conditions. *Water Resources Research, 39*, 1019. https://doi.org/10.1029/2002WR001370

Schaap, M. G., Porter, M. L., Christensen, B. S. B., & Wildenschild, D. (2007). Comparison of pressure-saturation characteristics derived from computed tomography and lattice Boltzmann simulations. *Water Resources Research, 43:W12S06*. https://doi.org/10.1029/2006WR005730

Schlüter, S., Sheppard, A., Brown, K., & Wildenschild, D. (2014). Image processing of multiphase images obtained via X-ray microtomography: A review. *Water Resources Research, 50*, 3615–3639. https://doi.org/10.1002/2014WR015256

Schlüter, S., Eickhorst, T., & Mueller, C. W. (2019). Correlative imaging reveals holistic view of soil microenvironments. *Environmental Science & Technology, 53*, 829–837. https://doi.org/10.1021/acs.est.8b05245

Schnepf, A., Black, C. K., Couvreur, V., Delory, B. M., Doussan, C., Koch, A., Koch, T., Javaux, M., Landl, M., Leitner, D., Lobet, G., Mai, T. H., Meunier, F., Petrich, L., Postma, J. A., Priesack, E., Schmidt, V., Vanderborght, J., Vereecken, H., & Weber, M. (2020). Call for participation: Collaborative benchmarking of functional-structural root architecture models. The case of root water uptake. *Frontiers in Plant Science, 11*(316). https://doi.org/10.3389/fpls.2020.00316

Scotson, C. P., Munoz-Hernando, M., Duncan, S. J., Ruiz, S. A., Keyes, S. D., van Veelen, A., Dunlop, I. E., & Roose, T. (2019). Stabilizing gold nanoparticles for use in X-ray computed tomography imaging of soil systems. *Royal Society Open Science, 6190769190769*. https://doi.org/10.1098/rsos.190769

Shen, C., Li, B., Wang, C., Huang, Y., & Jin, Y. (2011). Surface roughness effect on deposition of nano- and micro-sized colloids in saturated columns at different solution ionic strengths. *Vadose Zone Journal, 10*, 1071–1081. https://doi.org/10.2136/vzj2011.0011

Seaton, M. A., Anderson, R. L., Metz, S., & Smith, W. (2013). DL_MESO: Highly scalable mesoscale simulations. *Molecular Simulation, 39*(10), 796–821. https://doi.org/10.1080/08927022.2013.772297

Soto-Gómez, D., Pérez-Rodríguez, P., Vázquez-Juiz, L., López-Periago, E., & Paradelo, M. (2018). Linking pore network characteristics extracted from CT images to the transport of solute and colloid tracers in soils under different tillage managements. *Soil and Tillage Research, 177*, 145–154. https://doi.org/10.1016/j.still.2017.12.007

Spanne, P., Thovert, J. F., Jacquin, C. J., Lindquist, W. B., Jones, K. W., & Adler, P. M. (1994). Synchrotron computed microtomography of porous media: Topology and transports. *Physical Review Letters, 73*, 2001–2004. https://doi.org/10.1103/PhysRevLett.73.2001

Sukop, M. C., Huang, H., Lin, C. L., Deo, M. D., Oh, K., & Miller, J. D. (2008). Distribution of multiphase fluids in porous media: Comparison between lattice Boltzmann modeling and micro-x-ray tomography. *Physical Review E, 77*, 026710. https://doi.org/10.1103/PhysRevE.77.026710

Tecon, R., Ebrahimi, A., Kleyer, H., Levi, S. E., & Or, D. (2018). Cell-to-cell bacterial interactions promoted by drier conditions on soil surfaces. *PNAS, 115*, 9791–9796. https://doi.org/10.1073/pnas.1808274115

Torkzaban, S., & Bradford, S. A. (2016). Critical role of surface roughness on colloid retention and release in porous media. *Water Research, 88*, 274–284. https://doi.org/10.1016/j.watres.2015.10.022

Tötzke, C., Kardjilov, N., Hilger, A., Rudolph-Mohr, N., Manke, I., & Oswald, S. E. (2021). Three-dimensional in vivo analysis of water uptake and translocation in maize roots by fast neutron tomography. *Scientific Reports, 11*, 10578. https://doi.org/10.1038/s41598-021-90062-4

Tran, A. P., Yan, S., & Fang, Q. (2020). Improving model-based functional near-infrared spectroscopy analysis using mesh-based anatomical and light-transport models. *Neurophotonics, 7*, 015008. https://doi.org/10.1117/1.NPh.7.1.015008

Tuller, M., Or, D., & Dudley, L. M. (1999). Adsorption and capillary condensation in porous media: Liquid retention and interfacial configurations in angular pores. *Water Resources Research, 35*(7), 1949–1964. https://doi.org/10.1029/1999WR900098

Tuller, M., & Or, D. (2001). Hydraulic conductivity of variably saturated porous media: Film and corner flow in angular pore space. *Water Resources Research, 37*(5), 1257–1276. https://doi.org/10.1029/2000WR900328

Tyler, S. W., & Wheatcraft, S. W. (1990). Fractal processes in soil water retention. *Water Resources Research, 26*(5), 1047–1054. https://doi.org/10.1029/WR026i005p01047

Valvatne, P. H., & Blunt, M. J. (2004). Predictive pore-scale modeling of two-phase flow in mixed wet media. *Water Resources Research, 40*, W07406. https://doi.org/10.1029/2003WR002627

van Genuchten, M. (1980). A closed-form equation for predicting the hydraulic conductivity of unsaturated soils. *Soil Science Society of America Journal, 44*, 892–898. https://doi.org/10.2136/sssaj1980.03615995004400050002x

Van Loo, D., Bouckaert, L., Leroux, O., Pauwels, E., Dierick, M., Van Hoorebeke, L., Cnudde, V., De Neve, S., & Sleutel, S. (2014). Contrast agents for soil investigation with X-ray computed tomography. *Geoderma, 213*, 485–491. https://doi.org/10.1016/j.geoderma.2013.08.036

Vogel, H.-J. (2008). *QuantIm–some useful C/C++ functions for scientific image processing*. Helmholtz Center for Environmental Research.

Vogel, H.-J., & Roth, K. (2001). Quantitative morphology and network representation of soil pore structure. *Advances in Water Resources, 24*, 233–242. https://doi.org/10.1016/S0309-1708(00)00055-5

Vogel, H.-J., Tölke, J., Schulz, V. P., Krafczyk, M., & Roth, K. (2005). Comparison of a Lattice-Boltzmann model, a full-morphology model, and a pore network model for determining capillary pressure–saturation relationships. *Vadose Zone Journal, 4*, 380–388. https://doi.org/10.2136/vzj2004.0114

Vogel, L., Makowski, D., Garnier, P., Vieublé-Gonod, L., Coquet, Y., Raynaud, X., Nunan, N., Chenu, C., Falconer, R., & Pot, V. (2015). Modeling the effect of soil meso- and macropores topology on the biodegradation of a soluble carbon substrate. *Advances in Water Resources, 83*, 123–126. https://doi.org/10.1016/j.advwatres.2015.05.020

Vogel, L., Pot, V., Makowski, D., Garnier, P., & Baveye, P. C. (2018). To what extent do uncertainty and sensitivity analyses help unravel the influence of microscale physical and biological drivers in soil carbon dynamics models? *Ecological Modelling, 383*, 10–22. https://doi.org/10.1016/j.ecolmodel.2018.05.007

Wu, K., Nunan, N., Crawford, J. W., Young, I. M., & Ritz, K. (2004). An efficient Markov chain model for the simulation of heterogeneous soil structure. *Soil Science Society of America Journal, 68*, 346–351. https://doi.org/10.2136/sssaj2004.3460

Xiong, Q., Baychev, T. G., & Jivkov, A. P. (2016). Review of pore network modelling of porous media: Experimental characterisations, network constructions and applications to reactive transport. *Journal of Contaminant Hydrology, 192*, 101–117. https://doi.org/10.1016/j.jconhyd.2016.07.002

Xu, Y. (2004). Calculation of unsaturated hydraulic conductivity using a fractal model for the pore-size distribution. *Computers and Geotechnics, 31*, 549–557. https://doi.org/10.1016/j.compgeo.2004.07.003

Yu, B., & Liu, W. (2004). Fractal analysis of permeabilities for porous media. *American Institute of Chemical Engineers Journal, 50*, 46–57. https://doi.org/10.1002/aic.10004

Zarebanadkouki, M., Kim, Y., Moradi, A., Vogel, H.-J., Kaestner, A., & Carminati, A. (2012). Quantification and modeling of local root water uptake using neutron radiography and deuterated water. *Vadose Zone Journal, 11*(vzj2011), 0196. https://doi.org/10.2136/vzj2011.0196

Zarebanadkouki, M., Kroener, E., Kaestner, A., & Carminati, A. (2014). Visualization of root water uptake: Quantification of deuterated water transport in roots using neutron radiography and numerical modelling. *Plant Physiology, 166*, 487–499. https://doi.org/10.1104/pp.114.243212

Zhang, X., Crawford, J. W., Flavel, R. J., & Young, I. M. (2016). A multi-scale lattice Boltzmann model for simulating solute transport in 3-D X-ray micro-tomography images of aggregated porous materials. *Journal of Hydrology, 541*, 1020–1029. https://doi.org/10.1016/j.jhydrol.2016.08.013

Zhang, X., Neal, A. L., Crawford, J. W., Bacq-Labreuil, A., Akkari, E., & Rickard, W. (2021a). The effects of long-term fertilizations on soil hydraulic properties vary with scales. *Journal of Hydrology, 593*, 125890. https://doi.org/10.1016/j.jhydrol.2020.125890

Zhang, Y., Yang, Z., Wang, F., & Zhang, X. (2021b). Comparison of soil tortuosity calculated by different methods. *Geoderma, 402*, 115358. https://doi.org/10.1016/j.geoderma.2021.115358

Zheng, H., Kim, K., Kravchenko, A., Rivers, M., & Guber, A. (2020). Testing Os staining approach for visualizing soil organic matter patterns in intact samples via X-ray dual-energy tomography scanning. *Environmental Science & Technology, 54*, 8980–8989. https://doi.org/10.1021/acs.est.0c01028

Zhou, M., Lu, D., Dunsmuir, J., & Thomann, H. (2000). Irreducible water distribution in sandstone rock: Two phase flow simulations in CT-based pore network. *Physics and Chemistry of the Earth, Part A: Solid Earth and Geodesy, 25*(2), 169–174. https://doi.org/10.1016/S1464-1895(00)00027-2

Zhou, H., Yu, X., Chen, C., Zeng, L., Lu, S., & Wu, L. (2018). Evaluating hydraulic properties of biochar-amended soil aggregates by high-performance pore-scale simulations. *Soil Science Society of America Journal, 82*, 1–9. https://doi.org/10.2136/sssaj2017.02.0053

Zhou, H., Yu, X., Chen, C., Lu, S., Wu, L., & Zeng, L. (2019). Pore-scale lattice Boltzmann modeling of solute transport in saturated biochar amended soil aggregates. *Journal of Hydrology, 577*, 123933. https://doi.org/10.1016/j.jhydrol.2019.123933

Chapter 12
Future Perspectives

Sacha J. Mooney, Stephan Peth, Richard J. Heck, and Iain M. Young

12.1 Introduction

The adoption of X-ray CT imaging to the soil sciences has initiated a step-change in our understanding of how the design and complexity of the physical architecture of the soil porous network influences and regulates some of the most important soil functions, including water storage and flow, diffusion of gases, microbial behaviour and more. It has been well-appreciated, for decades, that soil structure is a vitally important soil property, but without an effective and rapid means to quantitatively assess it in 3-D, as it exists in the field, and how it changes dynamically over time, has been a too-frequently disregarded factor, often excluded from transport models and simplified excessively in the laboratory. Hounsfield's contribution to the development of X-ray CT in the 1970s (now over 50 years ago), best known for saving millions of lives each year through medical diagnostics, continues to support scientific advancement across a vast range of disciplines, especially those in the material and engineering sciences. However, one might argue that soil, the most

S. J. Mooney (✉)
School of Biosciences, University of Nottingham, Sutton Bonington Campus, Leics, UK
e-mail: Sacha.Mooney@nottingham.ac.uk

S. Peth
University of Hannover, Institute of Soil Science, Soil Biophysics Group, Hannover, Germany
e-mail: peth@ifbk.uni-hannover.de

R. J. Heck
School of Environmental Sciences, University of Guelph, Guelph, ON, Canada
e-mail: rheck@uoguelph.ca

I. M. Young
Biological and Environmental Science and Engineering Division, King Abdullah University of Science and Technology, Thuwal, Kingdom of Saudi Arabia
e-mail: iain.young@kaust.edu.sa

S. Jon Mooney et al. (eds.), *X-ray Imaging of the Soil Porous Architecture*,
https://doi.org/10.1007/978-3-031-12176-0_12

complex biomaterial on Earth, has been among those disciplines to benefit the most. Soil, with its highly complex yet fragile structure, that varies across many spatial and temporal scales, host to tremendous biodiversity, and upon which we are so dependent on for our food, our water, as well as the crucial role it has to play in mitigating the future changing climate, has long been considered something of a 'black box' when it comes to its structure. The ability to visualise the size, shape and connection of the microscopic pore structures (operating across multi scales), along with other soil material such as organic matter and those organisms that exist in soil such as roots and fauna, represents a huge advance in our current and future scientific capabilities. The key of course is to link such information to function. While in this book we provide a synthesis of the progress the community has made over the last c. 40 years, the next 40 years is likely to surpass these in terms of advancing our understanding of soil form and function. Considering the key chapter themes in this book, we here offer a future perspective on how we perceive the research fields may advance, highlighting where bottlenecks exist, where collective community action is needed and where greater interdisciplinary collaboration is necessary.

In most cases the cost of CT systems, following a broad period of plateau appears to be increasing, exacerbated no doubt by the recent component shortages associated with the Covid-19 pandemic. On this basis it seems unlikely that there will be significant changes in the near-term availability of CT systems to soil scientists, and in many cases, equipment sharing and scientific collaboration will continue to be necessary to support future scientific advances. Indeed, we would encourage such collaborative efforts, especially across disciplines. More positively, the future capability of instruments is likely to increase, though within the confines of current limitations, such as the sample size/resolution trade-off. The most likely future improvement to CT systems is around X-ray detector quality and automated feature detection. There continue to be regular but significant improvements in image quality and phase contrast alongside reductions in scan times (Chap. 2) and automated sampling is coming on stream. The soil science community stands to benefit considerably from this as it will advance our ability to undertake 4-D imagery, with high replication, important for studies considering water movement (Chap. 7) and root–soil interactions (Chap. 9) and improve the quantification of a wider range of soil components such as soil organic matter (Chap. 10). Such improvements will bring the current laboratory bench-top CT system closer to the synchrotron light source-based systems, but with the enhanced access that frequently impacts on the experimental plans for using the latter. One area that has not developed as much, in recent years, as might have been expected, is the integration of X-ray CT with other scientific instruments such as atomic absorption or mass spectrometry. This could facilitate the parallel 3-D quantification of chemical elements in soil, alongside the pore network data that is routinely collected from CT scanning, similar to what is possible via Nano-SIMs, albeit rather time consumingly and following resin impregnation and serial sectioning (e.g. Mueller et al., 2012). This would represent a major technological advance and remains an area in its infancy, although new applications and integration of X-ray CT with multi-modal correlative imaging are potentially very exciting new developments (e.g. Schlüter et al., 2019; Keyes et al., 2022).

Optimising the imaging process, both pre- and post-scan, remains a major challenge for the X-ray analysis of soil samples and, in many respects, it still remains a bit of a 'dark art' (Chap. 4). Though greater application and use of CT will increase the size of the size community and most likely bring in much needed new expertise. Most commercial CT systems are 'closed' in terms of operation, with little choice in terms of settings including use of filters, correction algorithms, etc. This can be advantageous as it means most systems are relatively simple to use, and the training requirement for the new user is generally light. However, there are disadvantages in that the options for users to make changes to processes, like reconstruction, are limited with many algorithms proprietary. Although most manufacturers seem receptive to the needs of users, and in the future, this could lead to greater options for certain operations, such as artefact correction.

One of the key areas where greater community action is needed is image processing and segmentation (Chaps. 5 and 6). Appropriate image thresholding remains the 'holy grail' for most analysis, and the tools/algorithms used often vary widely between researchers, which has negative impacts when it comes to comparing research outputs (see Baveye et al., 2010). Greater standardisation of methods and approaches would allow a much-improved transferability of research, something which is currently lacking. At present, soil scientists typically use a wide range of software/approaches which is often a result of software availability, though the freeware, ImageJ/FiJi (Schneider et al., 2012) and its associated libraries continue to be extensively utilised and developed by soil scientists (e.g. SoilJ (Koestel, 2018)). In the future there is potential for much greater application of more sophisticated algorithms, particularly those routinely employed in other disciplines such as the medical sciences and new statistical approaches which is likely to be a fruitful avenue to pursue. There is also potential for greater exploration of greyscale and blended (masked greyscale) data analysis rather than strictly binary, though there has been some development in this area recently (Martínez-Araya & Lark, 2018), it is certainly an area worthy of further research. Image quality inevitably varies widely across available instruments and scanner settings. Although we usually aim for the best possible quality scans, in most cases there is often a compromise between scanning time and image signal to noise ratios. This is particularly important when it comes to studies of dynamic processes where temporal resolution is in contradiction to spatial resolution (or better detail detectability) or where a very high numbers of samples are involved. What the best compromise between image quality and number of replicate samples to answer a specific research question will be has, so far, not been systematically studied. Certainly, this depends on the research question but the community could benefit from studies that develop protocols for benchmark tests. Many laboratories undertake this kind of work routinely before embarking on new experiments though it is rarely published. High throughput scanning of hundreds of samples is within reach and may be required when considering the considerable spatial heterogeneity of soil structure. At the same time, this calls also for automated image analysis routines and the employment of standards for quality control. Some recent advancements in machine learning approaches show potential to address these challenges, e.g. using lower number of projection images without loss of image quality or automated segmentation of soil pore networks, roots or organic matter (Wieland et al., 2021).

Quantification of the soil porous architecture has been the most common application of X-ray CT in soil science and the area which has advanced the discipline the most (Chap. 6). A significant development in the last 10 years has been for pore network analysis to move predominantly from 2-D to 3-D analysis. This has led to a greater understanding of how the soil pore network regulates key processes in soil like infiltration (Chap. 7). Many quantification approaches have become standardised, particularly those around measurement of the pore size distribution, facilitating a much greater comparison with other physical measurement methods such as the water release curve. Although soil pore connectivity, which is further emerging as a key property (e.g. Neal et al., 2020), is still highly varied in how it is measured. ImageJ/FiJi, with its many associated plugins, has become a very widely used platform which has facilitated the standardisation of approaches to an extent and this is likely to continue, particularly as many of the industrial standard imaging packages routinely sold with CT systems such as VG StudioMAX are very expensive (individual licenses typically costs > $10 K).

The application of X-ray CT specifically to issues concerning soil water physics is still in its infancy. While there are several examples in the literature that have sought to visualise water in pore spaces (Tracy et al., 2015), characterisation of water at the pore scale remains very complex and time consuming, whereas at the 'soil column' scale, the technical challenges of taking 'snapshots', i.e. repeated scans of a dynamic process remains a major issue. In this case, advances in scanner technology could have considerable impact, especially for 4-D imaging (with short acquisition times) and improving image/phase contrast (Chap. 7). Increasingly, CT imaging is poised to make major contributions to soil hydraulic models by parameterisation using real 3-D pore network information, especially if images of water-filled pores at tension in soils could be routinely captured and quantified. In comparison, the application of CT to other soil physical considerations, especially those related to soil deformation are more straightforward; as such this area has grown probably faster than most areas with CT measurement now routine for the characterisation of soil management, especially tillage impacts on soil structure (Chap. 8). However, there is still a lack of detailed studies on the rather complex localised stress-strain behaviour in soil (Peth et al., 2010) which would form the basis for the development of models describing soil structure dynamics. Ultimately, the coupled hydraulic-mechanical processes need to be quantified to foster our understanding of many practically relevant issues like shrink-swell behaviour or soil–root interactions. The development of in-situ testing stages, where samples can be manipulated during scanning, e.g. by loading, freezing-thawing or root growth could greatly facilitate the analysis of the micro-scale processes leading to structure dynamics. As the adoption of regenerative agriculture practices gathers pace globally, and the renewed necessity to conserve carbon and improve soil health is realised, the need for comprehensive understanding of soil structural dynamics, such as that available via X-ray CT, across a range of relevant spatial scales, will be even more prominent than now.

Our ability to visualise roots in soil, segment them from the soil matrix for quantification and assess the surrounding soil structure has been a 'game changer' for rhizosphere-related research (Chap. 9). This research has probably been the biggest growth area in the application of CT to soil sciences in the last 10 years' and has fostered much greater collaboration between plant and soil scientists. Great advances have been made in the 'root segmentation' challenge (difficult due to similarities in X-ray attenuation between roots and other organic soil fractions), and there is much promise in the application of machine learning/artificial intelligence (AI) approaches to assist with what is often a laborious process (Soltaninejad et al., 2020). However, there is still no universal approach for the segmentation of roots in soil CT images, and many of the available tools are either complex to use for a routine user or very time consuming, which becomes more problematic as sample numbers increase due to short scan times. Greater interdisciplinary collaboration is needed with imaging specialists and computer scientists to overcome current obstacles, which will also be supported by the enhanced phase contrast available on the newer detectors.

While the main focus in recent years, in the application of CT to soil sciences, has been on the soil physics and root–soil interactions area, there has been growing interest in the benefits of CT imaging to other aspects of soil sciences most notably soil biology (Chap. 10). Though the visualisation of micron-scale soil fauna is usually out of scope, due to attenuation or resolution limitations, many studies have used X-ray CT images to illustrate the feedbacks between soil fauna and their dynamic environment. Future advances in CT resolution, alongside novel staining approaches, could open new doors in this area. Another 'holy grail' has been the drive to visualise soil organic matter in-situ. Great advances have been made in this regard, and this is an area that is likely to be of considerable interest going forwards, given the importance of soil carbon to the climate emergency, food and soil security, etc. Even more challenging is the in-situ visualisation of soil microbes with new studies such as Keyes et al. (2022) which combined X-ray CT with synchrotron X-ray fluorescence/X-ray absorption near edge structure (XRF/XANES) element mapping showing promise. Straightforward solutions in that respect are currently not available but a compromise could be to combine 3-D scanning of structure followed by 2-D sectioning and mapping back the localised microbes into the 3-D structure. Here advances in image co-registration are particularly useful (Chap. 6). Developing microbial staining methods would be another promising but challenging path to go towards a full representation of the 3-D architecture of biotic and abiotic components.

The access to CT scanners to derive data concerning the multiscale 3-D structure of soil has opened up new possibilities for the discipline of pore network modelling (Chap. 11). No other technique can acquire the data needed for modelling approaches, such as Lattice-Boltzmann, so rapidly and across different scales. However, issues remain in handling the size of the data and the considerable complexity in the structural arrangement of soils, meaning that many models still work at a coarser resolution than the imaging (i.e. many modelling approaches convert fine resolution information to a coarser resolution to reduce data size), such that the important system complexity can be occluded or lost. There is great potential for AI approaches to help overcome some of these obstacles, along with enhanced computing capabilities. There remains, however, a problem with the application and future development of such pore scale models, as often this work is highly sophisticated, undertaken by a small number of people, and highly theoretical, which has led to a dearth of image-based models that are actually being utilised in prediction and decision making by key stakeholders which has to remain a future challenge. Model validation is another important issue requiring some physical data at the pore scale level which is difficult to obtain. The development of microsensors that can be located within a soil sample that locally measure state variables (e.g. oxygen concentration) or techniques that can measure processes such as enzyme activity by zymography could be very useful. The combination of imaging soil structure with pore scale modelling of soil processes and microsensing to validate such models is a very exciting future perspective with the potential to greatly enhance our understanding of structure and function in soils.

12.2 Conclusions

X-ray CT has made a very substantial contribution to soil sciences over the last 40 years, particularly so in the last 10 years, and we anticipate this to continue and increase in the future. Access to CT scanning for soil scientists has improved over time, and we are now seeing a greater emphasis on standardisation in research methods and protocols, and new technological developments. Though largely driven by other disciplines, new capabilities in CT systems will benefit researchers working across soil and plant sciences. While the inherent trade-offs associated with X-ray CT such as sample size vs resolution will remain, improved image quality especially in terms of phase contrast and resolution, alongside faster scan times, will offer huge benefits for multiscale imaging of dynamic processes in soil. Other issues related to data handling, such as appropriate image processing and measurement, might take longer to overcome, but developments in the areas of cloud computing, data science, AI/machine learning, in-situ soil sensors and so on, all offer promise. On this basis, we have reason to expect an even greater impact from the application of CT to soil sciences in the next 40 years, compared to the last 40 years.

In closing we offer a quote to the soil science community working with X-ray CT from Sir Godfrey Hounsfield, the instrument's inventor, that he delivered as part of a speech to children at his former school when he was invited back to open a new library named in his honour. Sir Godfrey said "*Each new discovery that is made brings with it the seeds of other, future inventions. There are many discoveries, probably just around the corner, waiting for someone to bring them to life, could this possibly be you?*" (Bates et al., 2012).

References

Bates, S., Beckmann, L., Thomas, A., & Waltham, R. (2012). *Godfrey Hounsfield: Intuitive genius of CT*. British Institute of Radiology.

Baveye, P. C., Laba, M., Otten, W., Bouckaert, L., Dello Sterpaio, P., Goswami, R. R., Grinev, D., Houston, A., Hu, Y., Liu, J., Mooney, S. J., Pajor, R., Sleutel, S., Tarquis, A., Wang, W., Wei, Q., & Sezgin, M. (2010). Observer-dependent variability of the thresholding step in the quantitative analysis of soil images and X-ray microtomography data. *Geoderma, 157*, 51–63. https://doi.org/10.1016/j.geoderma.2010.03.015

Keyes, S., van Veelen, A., McKay Fletcher, D., Scotson, C., Koebernick, N., Petroselli, C., Williams, K., Ruiz, S., Cooper, L., Mayon, R., Duncan, S., Dumont, M., Jakobsen, I., Oldroyd, G., Tkacz, A., Poole, P., Mosselmans, F., Borca, C., Huthwelker, T., Jones, D. L., & Roose, T. (2022). Multimodal correlative imaging and modelling of phosphorus uptake from soil by hyphae of mycorrhizal fungi. *The New Phytologist, 234*, 688–703. https://doi.org/10.1111/nph.17980

Koestel, J. (2018). SoilJ: An ImageJ plugin for the semiautomatic processing of three-dimensional X-ray images of soils. *Vadose Zone Journal, 17*, 170062.

Martínez-Araya M. and Lark R.M. 2018 Modelling of X-ray density (greyscale) signals from CT-scan images of rooted soil. 33rd international workshop on statistical modelling. Bristol, July 2018. In IWSM 2018 proceedings Volume I pp. 215–218.

Mueller, CW, Kölbl, A, Hoeschen, C, Hillion, F, Heister, K, Herrmann, A.M. & Kögel-Knabner, I (2012). Submicron scale imaging of soil organic matter dynamics using NanoSIMS–from single particles to intact aggregates, Organic Geochemistry, 42, 1476–1488, https://doi.org/10.1016/j.orggeochem.2011.06.003.

Neal, A. L., Bacq-Labreuil, A., Zhang, X., Clark, I. M., Coleman, K., Mooney, S. J., Ritz, K., & Crawford, J. W. (2020). Soil as an extended composite phenotype of the microbial metagenome. *Scientific Reports*. https://doi.org/10.1038/s41598-020-67631-0

Peth, S., Nellesen, J., Fischer, G., & Horn, R. (2010). Non-invasive 3-D analysis of local soil deformation under mechanical and hydraulic stresses by μCT and digital image correlation. *Soil and Tillage Research, 111*, 3–18.

Schneider, C. A., Rasband, W. S., & Eliceiri, K. W. (2012). NIH image to ImageJ: 25 years of image analysis. *Nature Methods, 9*(7), 671–675. https://doi.org/10.1038/nmeth.2089

Schlüter, S., Eickhorst, T., & Mueller, C. W. (2019). Correlative imaging reveals holistic view of soil microenvironments. *Environmental Science & Technology, 53*, 829–837. https://doi.org/10.1021/acs.est.8b05245

Soltaninejad, M., Sturrock, C. J., Griffiths, M., Pridmore, T. P., & Pound, M. P. (2020, May 12). Three dimensional root CT segmentation using multi-resolution encoder-decoder networks. *IEEE Transactions on Image Processing*. https://doi.org/10.1109/TIP.2020.2992893

Tracy, S. R., Daly, K. R., Mooney, S. J., Bennett, M. J., Crout, N. M. J., & Roose, T. (2015). Three dimensional quantification of soil hydraulic properties using X-ray computed tomography and image based modelling. *Water Resources Research, 51*, 1006–1022. https://doi.org/10.1002/2014WR016020

Wieland, R., Ukawa, C., Joschko, M., Krolczyk, A., Fritsch, G., Hildebrandt, T. B., Schmidt, O., Filser, J., & Jimenez, J. J. (2021). Use of deep learning for structural analysis of computer tomography images of soil samples. *Royal Society Open Science, 8*(3), 201275.